Ion Exchange and Solvent Extraction

Supramolecular Aspects of Solvent Extraction

Volume 21

ION EXCHANGE AND SOLVENT EXTRACTION SERIES

Series editors
Arup K. Sengupta
Bruce A. Moyer

Founding Editors
Jacob A Marinsky
Yizhak Marcus

Contents of Other Volumes

Ion Exchange and Solvent Extraction

Supramolecular Aspects of Solvent Extraction

Volume 21

Edited by

BRUCE A. MOYER

CRC Press
Taylor & Francis Group
Boca Raton London New York

CRC Press is an imprint of the
Taylor & Francis Group, an **informa** business

CRC Press
Taylor & Francis Group
6000 Broken Sound Parkway NW, Suite 300
Boca Raton, FL 33487-2742

First issued in paperback 2019

ISBN-13: 978-1-4822-0431-5 (hbk)
ISBN-13: 978-0-367-37906-3 (pbk)

Visit the Taylor & Francis Web site at
http://www.taylorandfrancis.com

and the CRC Press Web site at
http://www.crcpress.com

Contents

Preface

The theme of supramolecular chemistry (SC), entailing the organization of multiple species through noncovalent interactions, has permeated virtually all aspects of chemical endeavor over the past several decades. Given that the observed behavior of discrete molecular species depends upon their weak interactions with one another and with matrix components, one would have to conclude that SC must indeed form part of the fabric of chemistry itself. A vast literature now serves to categorize SC phenomena within a body of consistent terminology. The word *supramolecular* appears in the titles of dozens of books, several journals, and a dedicated encyclopedia. Not surprisingly, the theme of SC also permeates the field of solvent extraction (SX), inspiring the framework for this volume of *Ion Exchange and Solvent Extraction*. The six chapters of this volume attempt to identify both how supramolecular behavior occurs and is studied in the context of SX, and how SC is influencing the current direction of SX.

Researchers and practitioners have long dealt with supramolecular interactions in SX. Indeed, the use of polar extractant molecules in nonpolar media virtually ensures that aggregative interactions will dominate the solution behavior of SX. Analytical chemists working in the 1930s through to the 1950s with simple mono- and bidentate chelating ligands as extractants noted that extraction of metal ions obeyed complicated mass-action equilibria involving complex stoichiometries. As chemists and engineers developed processes for nuclear and hydrometallurgical applications in the 1950s and 1960s, the preference for aliphatic diluents only enhanced the complexity and supramolecular nature of extraction chemistry. Use of physical techniques such as light scattering and vapor-pressure measurements, together with various spectroscopic methods, revealed organic phase aggregates from well-defined dimers to small aggregates containing a few extractant molecules, to large inverse micelles swollen with water molecules. Extraction systems involving long-chain cations such as alkylammonium species or long-chain anions such as sulfonates or carboxylates proved especially prone to extensive aggregate formation. The related phenomenon of third-phase formation in SX systems, long misunderstood, now yields to spectroscopic and scattering techniques showing extensive long-range organization. Over the last 50 years, tools for studying the structure and thermodynamics of aggregation have grown increasingly sophisticated, leading to a rich and detailed understanding of what we can now recognize as SC phenomena in SX.

In the 1970s and 1980s, the rapid growth of SC elicited a paradigm shift in SX. The influence of SC principles had two major effects on the course of SX research: First, it provided a framework for understanding the supramolecular behavior that was already well appreciated in the field of SX, though earlier and without the SC terminology. Second, it provided the conceptual tools to control supramolecular behavior in SX, direct it for intended functionality, and simplify it. Extraction by designed reagents has been steadily progressing ever since, with commercial applications emerging to successfully validate this approach. With the discovery of crown ethers in the late 1960s, the advancement of extractant design has fruitfully employed the concept of inclusion. While considerable initial progress occurred with such molecules, especially because of their affinity and selectivity for alkali and alkaline earth metals, other molecular platforms, such as calixarenes, have proven more versatile. Multidentate receptors for partial to full inclusion of cations, anions, and ion pairs, as well as neutral species, have now become commonplace for selective extraction.

This volume of *Ion Exchange and Solvent Extraction* examines how the principles of SC are being employed both in advancing the design of new highly selective SX systems and in understanding aggregation phenomena in SX systems. Chapter 1 discusses the nature and definition of SC and how it is used generally in the design of novel SX reagents. Major approaches using SC principles are outlined and illustrated. Chapter 2 expands upon the theme of ion-pair recognition and introduces outer-sphere recognition of metal complexes, a novel idea with the potential for structural control of solvation, casting a new light on solvent modifiers. Chapter 3 reviews the large literature of calixarenes as extraction reagents for metal ions, where the synthetic versatility of this family of compounds has produced vast possibilities for inclusion and selective separations. Chapter 4 extends such chemistry to the extraction of biomolecules, where the potential for selective separations has only begun to be explored through site recognition in macromolecules. In Chapter 5, a detailed examination of the liquid–liquid interface as an expression of supramolecular phenomena in SX is presented. While most SX chemical research has focused on the complex interactions occurring in the bulk organic solvent, all transport in SX must occur through the liquid–liquid interface, where the sharp gradient of properties between the two immiscible phases serves as a remarkable force for the organization of amphiphilic species, also known as extractant molecules. Finally, Chapter 6 returns to the perennial problem of aggregation in SX and the progress that has been made recently within sight of the historical struggles to understand it.

Beyond this volume and into the future, SC will play a critical role in the future directions of SX. Molecular design will become more sophisticated, to the point that the selective binding of target ions and molecules by new extractants designed computationally will be assured, but a major question in fact concerns the quantitative prediction of equilibrium constants and distribution ratios. To approach this question will entail application of the principles of SC. These must be embodied in computational tools that take into account aggregative interactions, solvation, and hydration of organic phase species, which are only now beginning to be examined in present-day modeling. New extractants will behave more simply, which in effect means that supramolecular behavior will be understood to the point where unwanted aggregation can be avoided or predicted. Not only will the extraction complexes be well behaved, but free extractant molecules will be designed to either not aggregate or aggregate in a controlled fashion so that aggregated structures will be compatible with the solvent matrix. Solvent modifiers, presently used in the form of branched alkanols, alkylamides, or other monofunctional polar molecules, will also yield to computational design so that they will effectively function as selective solvating agents. That is, only the target metal complexes will be solvated and therefore receive selectively enhanced distribution. Self-assembly will become a refined concept for extractant design from small molecular units. Whereas extractants such as cryptands that fully encapsulate target ions have proven to involve difficult organic synthesis, the practicality of extractants that function by encapsulating guest species can be expected to improve if they can be designed to self-assemble from simple units. Self-assembly can be viewed as controlled aggregation, a logical outgrowth of the current body of research aimed at understanding the structural basis of aggregation phenomena in SX. Taking the above points dealing with control of complexation, solvation, aggregation, and phase behavior together, it would appear that the future progress and success in the field of SX in large part will grow out of the principles of supramolecular chemistry.

About the Editor

Bruce A. Moyer, PhD, has built a career in chemical sciences at Oak Ridge National Laboratory, specializing in both fundamental and applied aspects of solvent extraction and ion exchange, spanning 34 years. His interests in fundamental aspects include thermodynamics, equilibrium modeling, and interfacial phenomena from third-phase formation to phase disengagement to crud. He has investigated diverse extractant classes, including amines, sulfoxides, sulfonic acids, carboxylic acids, crown ethers, calixarenes, and various anion receptors such as calixpyrroles, macrocyclic amides, and sulfonamides. Extracted ions examined include alkali metals, alkaline earths, various transition metals, actinides, lanthanides, and a variety of inorganic anions including oxyanions and halides. Dr. Moyer's applied interests have led to studies of problems in uranium milling, cleanup of legacy nuclear waste, and nuclear fuel-cycle separations. His most successful application is the development of the Caustic-Side Solvent Extraction (CSSX) process, which has been operating successfully in the Modular CSSX Unit at the Savannah River Site, processing more than 4 million gallons of high-level waste.

Dr. Moyer has published over 180 open-literature articles, book chapters, proceedings papers, and reports. His patents range from solvent extraction of cesium for nuclear-waste cleanup to supported liquid–membrane systems and novel anion-exchange resins. He has received a number of awards: 2011 Council of Chemical Research Collaboration Award for Development and Implementation of High-Level Salt-Waste Processing Technology (team award); IR-100 Award in 2004 for a highly selective, regenerable perchlorate treatment system; UT-Battelle Technical Achievement Award in 2000 for contributions to the development of novel resin regeneration techniques; three Lockheed Martin Research Corporation Achievement awards in 1999—leadership award, development award (novel bifunctional anion exchange resin), and development award (novel process for cesium separation from waste). Dr. Moyer received his BS in chemistry, summa cum laude, from Duke University in 1974 and his PhD in inorganic chemistry from the University of North Carolina at Chapel Hill in 1979. Currently he serves as co-editor of the journal *Solvent Extraction and Ion Exchange* and this book series, Ion Exchange and Solvent Extraction. In 2008, he served as the technical chair of ISEC 2008 and editor-in-chief of the proceedings, and in 2011 he served as member of the advisory committee, program chair for nuclear separations, and co-editor of the *Proceedings of ISEC 2011*. In addition to his duties as a group leader at Oak Ridge National Laboratory, Dr. Moyer leads the Sigma Team for Minor Actinide Separations for the Fuel Cycle Research and Development program of the United States Department of Energy (USDOE), Office of Nuclear Energy. Recently, he led the diversifying supply focus area of the Critical Materials Institute, a USDOE Energy Innovation Hub led by Ames Lab.

List of Contributors

Mark R. Antonio
Chemical Sciences and Engineering
   Division
Argonne National Laboratory
Argonne, Illinois
antonio@anl.gov

Karsten Gloe
Department of Chemistry and Food
   Chemistry
Technischen Universität Dresden
Dresden, Germany
Karsten.Gloe@chemie.tu-dresden.de

Kerstin Gloe
Department of Chemistry and Food
   Chemistry
Technischen Universität Dresden
Dresden, Germany
kerstin.gloe@chemie.tu-dresden.de

Feng Li
School of Science and Health
University of Western Sydney
Penrith, Australia
Feng.Li@uws.edu.au

Leonard F. Lindoy
School of Chemistry
University of Sydney
Sydney, Australia
L.Lindoy@chem.usyd.edu.au

Mikael Nilsson
Chemical Engineering and Materials
   Science
University of California, Irvine
Irvine, California
nilssonm@uci.edu

Keisuke Ohto
Department of Chemistry and Applied
   Chemistry
Faculty of Science and Engineering
Saga University
Saga, Japan
ohtok@cc.saga-u.ac.jp

Tatsuya Oshima
Department of Applied Chemistry
Faculty of Engineering
University of Miyazaki
Miyazaki, Japan
oshimat@cc.miyazaki-u.ac.jp

Benjamin D. Roach
Chemical Sciences Division
Oak Ridge National Laboratory
Oak Ridge, Tennessee

Peter A. Tasker
School of Chemistry
University of Edinburgh
Edinburgh, United Kingdom
peter.tasker@ed.ac.uk

Hitoshi Watarai
Institute for NanoScience Design
Osaka University
Osaka, Japan
watarai@chem.sci.osaka-u.ac.jp

Marco Wenzel
Department of Chemistry and Food
   Chemistry
Technischen Universität Dresden
Dresden, Germany
marco.wenzel@ikts.fraunhofer.de

Peter R. Zalupski
Aqueous Separations and
   Radiochemistry Department
Idaho National Laboratory
Idaho Falls, Idaho
peter.zalupski@inl.gov

1 Supramolecular Chemistry in Solvent Extraction: Toward Highly Selective Extractants and a Better Understanding of Phase-Transfer Phenomena

Karsten Gloe, Kerstin Gloe, and Marco Wenzel
Department of Chemistry and Food Chemistry,
Technische Universität Dresden,
Dresden, Germany

Leonard F. Lindoy
School of Chemistry, University of Sydney,
Sydney, Australia

Feng Li
School of Science and Health, University of Western Sydney,
Penrith, Australia

CONTENTS

1.1   INTRODUCTION

Solvent extraction provides the basis for important separation and concentration processes that are widely used in hydrometallurgy, waste treatment, effluent purification, and material preparation[1-3]. In a typical extraction process, the complexing agent, the extractant, plays a key role and is responsible for the phase transfer of one or more species of interest between the particular liquid–liquid phases employed. At present, four categories of metal extractants are used in industry: acidic and chelating cation exchanging reagents as well as solvating and anion exchanging reagents. Typical structure examples are shown in Figure 1.1.

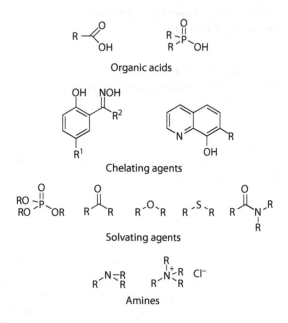

FIGURE 1.1   Examples of extractant types used in industry.

The use of these four reagent types is based on coordination chemistry principles that involve mainly electrostatic interactions between the ionic substrate and the extractant. In the case of chelating agents, both stability and selectivity of the system are often improved, arising from the operation of the entropy-controlled chelate effect. Despite the long-standing successful operation of numerous extraction plants, a number of problems remain to be solved, because of both the emergence of new types of challenging separation problems and expectations for better, more competitive performance by industry. In particular, individual extractants often display less than ideal selectivity for a desired species, and there tends to be limited understanding of structure-property relationships in multicomponent extraction systems. The development of supramolecular chemistry, the chemistry of weak noncovalent interactions, has opened new approaches for understanding the subtleties of solvent extraction. Such understanding can contribute both to the development of new tailor-made extraction systems with improved selectivity, and to the better understanding of the crucial role of weak noncovalent interactions in phase-transfer reactions. This has direct implications for the continuing need in solvent extraction for a rational approach toward the development of selective reagents, reflecting the requirements of a wide range of industrial, environmental, and mineral processing applications. However, the design of such reagents is often far from straightforward given the complexity of a typical two-phase system. Besides the inherent affinity of the extractant (or extractants) for the substrate species of interest, a host of other factors, many of them supramolecular in nature, can influence the outcome of a given extraction process. These include solvation effects, the relative lipophilicities of the various species present, the nature of the diluent and modifier, interfacial phenomena, the ionic strength of the aqueous phase, and the presence or absence of additional interactions (including ion pairing and the association of other species).

An aim of this review is to show how supramolecular concepts may play a role in influencing extraction processes with the focus placed on the more recent work. Since both calixarene-derived extractants and metal salt extractants are treated in other chapters of this book, these areas have been given less emphasis in the discussion that follows.

1.2 SUPRAMOLECULAR CHEMISTRY IN RELATION TO SOLVENT EXTRACTION

1.2.1 What Is Supramolecular Chemistry?

Supramolecular chemistry can be considered to be the chemistry of di- or multicomponent molecular or ionic assemblies in which the ionic and molecular building blocks are held together by weak noncovalent interactions.[4,5] The rise of supramolecular chemistry over the past three decades or so has thus led to a steady appreciation of the role of such noncovalent interactions in forming larger assemblies. These weak interactions include hydrogen bonding, van der Waals forces, π-π stacking, π-cation and π-anion interactions, as well as the full complement of electrostatic (ion-ion,

---

TABLE 1.1

Strengths of Supramolecular Interactions in kJ/mol

Coordinate bonding	40–120
Electrostatic interactions (ion–ion, ion–dipole, dipole–dipole)	5–300
Hydrogen bonding	4–60
π–π interactions	5–50
van der Waals forces	1–5
Solvophobic effects	2–15

---

ion-dipole, dipole-dipole) interactions. Hydrophobic-hydrophilic (solvophobic) inter-actions can also play their part in a given supramolecular assembly process. Usually, coordinate bonds, which characteristically occur in many metal complexes, are also included in the supramolecular realm, even though their properties and bond energies are normally quite different from the other (weak) interactions mentioned above. The respective interaction energies are listed in Table 1.1.[5–7]

Overall, these are the same interactions that nature uses to assemble its vast array of self-assembled biostructures. In forming such assemblies, a degree of both steric and electronic complementarity is necessary between the component building blocks in order that initially recognition and then assembly can occur. A further important common feature of supramolecular systems is that their formation is reversible, reflecting the presence of the weak binding interactions just mentioned. This results in the ability to undergo error correction during an assembly process so that the most stable thermodynamic product is often gener-ated in high yield.[5,7,8]

1.2.2 SUPRAMOLECULAR ASPECTS OF SOLVENT EXTRACTION SYSTEMS

In the remainder of this chapter, we examine examples of selected two-phase solvent extraction systems, with emphasis on supramolecular aspects of their behavior. The treatment is illustrative rather than comprehensive. However, it needs to be stated that it is often difficult to define exactly what constitutes "supramolecular" behavior in this context, and the meaning emerges mainly by example. It is fair to point out that solvent extraction has always been dominated by supramolecular interactions, as simple lipophilic extractant molecules were used initially to extract ionic species through classical complexation equilibria. Since these processes often entailed complex and uncontrolled behavior, with resulting unpredictable affinity and selectivity, the object of deliberate supramo-lecular approaches in more recent years has been to simplify behavior and gain control through appropriately designed multidentate extractants.

Many supramolecular extractants possess a number of donor functions in a preorganized arrangement for selectively binding a substrate. Some systems of this type are shown in Figure 1.2. Early examples were the macrocyclic crown compounds and the related macrobicyclic cryptands, which normally show strong

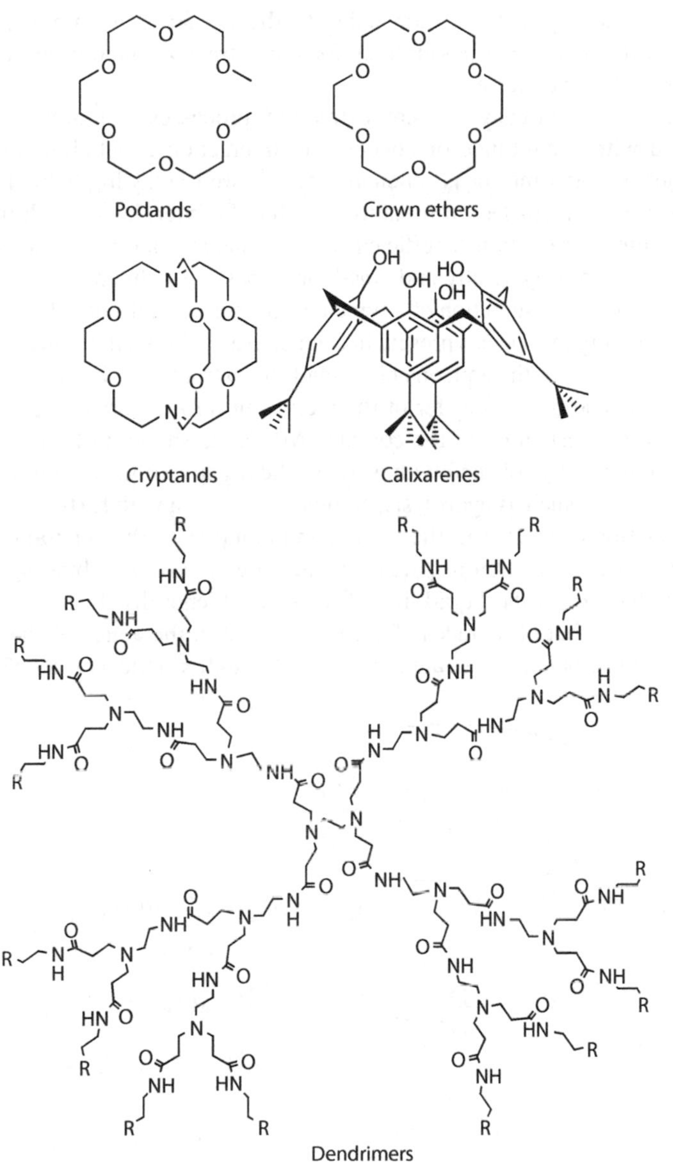

Podands

Crown ethers

Cryptands

Calixarenes

Dendrimers

FIGURE 1.2 Chemical structures of some typical supramolecular extractants representing different extractant categories.

binding and efficient extraction properties toward alkali and alkaline earth metal ions.[9–11] A further development was directed toward the synthesis and use of related open-chain counterparts (podands); however, these are mostly weaker ligands and bind cations with less selectivity than corresponding macrocyclic systems.[12] A range of other complexing agent types, such as calixarene derivatives[13–15] and dendrimer-based systems, are amenable to diverse structure modifications and open up wide application possibilities for the extractive separation of particular ionic and molecular species.

As mentioned already, solvent extraction processes are almost invariably associated with a multitude of noncovalent interactions, including solvation,[16,17] association,[18–21] and interfacial phenomena,[22] as well as hydrophobic-hydrophilic interactions.[23] Nevertheless, it remains true that the interplay of such interactions on, for example, phase-transfer efficiency in the majority of reported solvent extraction studies remains generally undefined (or, at best, ill-defined).

A schematic representation of a typical two-phase metal salt solvent extraction process, showing the species present in each phase, is shown in Figure 1.3. Initially the metal salt MX in the aqueous phase and the extractant L in the organic phase are solvated. The phase transfer of the metal salt into the organic phase is associated with the formation of the complex ML_sX, as shown in Figure 1.3. These species are normally solvated with water or the organic diluent. Additionally, further interactions, such as hydrolysis, complex formation with further components, and association can occur in the system and complicate the general description. As shown, L can partition in part to the aqueous phase, and binding can occur in either phase or at the liquid–liquid interface. Generally, L is designed to be hydrophobic so that only traces of L can be found in the aqueous phase, thereby minimizing losses in process applications. Although kinetic aspects of extraction

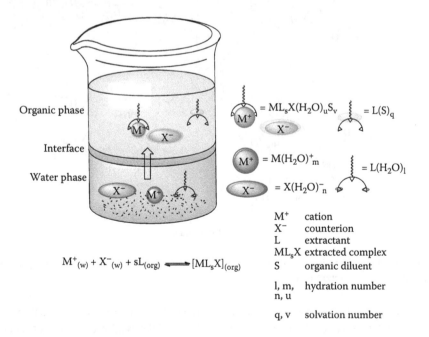

Organic phase M^+ $= ML_sX(H_2O)_uS_v$ $= L(S)_q$

Interface M^+ $= M(H_2O)^+_m$ $= L(H_2O)_l$

Water phase X^- $= X(H_2O)^-_n$

$$M^+_{(w)} + X^-_{(w)} + sL_{(org)} \rightleftharpoons [ML_sX]_{(org)}$$

M^+	cation
X^-	counterion
L	extractant
ML_sX	extracted complex
S	organic diluent
l, m, n, u	hydration number
q, v	solvation number

FIGURE 1.3 Interactions in a solvent extraction system employing a solvating podand as extractant, illustrating the species in the aqueous and organic phases after equilibrium is achieved.

are beyond the scope of this review, it is generally accepted that as the extractant is made sufficiently hydrophobic, it becomes negligibly partitioned to the aqueous phase, and the binding step occurs mainly at the interface.

In the following discussion, we will concentrate on examples of solvent extraction where supramolecular receptors were employed or supramolecular influences have been either identified or postulated to occur.

Clearly both the use of rationally designed receptors and an appreciation of the interactions involved, coupled with the ability to moderate supramolecular influences, show much potential for achieving more favorable control of the selectivity and efficiency of separation processes.[24–27]

1.3   SELECTED EXAMPLES ILLUSTRATING SUPRAMOLECULAR CONCEPTS IN ACTION

A classification of supramolecular cation, anion, and metal salt extraction processes due to the authors is shown in Figures 1.4 and 1.5.[28,29] Nine different approaches have been identified for supramolecular single- (cation or anion) and

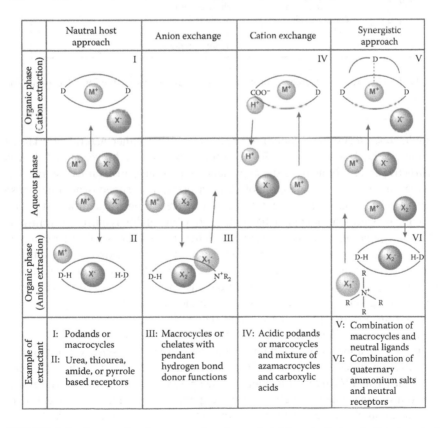

FIGURE 1.4   Supramolecular approaches for undertaking single-ion (cation or anion) extraction.

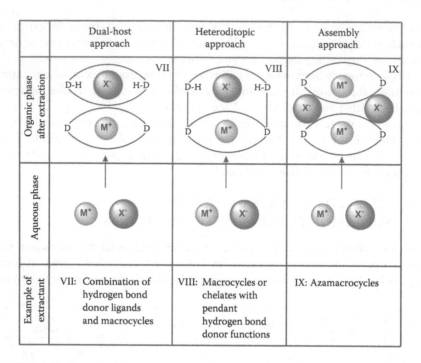

FIGURE 1.5 Supramolecular approaches for simultaneous extraction of cations and anions.

simultaneous (cation and anion) ion extraction using either neutral or charged ligands. This classification is similar to that presented by Moyer et al.[30,31]. It is instructive to briefly analyze the approaches represented in Figures 1.4 and 1.5, giving emphasis to their supramolecular aspects. For aqueous-organic phase transfer, charge neutrality needs to be maintained. Thus, a neutral extractant system, as is present in approaches **I**, **II**, and **V**, inevitably requires co-transport of a corresponding counterion together with the charged supramolecular complex into the organic phase. As a consequence, an efficacious extraction of either a cation or an anion will depend strongly on the lipophilicity of the counterion. In contrast, this is not a requirement in the case of the cation or anion exchange approaches given by **IV**, **III**, and **VI**, where the extraction is accompanied by a countertransport of equally charged ions, H^+ or X^-.

The approach **IVb** illustrates a synergistic cation exchange system based on preorganization of an optimum metal ion extractant assembly using a self-organization process between an azacrown macrocycle and a carboxylic acid. Examples **V** and **VI** are modifications of the approaches given by **I** and **II**, respectively, involving synergistic effects that are generated through the introduction of a second complexing agent in the organic phase. Three supramolecular extraction approaches concerned with the simultaneous extraction of cations and anions (**VII**, **VIII**, and **IX**) are illustrated in Figure 1.5. For the dual-host approach, **VII**, two ligands are present for binding each ion type, whereas the approach given by **VIII** is based on the presence of two different binding sites in one heteroditopic ligand molecule. Here, application of the principle exemplified by **VII** may involve less synthetic effort, because it is based on the use of single-function complexing agents that are often more readily available. In contrast, approach **VIII** may lead to enhanced extraction selectivity and efficiency through the presence of direct electrostatic interaction (contact ion pairing) between the bound cation and anion. Furthermore, specific self-organization processes involving defined metal binding sites may favor the effective phase transfer of assemblies incorporating metal salts, as shown in the approach illustrated by **IX**.

1.3.1 CATION BINDING AND EXTRACTION

1.3.1.1 Solvent Extraction Employing Mixed Extractants

As has been well established over many decades, synergistic effects arising from combinations of extractants in the organic phase may lead to both enhanced extraction selectivity *and* more efficient metal extraction.[1-3] In early studies, the use of a combination of extractants to ensure the saturation of the coordination sphere of an extracted metal ion (leading in turn to increased lipophilicity of the extracted complex) was proposed to promote synergistic extraction by O- or S-macrocycles in the presence of carboxylic, phosphoric, or sulfonic acids.[32-34] Such solution behavior was supported to some degree by the solid-state X-ray structures of corresponding metal complexes in particular cases. Although this concept of "synergized cation exchange" was successful in expanding the potential uses of neutral crown ethers in solvent extraction, it proved to be a complicated matter to understand and control the resulting complicated extraction equilibria, which tend to involve unpredictable organic-phase aggregation phenomena.

In other studies, lipophilic tetraalkylammonium salts of organic acids, termed acid-base couples or binary extractants, have been shown to act as efficient extractants for selected metal ions.[35-40] Such binary extraction systems have been employed for the recovery and separation of individual nonferrous, rare earth, noble, and associated metals. In one such application, the binary extractant, tetraoctylammonium dialkyldithiophosphate, was employed for the extraction of copper from leach solutions of oxidized ores. The extraction was characterized by a large separation factor for this metal relative to iron, as well as exhibiting fast extraction and stripping kinetics. However, while ion pairing is undoubtedly involved, the overall rationale for the enhanced performance for this and related binary extraction systems remains somewhat unclear. Nevertheless, the possibility that something akin to an assembly effect is contributing to the observed enhanced behavior cannot be ruled out.

The formation of supramolecular assemblies has long been recognized as a potentially useful approach for controlling both selectivity and efficiency in metal separation processes.[41,42] In this vein, it was realized that supramolecular host-guest assembly between ligand moieties in the required stoichiometry for binding the metal ion of interest can lead to enhanced metal ion binding. This led to the concept of an assembly effect.[43–48] Namely, the use of a discrete ligand assembly of this type has the potential to result in enhanced complex stability provided the back-equilibrium to the free (separated) ligands is unfavorable. In one sense, the operation of the assembly effect can be considered to be associated with a degree of preorganization of the coordination sphere, and thus there is less loss of disorder on metal complexation relative to the (hypothetical) situation occurring if the component ligands were not assembled. Ligand assembly is thus expected to result in a positive contribution to the overall entropy of complexation.

In a number of cases, it has been possible to characterize metal-free ligand assemblies in both solution and the solid state that show stoichiometries corresponding to potential metal coordination spheres.[43–48] In particular, ^{1}H nuclear magnetic resonance (NMR) titration studies have been employed to detect host-guest formation between potential ligands in $CDCl_3$, as well as for defining the nature of the host-guest interaction sites; it has also enabled the stoichiometry of the resulting assemblies to be determined. In a series of studies, a range of both polyamine and mixed-donor (O_mN_n) macrocycles were shown to form 1:1 and 2:1 adducts with *tert*-butylbenzoic or palmitic (hexadecanoic) acid under the conditions employed.[43,44,48,49] Interestingly, the maximum observed stoichiometry of the respective adducts corresponded to the number of amine sites in the host ligand that have (log) protonation constants equal to or greater than 6–7 in aqueous or aqueous-methanol media.

The application of the assembly effect to the solvent extraction of metal ions in two-phase systems has characteristically been associated with the formation of a ligand assembly in the organic phase via proton transfer from an organic acid to an amine-containing ligand (often a macrocycle), resulting in charge-separated hydrogen bond interactions (see also later) between these ligand components:

$$L_{(org)} + nRXH_{(org)} \leftrightharpoons [LH_n(RX)_n]_{(org)}$$

Metal ion extraction then occurs via phase transfer accompanied by proton loss from the ligand assembly:

$$[LH_n(RX)_n]_{(org)} + M^{n+}_{(w)} \leftrightharpoons [LM(RX)_n]_{(org)} + nH^+_{(w)}$$

For example, when the nonacid ligand component is a macrocycle, the process may be illustrated as shown in Figure 1.6.

Namely, the formation of an assembled extractant "package" of appropriate stoichiometry is postulated to occur in the organic phase, resulting in synergistic solvent extraction being engendered. The stabilities of the species involved in these equilibria depend upon such factors as the nature of the acid (lipophilicity, pK_a), the nature (including the basicity and hydrophobicity) of the amine ligand, the coordination requirements of the metal ion, the intrinsic affinities of the metal ion for the ligands, the pH of the aqueous phase, and the nature of the solvent employed for the organic phase.

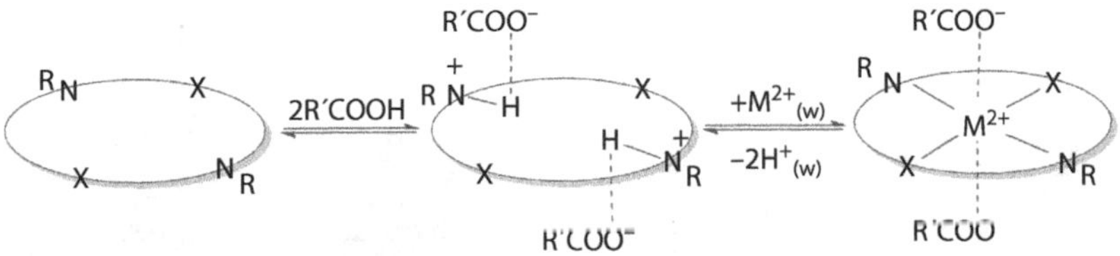

FIGURE 1.6 Operation of an assembly effect in the extraction of a metal cation by an azamacrocycle–carboxylic acid ligand combination.

Besides offering a basis for the rational design of improved metal uptake systems, the assembly concept also yields the possibility of rationalizing a range of previously reported metal ion extraction behavior. Thus, a significant number of examples of the use of synergistic combinations of amine macrocycles and organic acid extractants have been reported over many years; at least in some instances, these are very likely associated with the operation of an assembly effect.[39,50–56]

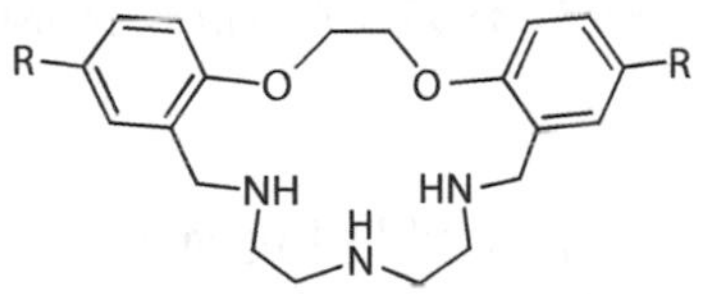

1, R = H
2, R – C(CH₃)₃
3, R = C₉H₁₉

In representative studies, ¹H and ¹³C NMR titrations have demonstrated the formation of ligand assemblies in CDCl₃ between the macrocycles **1–3** and *tert*-butylbenzoic, palmitic, phenylphosphinic, diphenylphosphinic, and salicylic acids.[45,48] For example, the addition of palmitic (hexadecanoic) acid to **1** led to induced shifts for the signals corresponding to the methylene protons adjacent to the nitrogen donors in keeping with the presence of significant interaction between the NH groups and palmitic acid. Similar behavior was observed when 4-*tert*-butylbenzoic acid was substituted for palmitic acid. The solution behavior in this latter case parallels the situation confirmed in the solid state by an X-ray structure determination of the corresponding crystalline 2:1 assembly between 4-*tert*-butylbenzoic acid and **1** (R = H). The carboxylic groups interact directly via hydrogen bonds with two amine nitrogens of the macrocycle as well as with an included water molecule.[45] In this adduct, based on NMR studies and precedence from the parallel study involving cyclam mentioned above, carboxylic protons from the 4-*tert*-butylbenzoic acid moieties are postulated to be essentially transferred to the two nonadjacent amine sites on the N₄-cyclam

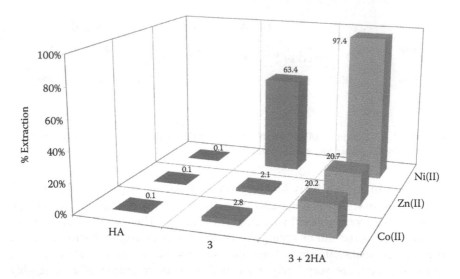

FIGURE 1.7 Comparison of the percent metal extraction into a chloroform organic phase by the ligand assembly (**3** + 2HA) (see text), where HA = 4-*tert*-butylbenzoic acid; $[M(NO_3)_2] = 1 \times 10^{-4}$ M; pH = 5.5 (MES/NaOH buffer); $[NaNO_3] = 1$ M; $[HA] = 2 \times 10^{-3}$; and **3** = 1×10^{-3} M.

host such that two charge separated hydrogen bonds are present between the protonated amines and the 4-*tert*-benzoate anions.

Following confirmation that 2:1 (carboxylic acid:macrocycle) ligand assemblies of the above type based on ligands **1–3** readily form in CDCl$_3$, studies were undertaken to assess the efficacy of the individual assemblies (water-chloroform) for the synergistic extraction of Co(II), Ni(II), Cu(II), and Zn(II) from an aqueous phase into a chloroform phase.[48] The investigation involving the lipophilic derivative **3** and 4-*tert*-butylbenzoic acid is representative.

Extraction results have been reported for Co(II), Ni(II), Cu(II), and Zn(II) for the system $M(NO_3)_2$-$NaNO_3$-MES/NaOH buffer/**3**-HA-CHCl$_3$ (where MES = 2-[*N*-morpholino]ethanesulfonic acid and HA = 4-*tert*-butylbenzoic acid). The results are presented in Figure 1.7 and provide clear evidence for the existence of strong synergistic effects in all cases. The order of increasing extraction for this ligand assembly is Ni(II) > Co(II) ~ Zn(II).

A subsequent solvent extraction experiment for Cu(II) in which the ionic strength of the aqueous phase was not adjusted (but otherwise under similar conditions to those presented above) also gave clear evidence for strong synergistic extraction.

1.3.1.2 Dendritic Cation Extractants

Dendrimers are defined, highly branched polymers with a spherical globular shape in solution. The structural similarity to natural proteins offers a comparable host-guest chemistry that can be modified by additional internal or peripheral end groups. Such multifunctional systems can be used for different binding processes of cationic and anionic species. While the extraction behavior of dendritic systems can be readily documented experimentally, a quantitative understanding of the extraction process in such systems has proven to be less than straightforward. For example, Fréchet-type poly(benzyl ether) dendrons functionalized with crown ether receptor sites have been employed as extractants for Li^+, Na^+, K^+, Rb^+, and Cs^+ picrates from alkaline solution. In all, nine different dendrimers (**4–12** in Figure 1.8) were investigated.[57]

Parallel extraction experiments employing the free crown ethers, benzo-12-crown-4, benzo-15-crown-5, and benzo-18-crown-6, were also performed. In general, dendrimer extraction of the alkali metal ions showed an increase in efficiency on moving from the parent crown ethers to the first-generation dendrimers, but a gradual decrease was observed on moving from the first generation to the second and third generations. Thus, both positive and negative dendritic effects were in evidence. The reasons for such behavior remain somewhat unclear but could reflect statistical and proximity (including charge

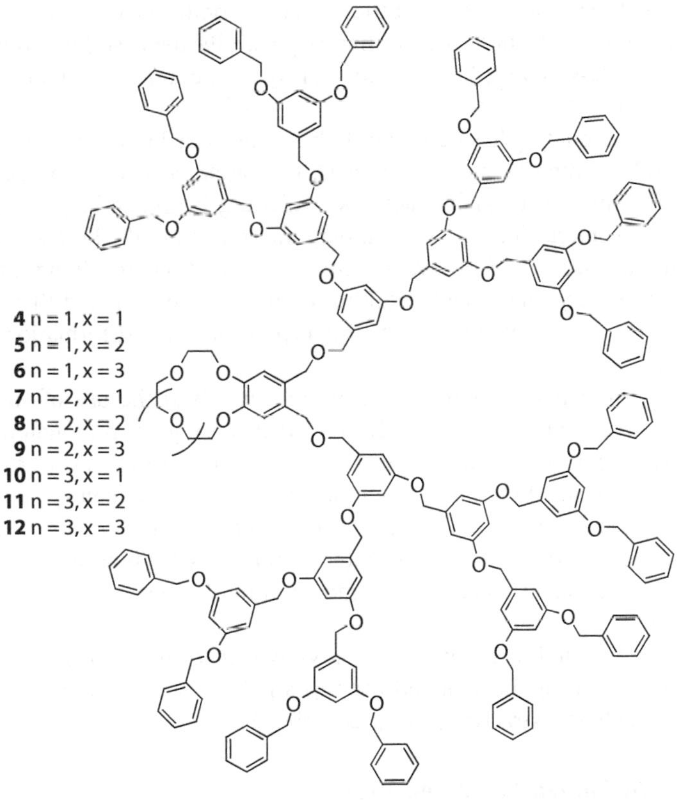

4 n = 1, x = 1
5 n = 1, x = 2
6 n = 1, x = 3
7 n = 2, x = 1
8 n = 2, x = 2
9 n = 2, x = 3
10 n = 3, x = 1
11 n = 3, x = 2
12 n = 3, x = 3

FIGURE 1.8 The crown ether functionalized dendrimers showing the first-, second-, and third-generation structures.

buildup) effects. The authors suggest that in the case of the first-generation den-drimers, the donor atoms located in the dendritic "arms" are able to bind with the alkali metal centers, while this does not occur in the second- and third-gen-eration systems due to steric crowding. Also, it was suggested that in the latter systems, steric crowding may also inhibit access to the crown ether cavity, thus effectively lowering its metal binding ability. It is also possible the environment for the large picrate counterion will be changed with increasing generation of the dendrimer.

1.3.2 ANION BINDING AND EXTRACTION

1.3.2.1 General Considerations

It is noted that spherical anions are generally larger than the corresponding iso-electronic cations [Na^+ (116)/F^- (119), K^+ (152)/Cl^- (167), Rb^+ (166)/Br^- (182), Cs^+ (181)/I^- (206 pm)], and thus have a lower charge-to-radius ratio. Also, many anions have well-defined nonspherical shapes and typically are associated with high solvation energies. In some cases they can undergo acid-base reactions, giv-ing rise to pH dependence. Each of these factors combines to make simple anion extractants more difficult to design than is generally the case for cation extract-ants. Nevertheless, a very large number of synthetic anion receptors have now been reported.[58–65]

While anion binding has traditionally been discussed in terms of electrostatic (ion-ion and ion-dipole) interactions, it is, of course, also well appreciated that other weak supramolecular interactions, including hydrogen bonding and other van der Waals interactions, very often play major roles in the host-guest chemis-try of anionic species.[66–68] Favorable anion-π interactions involving unsaturated receptors may also contribute[69–78] (although it needs to be noted that the wide-spread significance of interactions of this type in host-guest formation has been contested).[79]

Much of the early work on anion binders involved the use of simple neu-tral organic receptors in organic solvents, while a range of protonated polyaza-containing receptors, such as protonated azacrown macrocycles, were typically employed for anion binding in aqueous media.[80] Subsequently, new classes of anion binders were developed. These include "metal complex receptors" incorporating coordinatively unsaturated metal sites for binding the anion,[81,82] interlocked (catenated) receptors,[83] and supramolecular self-assembled cage receptors.[84–90]

Finally, it is noted that a number of review articles have appeared that cover aspects of both the application and the supramolecular aspects of anion binding receptors in solvent extraction processes.[29,80,91–95]

1.3.2.2 Uncharged Anion Receptors

Over recent years a variety of receptors containing neutral hydrogen bond donors capable of selectively complexing anions in *aqueous* solution have been reported. Such species are hence capable of overcoming the considerable enthalpic cost associated with dehydration of the strongly hydrated anions (see later). In general, this has been very often achieved by receptor design that results in strong anion bonding such that the receptor (or receptors) effectively encapsulates the anion, with the latter being surrounded by sufficient donor sites to cause coordination saturation. In one earlier example, Kubik et al.[96] reported receptors based on linked pairs of cyclic peptide subunits (see **13**) that fold to encapsulate and strongly complex a sulfate ion. In the case of **13**, a binding constant for this ion of 3.5×10^5 in 1:1 methanol/water was determined.

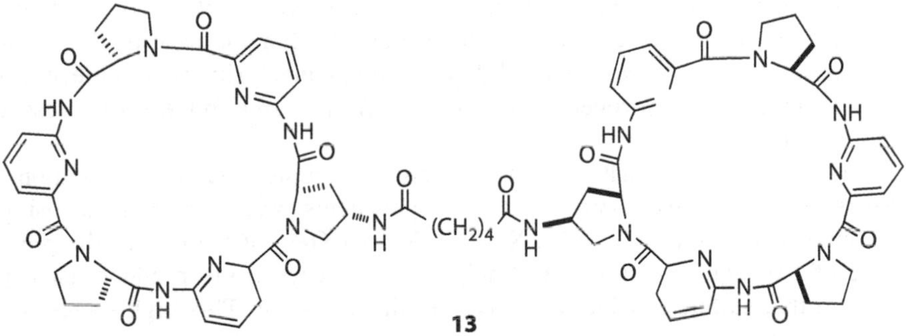

13

Uncharged receptors for HSO_4^- and $H_2PO_4^-$ have been reported over recent years, typically incorporating binding sites consisting of N–H groups[97,98] or urea derivatives.[99] Similarly, other neutral amine-containing systems have been demonstrated to bind SO_4^{2-}.[100–105] For example, a number of tripodal tris-urea derivatives derived from tris(2-aminoethyl)amine of general type **14** (where R can be a number of substituent types) have been demonstrated to bind this dianion, with two molecules of **14** essentially encapsulating the sulfate ion in each case.[100–104] At least in some of these systems, the SO_4^{2-} anion is surrounded by its maximum of 12 hydrogen bonds—even though in particular instances the orientation of such bonds is less than ideal. Nevertheless, the ability to essentially saturate the anion in terms of the number of hydrogen bond linkages is clearly important[62] in maximizing host-guest binding, and when steric (directional) complementarity also occurs, binding strength will be further enhanced.

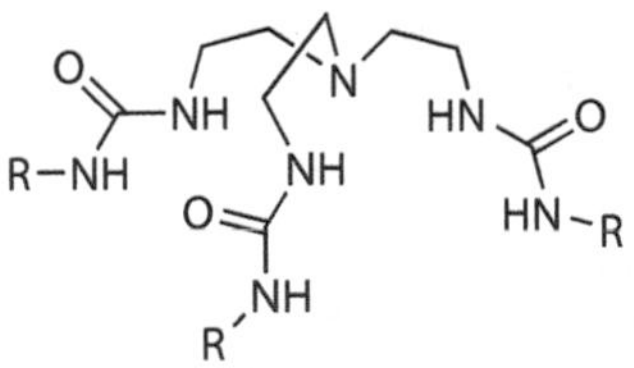

14

In a further study, the degree of anionic guest complementarity in the crystalline state has been employed as a guide to anion selectivity with respect to SO_4^{2-} over SO_3^{2-}, CO_3^{2-}, and SeO_4^{2-}.[101] The crystallization of the neutral receptor with various MSO_4 salts from 1:1 H_2O/MeOH solutions resulted in single crystals with a stoichiometry of ML_2(dianion)$(H_2O)_6$, where L = **14** (R = 3-pyridyl). The X-ray structures of these products showed that they are essentially isostructural in that the resulting similar framework persists in the presence of the different-shaped (or different-sized in the case of SeO_4^{2-}) anions. In each structure, one anion exists inside a cage-like structure composed of two **14** (R = 3-pyridyl) units, with the included anion surrounded by an array of hydrogen bonds. Interestingly, no crystalline products were obtained when similar crystallization experiments were performed using singly charged anions instead of the above dianions. For the case of the tetrahedral sulfate guest, excellent solid-state complementarity was achieved; a regular array of 12 hydrogen bonds to SO_4^{2-} from 6 urea functions is present in this structure.

In order to investigate sulfate selectivity against the above dianions, pairwise competitive crystallization experiments were undertaken using 1:1 mixtures of $MgSO_4$ and Na_2X (X = SO_3^{2-}, CO_3^{2-}, or SeO_4^{2-}) in the presence of two equivalents of **14**. Analysis of the crystalline products clearly showed that sulfate is selectively taken up in each case. This high selectivity for SO_4^{2-} was attributed to its near ideal complementarity for the available urea binding sites.

In an extension to the above studies, the use of the tripodal hexa-urea host **15** (see Figure 1.9) has been demonstrated to be an efficient extractant for SO_4^{2-}.[106] In this case, the results are in accord with the *ortho*-substituted phenyl bridge serving as a "corner" in a tetrahedral coordination shell that is

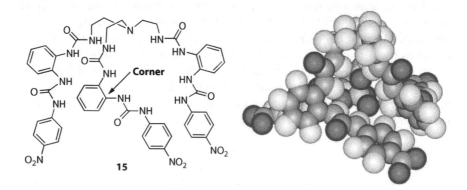

FIGURE 1.9 X-ray structure of the SO_4^{2-} complex **15** showing how **15** wraps around a SO_4^{2-} anion in a tetrahedral fashion such that 12 urea (NH) donors are directed inwards toward this ion. (From Jia, C. D. et al., *Angew. Chem. Int. Ed.,* 50, 486–490, 2011.)

capable of surrounding either a single SO_4^{2-} or PO_4^{3-} ion. It was assumed that the tris-chelating nature of **15** might enhance binding due to the operation of a chelate effect in a parallel manner to that observed in metal ion receptor chemistry involving chelating ligand systems. In this study, it proved possible to crystallize the 1:1 complex $(NBu_4)_2[(15)(SO_4)]\cdot DMSO$ (where DMSO is dimethyl-sulfoxide) by adding **15** to excess $(NBu_4)_2SO_4$ in a water-DMSO mixture. An X-ray crystal structure (Figure 1.9) confirmed the tetrahedral arrangement of bound **15**; all 6 urea functions are bound to the SO_4^{2-} ion such that 12 hydrogen bonds are again formed, with the N–O distances ranging from 2.90 to 3.16 Å. The bound anion is thus locked in a cavity that is surrounded by aryl rings. Three T-shaped CH···π interactions between the terminal *p*-nitrophenyl and nearby 1,2-substituted aryl groups also act to stabilize this structure. The "outside" of the bound ligand is somewhat hydrophilic due to the exo-orientation of the urea and nitro oxygen atoms.

^{1}H NMR studies of the above system have been carried out in DMSO/0.5% D_2O. The results are in agreement with the presence of a two-step sulfate binding mechanism that involves 1:1 binding followed by 1:2 (host:guest) binding of the anion. In extraction studies, sulfate was almost quantitatively transferred from an aqueous phase into an organic phase. Competitive extractions employing other anions were also undertaken in the presence of 25% D_2O, with a selectivity sequence of SO_4^{2-} > $H_2PO_4^-$ >> other anions being established. Overall, this study exemplifies the manner by which a combination of host-guest complementarity, coupled with the use of both the chelate and hydrophobic effects, provides a strategy for achieving enhanced anion binding.

1.3.2.3 Charged Anion Receptors

As discussed above, there has been much interest in the design and synthesis of sulfate binding receptors—due, in part, to the considerable importance of this dianion in particular environmental and biological systems.[59,66,83,107-114] For example, one application relates to the need for a selective anion binding extractant for the removal of sulfate from nitrate-containing solutions for the remediation of nuclear waste.[106] However, such a separation is not without considerable difficulty, since there is a higher concentration of NO_3^- in nuclear waste, and a successful separation must overcome sulfate's extremely large hydration energy (-1080 kJ mol^{-1}) relative to NO_3^- (-300 kJ mol^{-1}).[115] That is, the Hofmeister bias[116-118] (which, of course, significantly favors the extraction of the less hydrophilic NO_3^- over SO_4^{2-} from water) needs to be overcome.

A range of examples of positively charged SO_4^{2-} binding receptors have been reported, very often incorporating binding sites consisting of N–H$^+$ groups (see below). Positively charged receptors, of course, provide the opportunity for direct charge neutralization without the need for the presence of separate counterion(s).

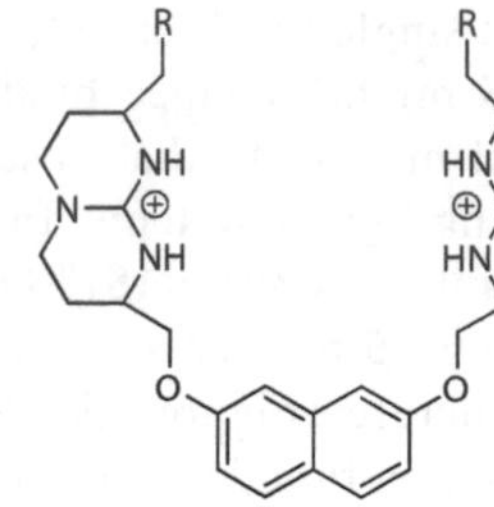

16

In an early study, it was demonstrated that a range of lipophilic open-chain bis-guanidinium receptors that included ditopic species of type **16**, incorporating bicyclic guanidines as the receptor unit, act as efficient 1:1 extractants for the strongly hydrated tetrahedral anions HPO_4^{2-} and SO_4^{2-} over a wide pH range,[119] despite both these anions lying at the "hydrophilic" end of the Hofmeister series.[116,117] The study confirmed that such appropriately designed, positively charged receptors result in strong anion binding reflecting both overall electrostatic and spatial complementarity. Indeed, for **16** with R = −OSi(*tert*-Bu) $(C_6H_5)_2$, surprisingly high selectivity for SO_4^{2-} over HPO_4^{2-} was observed. While these are relatively similar anions with respect to their size and Lewis basicity, it was suggested that the observed extraction selectivity for SO_4^{2-} arises from the lower hydration free energy for this ion,[115] coupled with the fact that HPO_4^{2-} has an additional proton that is available for forming further hydrogen bonds to water, hence increasing the effective hydrophilicity of its host-guest complex compared to that of sulfate.

A further doubly positively charged receptor of the above general type incorporating two guanidinium binding sites has also been demonstrated to be an efficient extractant of SO_4^{2-} from water into chloroform. Once again, the stoichiometry of the resulting sulfate complex is 1:1 (see **17**). This receptor is also a highly efficient extractant of I⁻ and adenosine triphosphate (ATP).[91,120]

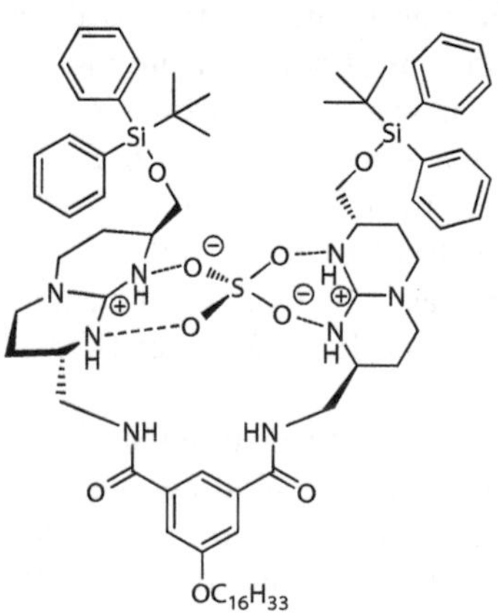

17

More recently, the diprotonated lipophilic macrocyclic cyclo[8]pyrrole derivative **18**, bearing eight undecyl chains on its pyrrolic backbone, was demonstrated to be an efficient extractant for SO_4^{2-} from a neutral aqueous phase into a toluene phase. This macrocyclic receptor thus has eight hydrogen bond donors pointing into its cavity and, coupled with its double positive charge, appeared ideal for binding a polyoxo dianion of suitable size. Indeed, **18** was claimed to be the first synthetic receptor to show high selectivity for sulfate in the presence of excess nitrate under conditions suitable for their separation by solvent extraction.[121] Initially, the kinetics of SO_4^{2-} transfer between the two phases was observed to be very slow. However, in the presence of the commercial phase-transfer catalyst, Aliquat 336-nitrate (A336N), a lipophilic quaternary ammonium salt, the exchange kinetics was greatly improved. Subjecting the toluene phase containing bound sulfate (0.5 mM) and 0.1 mM trioctylamine (TOA) to successive equilibrations with aqueous 0.1 M HNO_3 (until sulfate was no longer detected in the aqueous phase) led to generation of the bis-nitrate complex of **18**. This bis-nitrate complex in toluene in the presence of $(TOAH)NO_3$ is itself an effective extractant for sulfate, confirming the selectivity of **18** for this ion.

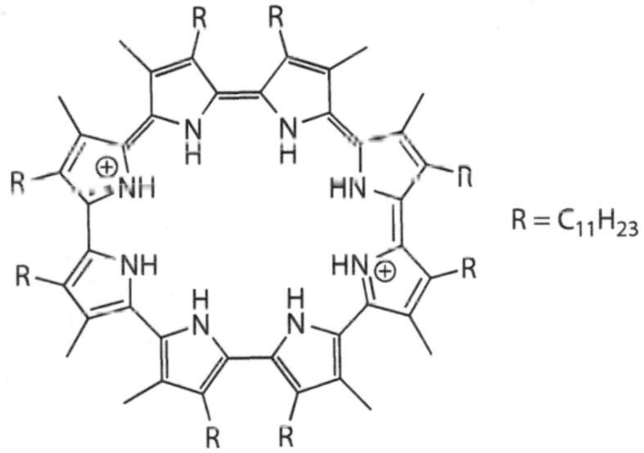

$R = C_{11}H_{23}$

18

Anion exchange is now an established industrial process for achieving anion separation,[122,123] and the use of anion exchange to achieve selective sulfate extraction, as well as for overcoming the Hofmeister bias, has been discussed.[110,124,125] In one study of this type, the neutral hydrogen bonding anion receptors of types **19–21** were employed in conjunction with the quaternary ammonium extractant Aliquat 336-nitrate {$CH_3N[CH_2)_{7-9}(CH_3)]_3NO_3$; A336N}. Each of the fluorinated calixpyrroles **19** and **20** and the tetraamide macrocycles **21** (with R^1 = H, R^2 = dansyl or R^1 = tert-butyl, R^2 = CH_3), but not the less lipophilic derivative **21** (with R^1 = H, R^2 = CH_3), were demonstrated to enhance the solvent extraction of sulfate over nitrate in conjunction with the above quaternary cation.[113]

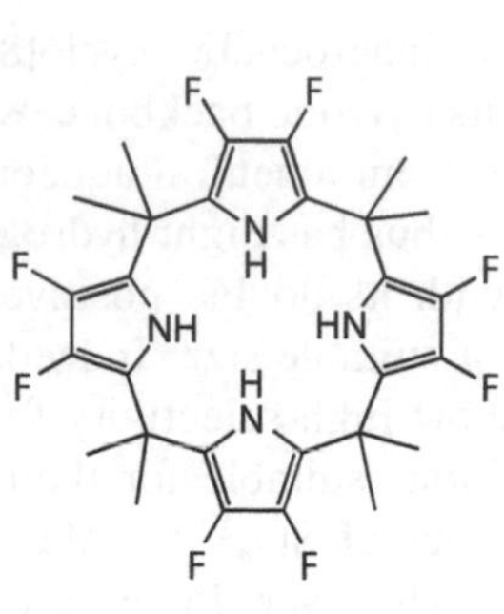

19

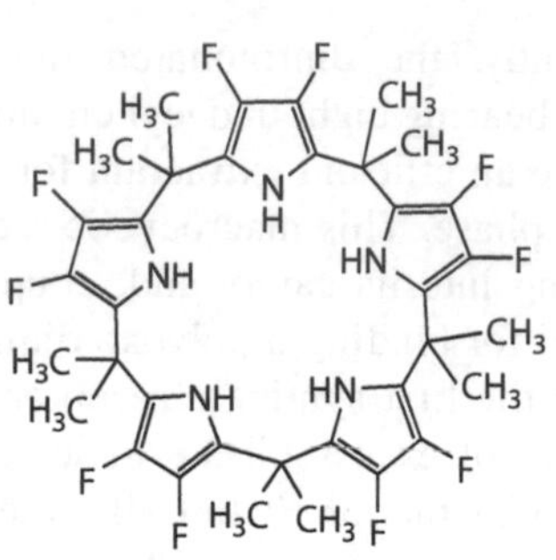

20

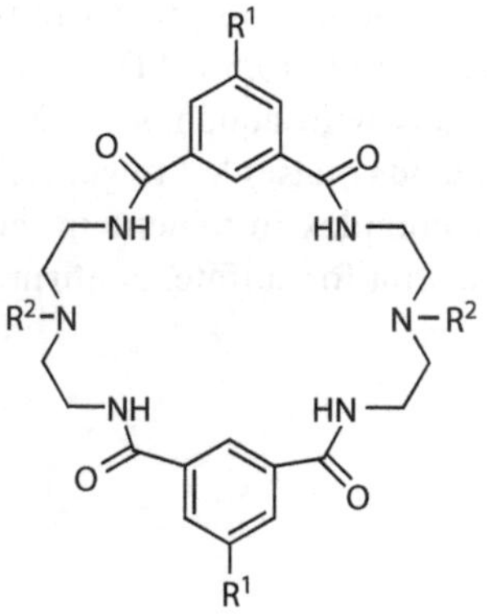

21

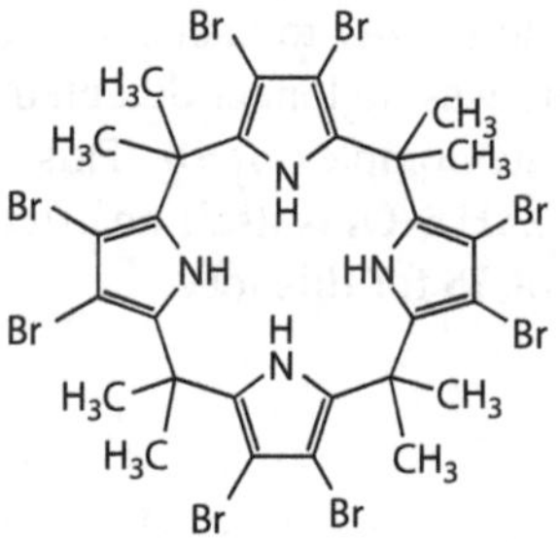

22

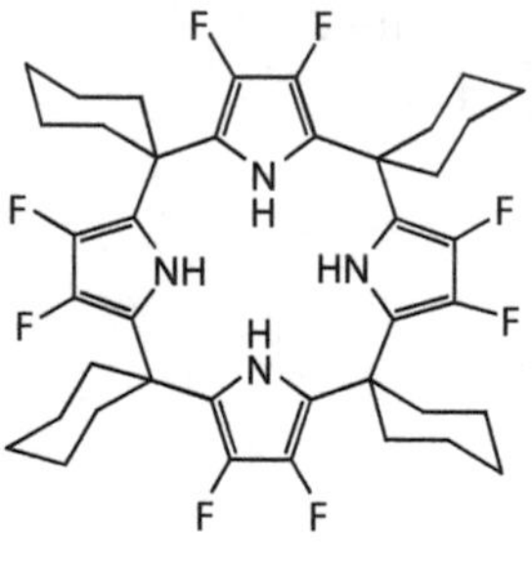

23

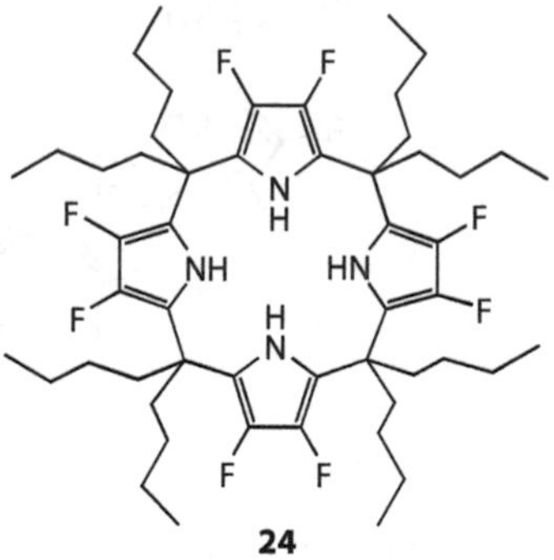

24

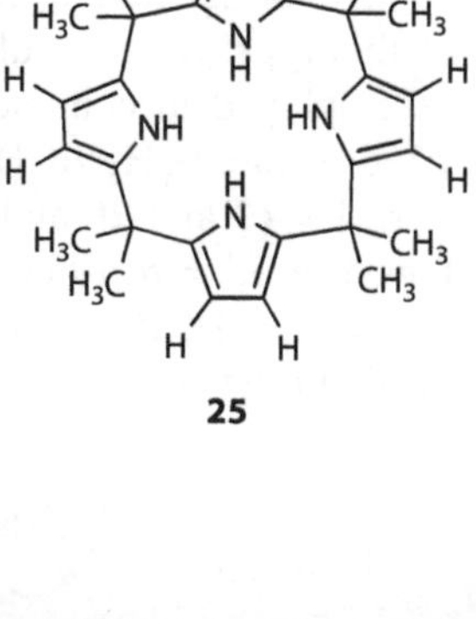

25

In the case of **19** and **20**, both toluene and chloroform were employed as the organic-phase diluents in order to minimize competing anion solvent interactions. Even so, significant differences in extraction behavior were observed: in toluene the extraction efficiency of these two receptors was almost identical, while in chloroform the larger ring **20** was 10 times more effective than **19**. Such a difference presumably reflects subtle supramolecular sulfate influences resulting in different speciation in the respective diluents. Nevertheless, where observed, the extraction enhancement was ascribed to specific binding between these fluorinated calixpyrroles and the sulfate anion, although in this study the exact speciation present in the respective organic phases was not elucidated.

Subsequently, this study was extended to the use of other neutral macrocyclic receptors that included two tetraamide derivatives **21** (with R^1 = H, R^2 = dansyl and R^1 = *tert*-butyl, R^2 = CH_3) and a further four sulf[4]pyrrole derivatives **22–25**. These, as well as **19** and **20**, were compared in liquid–liquid extraction studies. Once again, individual members of these cyclic species (at 0.1–10 mM), together with the nitrate-containing Aliquat 336N (at 10 mM), were each present in the respective organic (chloroform) phases;[126] the aqueous phase contained 10 mM of sodium nitrate and 0.1 mM of sodium sulfate in each case. In the presence of Aliquat 336N, the extraction of SO_4^{2-} from the aqueous phase containing excess sodium nitrate was clearly enhanced. The observed increase was attributed to sulfate binding by the chosen macrocyclic receptor in the organic phase through anion exchange with the initially bound nitrate. Namely, anion binding may essentially be considered to occur in the water-immiscible organic phase (with the latter contacting the aqueous phase containing the SO_4^{2-}/NO_3^- mixture), with the overall result being a redistribution of the anions between the two phases.

Distribution data for each of the above macrocyclic systems were in accord with both 1:1 and 1:2 (receptor: SO_4^{2-}) formation in the chloroform phase, with the log K for 1:1 binding falling in the range 2.1–4.8. The extraction data show that the macrocycle substituents have a strong influence on the degree of sulfate extraction, but initial attempts to correlate macrocyclic ligand structure with extraction efficiencies gave no clear relationship in terms of simple chemical expectations. This is in keeping with the inherent complexity of these multispecies systems, as well as the difficulty in estimating the overall contribution of a number of weak supramolecular interactions contributing to a given extraction outcome.

A further study focused on the "parent" octamethyl-substituted ring system, *meso*-octamethylcalix[4]pyrrole (**25**), was more successful in unraveling details of the observed anion exchange behavior.[127] For this system, it was demonstrated that the structure of the quaternary cation employed plays a key role in influencing sulfate selectivity. In part, based on the results of previous studies,[128–132] the enhanced extraction was postulated to reflect the supramolecular binding of the methyl group of the Aliquat cation in the cavity of **25** (in its cone configuration), with **25** also showing concomitant hydrogen bond formation of its four N–H donor groups with the sulfate anion on the opposite side of this receptor. Indeed, comparative experiments in which the influence of the methyl-containing Aliquat was compared with symmetrical (solely) long-chain quaternary ammonium cations showed that extraction of SO_4^{2-} was "switched off" in each of the latter cases. Further evidence for such an ion-pair receptor mode of action was obtained from an X-ray structure of the solid-state adduct between **25** and tetramethylammonium sulfate; this structure shows an analogous supramolecular host-guest arrangement involving simultaneous cation and anion binding arrangements to those proposed to occur in solution.

Eight new uncharged anion receptors based on the rigid steroidal cholapod skeleton (derived from chloric acid) incorporating up to six hydrogen bond donors have been employed as extractants in solvent extraction (H_2O/CHCl$_3$) experiments for a range of seven monovalent anions initially added as their tetraethylammonium salts to the respective aqueous phases.[133] Three representative examples of such receptors, each incorporating six hydrogen donor sites, are given by **26–28**.

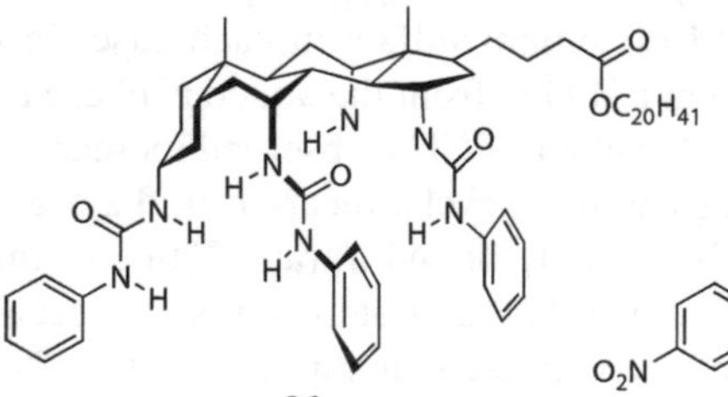

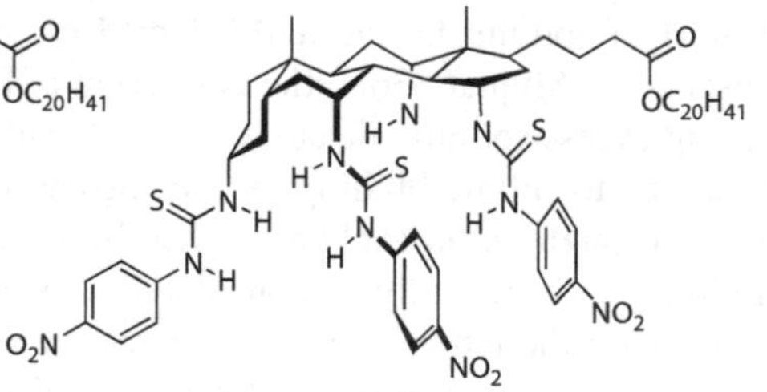

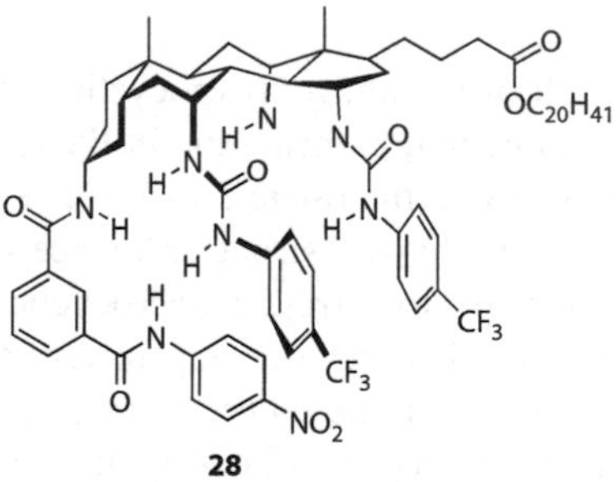

For this receptor category, the spacing between the three pendant arms limits the degree of intramolecular contact possible, with the lipophilic nature of the cholapod backbone favoring the use of such receptors in nonpolar media. As demonstrated in part by **26–28**, the nature of the functional groups attached to each arm can be readily varied to incorporate urea, thiourea, sulfonamide, carbamate, trifluoroacetamide, and isophthalamide binding units, thus allowing a variety of hydrogen donor patterns to be generated.

High anion binding constants ($>10^{10}$ M^{-1}) were observed for particular receptor-anion systems; for example, the binding constant for **27** toward Cl$^-$ is 1.8×10^{11} in chloroform. While, as expected, the constants vary with receptor geometry; for systems with similar geometries, the relative anion *selectivities* between receptors were shown to be dependent on the respective intrinsic binding strengths toward Cl⁻, with enhanced selectivities being engendered for those systems displaying overall their strongest binding toward this anion. Thus, it was observed that as the constants for Cl⁻ rose, the selectivity figures for Br⁻, I⁻, NO_3^-, ClO_4^-, and $EtSO_3^-$ all decreased (although those for AcO⁻ rose). Such behaviour was termed the *affinity-selectivity* effect, and the possibility that this effect might operate more widely in supramolecular host-guest chemistry was raised by the authors.

1.3.2.4 Dendritic Anion Extractants

Dendrimer-based ligands have been demonstrated to bind and transport a range of guest types that include anionic species.[94,134] Early studies were carried out using dendrimers based on different generations of POPAM (oligoethyleneoxy-modified polypropyleneamine) precursors such as **29** with X = –C(O)NHR. These are readily protonated at their tertiary nitrogen amine functions (and are expected to be at least half protonated around pH 5–6) and act as efficient phase-transfer reagents (water-chloroform) for pertechnetate, perrhenate, and ATP anions.[135]

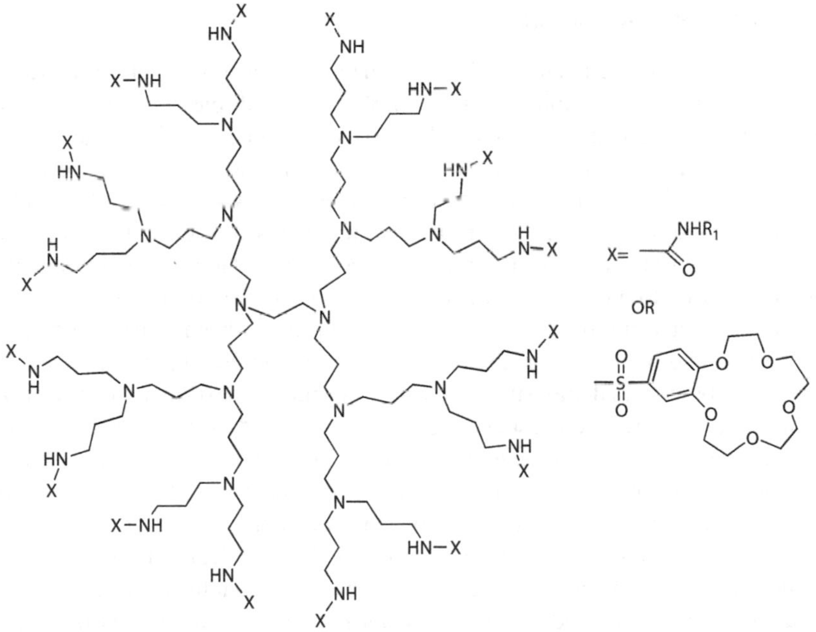

29

The presence of multiurea domains that act as hydrogen bonding donor sites for these anions[66,136,137] is a key aspect underpinning the efficiency of extraction of this dendrimer type. Further, the presence of lipophilic alkyl groups (R = C_6H_{13}, C_8H_{17}) on the periphery of **29** [with X = –C(O)NHR] is clearly advantageous in assisting phase transfer, while the pH dependency of anion binding enables the release of the anion under pH control.

In a different approach, a series of dendrimer extractants, in this case based on four generations of POPAM dendrimers with, respectively 4, 8, 16, and 32 sulfonyl-substituted benzo-15-crown-5 units appended at their peripheries, have been employed in solvent extraction (water-chloroform) studies.[138] The third-generation derivative is illustrated by **29** (where X is the sulfonyl-substituted benzo-crown derivative mentioned above). These dendrimeric species are soluble in common organic solvents but insoluble in water. The third- and fourth-generation dendrimer derivatives, besides extracting mercury(II) chloride efficiently, were also shown to act as anion extractants for pertechnetate (TcO_4^-). In this case, this anion is bound to protonated amine groups in the "interior" of the dendrimer; analysis of the organic phase showed that no Na^+ ions ($NaTcO_4$ was employed for the study) were extracted into the organic phase. Thus, in this system, it appears that it is supramolecular hydrogen bonding between the protonated amine sites on the dendrimer's backbone and the oxygens of the pertechnetate anion that dominate anion uptake into the organic phase.

1.3.3 ION-PAIR RECEPTORS

It has long been recognized that the counterion employed in a solvent extraction process involving a neutral extractant can play a crucial role in determining both extraction efficiency and selectivity. The roles of counterions in supramolecular systems in nonaqueous media have been reviewed, with aspects of the discussion being of relevance to the behavior of ions in the organic phase of two-phase solvent extraction systems.[139] Supramolecular aspects of the influence of steric, charge, and polarization effects, as well as the nature of ion pairs and the ability for a counterion to interact with further species, were also discussed.

A considerable number of ion-pair receptors have now been investigated as both recognition and binding reagents for use in solvent extraction, and such reagents have been sometimes demonstrated to result in enhanced extraction behavior over the use of a mixture of corresponding single-cation and -anion extractants. Such enhancement can reflect binding cooperativity occurring in the dual system, for example, due to favorable electrostatic interaction occurring between the cation and anion in the binding site(s) of the ion-pair receptor.

The simultaneous extraction of a cation and anion from an aqueous to an organic phase by an uncharged extractant has traditionally been described as occurring by an associated ion pair or by a separated ion pair. The associated ion pairs and separated ion pairs were referred to by Kim and Sessler[95] as "ion pairs" and "pairs of ions," respectively. However, the distinction between these is somewhat arbitrary in that "intermediate" ion pairing may also occur, and indeed, a range of extractants incorporating both adjacent and variously separated ion-pair binding sites have now been investigated. Three limiting cases are illustrated in Figure 1.10 and correspond to (a) a contact ion pair, (b) a solvent-separated ion pair, and (c) a host-separated ion pair. Of these, it is noted that contact ion-pair

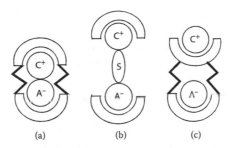

(a)   (b)   (c)

FIGURE 1.10   Schematic representation of (a) a contact ion pair, (b) a solvent-separated ion pair (where S represents a single solvent molecule or multiple solvent molecules), and (c) a host-separated ion pair. (From Kim, S. K., and Sessler, J. L., *Chem. Soc. Rev.*, 39, 3784–3809, 2010.)

formation avoids any "columbic penalty" associated with charge separation of the respective ions before binding, as may occur in the cases of (b) and (c).

In 2010, Kim and Sessler reviewed the state of the art with respect to synthetic ion-pair receptors, and a range of ion-pair receptor types were discussed.[95,140] These include such receptors employing Lewis acid, urea, amide, pyrrole, indole, hydroxyl, and positively charged groups for anion recognition; phosphine oxide and sulfoxide-based ion-pair receptors, along with G-quartets and mixed calixarene-calixpyrroles receptors were all discussed.

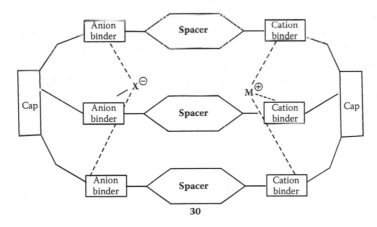

30

Computational (Hartree–Fock, density functional theory [DFT], and Moller–Plesset perturbation theory [MP2]) studies have been applied to the design of cryptand-like C_3-symmetric cages incorporating adjacent cation-anion binding sites and have enabled evaluation of a number of aspects of the supramolecular interactions applicable to such systems.[141] An outcome of the study, which included modeling of structures of type **30**, was that several promising candidate host molecules for ion-pair binding of NaCl were identified. The strategy employed was to maximize the NaCl binding through (1) optimizing the cavity size and shape for both the cation and anion binding sites, (2) variation of the donor sets employed for cation and anion binding, and (3) optimizing the spacer groups separating the cation and anion sites. Comparative analysis of the structures and charge distributions in the generated ion-pair-loaded receptors yielded insight into the factors contributing to strong NaCl binding. The best candidate incorporated a tripodal triether-substituted amine, $N(CH_2CH_2OR)_3$, for binding the sodium ion and three thiourea groups for binding the chloride ion. In this optimized structure, the thiourea groups are arranged in a tripodal fashion connected to a 1,3,5-trisubstituted benzyl head group to yield a $C_6H_3(CH_2NHC(=S)NHR)_3$ arrangement, with three $-CH_2CH_2-$ groups optimum for linking the cation and anion receptor sites. For the systems investigated, thioamide and thiourea derivatives were predicted to bind Cl⁻ more strongly than the corresponding amide and urea derivatives.

An outcome of the study is that such C_3-symmetric cages with adjacent cation and anion binding sites can lead to very strong ion-pair binding (in excess of 300 kJ mol⁻¹ for bound NaCl in a vacuum). The study also demonstrated the importance of the nature of the spacer groups between the binding sites for maximizing the ion-pair interaction. More generally, the study also gives insight into the supramolecular factors influencing the recognition of the NaCl ion pair and points the way for the design of further ditopic receptors that exhibit cooperativity.

Smith et al. have synthesized the ditopic ion-pair receptors 31[142] and 32,[143] each incorporating two amide functions for binding an anion and a dibenzo-18-crown-6 site in the case of 31, or a diaza-18-crown-6 site in the case of 32, for binding an alkali metal cation. In the case of 31, ^1H NMR studies in both DMSO-d^6 and a DMSO-d^6/CD$_3$CN mixture coupled with X-ray structure evidence indicated that 31 binds both sodium chloride and potassium chloride as ion pairs cooperatively. Thus, when one equivalent of Na⁺ or K⁺ (as their tetraphenylborate salts) was initially added to 31, followed by chloride ion, binding of the latter was observed to be enhanced by a factor of ~10 relative to the situation where the bound alkali metal cation was absent.

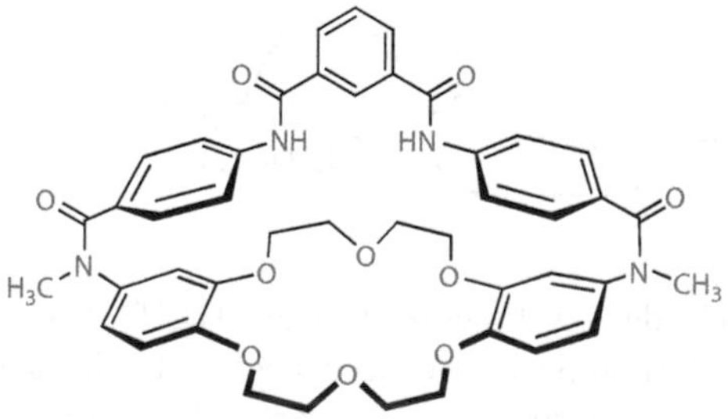

31 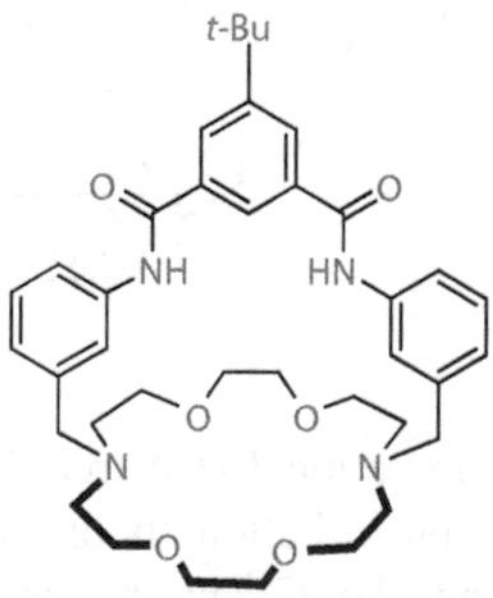   32

The X-ray structure of the sodium chloride complex of 31 (Figure 1.11(a)) showed that this salt is bound as a solvent-separated ion pair in the solid state.[142] As predicted, the Na⁺ binds in the cavity of the dibenzo-18-crown-6 site with an

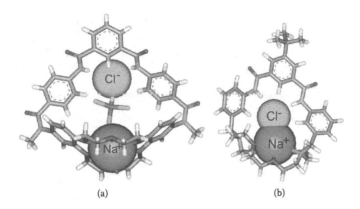

(a)   (b)

FIGURE 1.11 X-ray structures of (a) the chloroform-separated Na⁺Cl⁻ ion pair in the ditopic receptor **31** and (b) the contact Na⁺Cl⁻ ion pair in receptor **32**; solvate molecules are not shown. (From Deetz, M. J. et al., *J. Am. Chem. Soc.,* 122, 6201–6207, 2000; Mahoney, J. M. et al., *J. Am. Chem. Soc.,* 123, 5847–5858, 2001.)

axial bound water molecule pointing away from the cavity also forming part of the coordination sphere. The Cl⁻ anion has hydrogen bonds to each of the amide N–H groups as well as to the C–H of the adjacent aromatic ring and the C–H of a chloroform molecule, with the latter effectively separating the individual ions in the Na⁺Cl⁻ ion pair.

Based on the observed ion-pair binding behavior of **31**, receptor **32** was synthesized with a shorter distance between the potential cation and anion binding sites in an attempt to promote binding of NaCl or KCl as contact ion pairs.[143] The X-ray structures of the complexes of both these salts revealed that this expectation was realized in the solid state in each case; the structure of the 1:1 NaCl complex of **32** (Figure 1.11(b)) thus contrasts with that of the corresponding 1:1 complex of **31** (Figure 1.11(a)) with its solvent-separated ion pair. Cooperative binding of the chloride ion was observed for the potassium complex; however, for the sodium complex, little cooperative effect was observed. This observed different behavior was assigned to differences in the coordination geometries adopted by these two alkali cations.

Although no liquid–liquid extraction experiments are reported, solid-liquid extraction studies were undertaken in which an excess of powdered NaCl, NaBr, NaI, KCl, KBr, or KI was exposed to a solution of **32** in CDCl₃, with the outcome of the extraction monitored by ¹H NMR and mass spectrometry.[144] The results were in accord with each of these alkali metal salts being bound in a 1:1 fashion within the cavity of **32**. In further studies, it was shown that **32** promotes the transfer of individual alkali metal salts across both a phospholipid bilayer and, in the case of KCl, a supported liquid–membrane consisting of receptor **32** dissolved in 2-nitrophenyl-octyl ether.

In a further study, **32** was also shown to extract LiCl and LiBr from the solid state to produce inclusion complexes in which each of these salts is bound as a water-separated ion pair, contrasting with the above contact ion pairing that occurs with the larger Na^+ and K^+ alkali metal salts.[145] The Smith group has also demonstrated that **32** forms strong complexes with trigonal anions such as NO_3^- and OAc^-, with this receptor solubilizing both solid $NaNO_3$ and $KOAc$ in $CHCl_3$.[146]

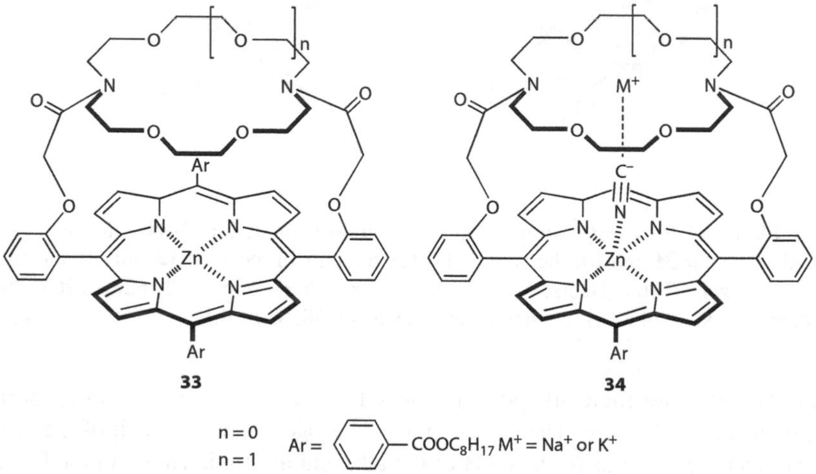

33   **34**

$n = 0$
$n = 1$ $Ar = $ ⬡—$COOC_8H_{17}$ $M^+ = Na^+$ or K^+

Two ditopic receptors involving a porphyrin-bound zinc(II) site for anion binding connected to a "capping" azacrown site for cation binding are given by **33** ($n = 0$ or 1).[147] Based on UV-vis and ^1H NMR evidence, it was demonstrated that both these receptors bind NaCN and KCN in methanol, with **33** ($n = 0$) binding NaCN about 56 times more strongly than KCN, whereas for **33** ($n = 1$) KCN is bound approximately 13 times more strongly than NaCN. The proposed binding of the respective metal cyanide ion pairs is illustrated in **34**.

The facile supramolecular self-assembly of a redox-responsive receptor in the form of the trinuclear ruthenium(II) metallacycle **35** (Figure 1.12) has been reported.[148] The product displayed good solubility in a variety of organic solvents, including chloroform, and thus proved suitable for use in $H_2O/CHCl_3$ solvent extraction experiments. This product was found to be remarkably selective for Li^+ (as LiCl) over other alkali and alkaline earth metal ions present in the aqueous phase, despite the very high enthalpy of hydration of Li^+ (-530 kJ mol^{-1}) and Cl^- (-365 kJ mol^{-1}).[149] For example, lithium selectivity was demonstrated to occur in the presence of large excesses of Na^+, K^+, Cs^+, Ca^{2+}, and Mg^{2+} ions, with the Li^+/Na^+ selectivity being greater than 1000:1 under the conditions employed (even though the hydration energy for Na^+ is of course significantly lower). Such high selectivity is of potential use in a number of applications, including the construction of a sensor (see below) for monitoring the lithium concentration in the blood of patients with bipolar disease, a condition treated with lithium carbonate for which the toxic dose is close to the therapeutic dose.

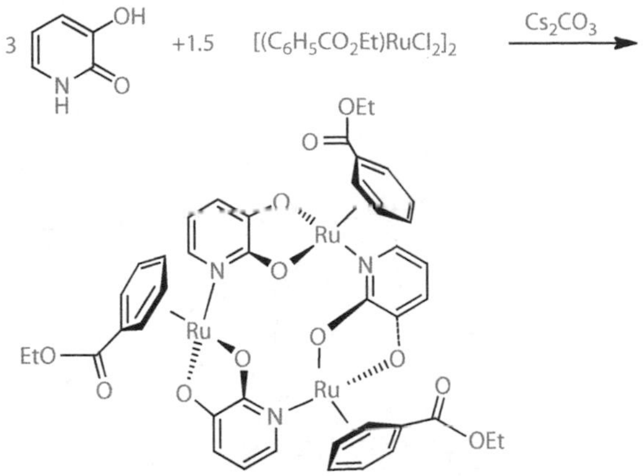

FIGURE 1.12 Supramolecular assembly of the trinuclear ruthenium(II) receptor **35** showing the central three-oxygen site where a lithium chloride (contact) ion pair binds. (From Piotrowski, H., and Severin, K., *Proc. Natl. Acad. Sci.,* 99, 4997–5000, 2002.)

The facile (one-step) assembly of receptor **35** in high yield from simple starting materials is an especially attractive feature of this system, with the presence of three ruthenium(II) centers resulting in it being electrochemically active, and hence of potential use as the basis of an electrochemical sensor. Reaction of **35** with excess lithium chloride in ethanol, followed by removal of the solvent and extraction of the residue into chloroform led to isolation of the adduct **35·LiCl·H₂O**, whose X-ray structure showed that the Li⁺ ion adopts a tetrahedral coordination geometry with all three oxygens of the cyclic receptor coordinating and the remaining site is occupied by a terminal chloride anion. The chloro ligand is closely approached by three arene groups that give rise to three short CH···Cl distances (2.70, 2.73, and 2.83 Å) that appear to assist in stabilizing the structure. Interestingly, **35·NaCl** can also be isolated and was shown to adopt a related structure to that of its lithium analog.

The high affinity of **35** for LiCl was ascribed to, first, the perfectly preorganized arrangement of the three oxygen atoms for binding to lithium (see Figure 1.13). Second, the energetic penalty for desolvation of the lithium-free cavity is low, since it is only capable of including one water molecule before complexation (confirmed in the solid state by an X-ray structure determination). Third, the LiCl is present as an ion pair so that unfavorable charge separation does not occur, with the chloride anion encapsulated by the arene ligands.

The redox properties of **35** and **35·LiCl** were compared using cyclic voltammetry. In the former case, three irreversible oxidations were observed, while in the presence of LiCl the potential of the first peak is shifted anodicly more than 350 mV, thus confirming the potential for this system to act as a chemosensor for detection of Li⁺.

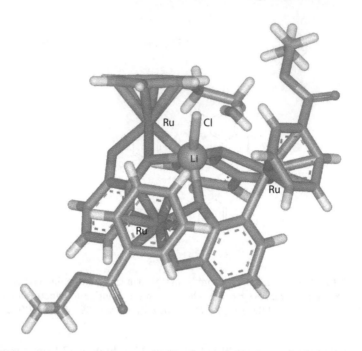

FIGURE 1.13 X-ray structure of the **35**·LiCl·H$_2$O complex. (From Piotrowski, H., and Severin, K., *Proc. Natl. Acad. Sci.*, 99, 4997–5000, 2002.)

Supramolecular assembly has resulted in the formation of a large heterotopic (anion-cation) ion-pair receptor involving the lipophilic 1,3-alternate calix[4] arene-guanosine conjugate **36**. This species was shown to dimerize in water-saturated CDCl$_3$ to yield (**36**)$_2$·(H$_2$O)$_n$. In this supramolecular assembly, the water molecules were shown to assist in stabilizing the dimeric structure, presumably by aiding the formation of an intermolecular hydrogen-bonded G-quartet located in the plane between the two component calix[4]arene derivative moieties.[150] While water usually inhibits the formation of hydrogen-bonded structures in organic solvents (through competing hydrogen bond formation), there are a range of examples known where hydrogen-bonded, usually bridged, water molecules assist in stabilizing a particular supramolecular structure.[151–159]

In the present case, the use of wet nonpolar organic solvents enhances both the solubility of (**36**)$_2$·(H$_2$O)$_n$ in the organic solvent and the promotion of its effectiveness as an ionophore, presumably through assisting in the stabilization of the G-quartet as mentioned above. Thus, this assembly was found to extract particular Na$^+$ and K$^+$ salts from water into nonpolar media such as chloroform or dichloromethane, with the evidence strongly suggesting that a salt-bound planar dimer is generated of the type depicted in Figure 1.14.

With respect to this, G-quartets have long been known to exhibit affinity for alkali metal cations.[160,161] The presence of water in CDCl$_3$ was also found to be essential for the stability of the salt-bound dimer, (**36**)$_2$·MX·(H$_2$O)$_n$ (Figure 1.14). Both ^{1}H

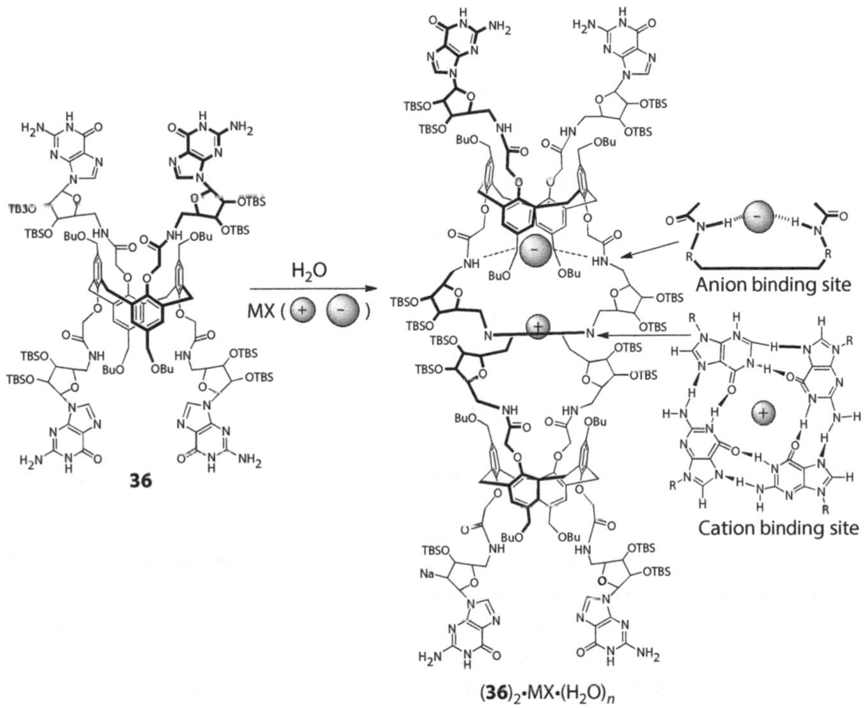

FIGURE 1.14 Binding of alkali metal salts (MX) by the water-stabilized G-quartet $(36)_2 \cdot (H_2O)_n$. (From Kotch, F. W. et al., *J. Am. Chem. Soc.*, 125, 15140–15150, 2003.)

NMR and chromatography were employed for investigating the use of $(36)_2(H_2O)_n$ in chloroform as an extractant for NaCl and NaBr from water. The studies revealed that $(36)_2(H_2O)_n$ shows only moderate selectivity (2:1) for K^+ over Na^+, and this ratio is essentially unaffected by whether Cl^- or Br^- is the counterion. Such a selectivity order is not unexpected, given the lower desolvation energy for K^+ relative to Na^+ (295 kJ mol^{-1} against 365 kJ mol^{-1}). In a further experiment, when NaCl was compared with NaBr, a 2:1 Br^-:Cl^- selectivity was observed. Nevertheless, based on these values alone, one would expect greater discrimination for K^+ than was observed, suggesting that only partial dehydration of these ions is likely to be required for host-guest formation. The evidence strongly indicates that the alkali cation resides in the G-quartet, while two neighboring 5′-amide groups act as a receptor for the halide ion via hydrogen bond interactions (Figure 1.14). The formation of the resulting ion pair helps to further stabilize the overall supramolecular assembly. Interestingly, the structure is anion dependent; use of a noncoordinating tetraphenylborate anion switches the discrete dimeric species to a supramolecular (that is, noncovalent) polymer.

A further category of metal salt extractants involving the use of zwitterion reagents for metal salt extraction has been reviewed by Tasker et al.[162,163] Studies of this type are presented elsewhere in this volume, and only a brief discussion of the use of this ligand category is given here.

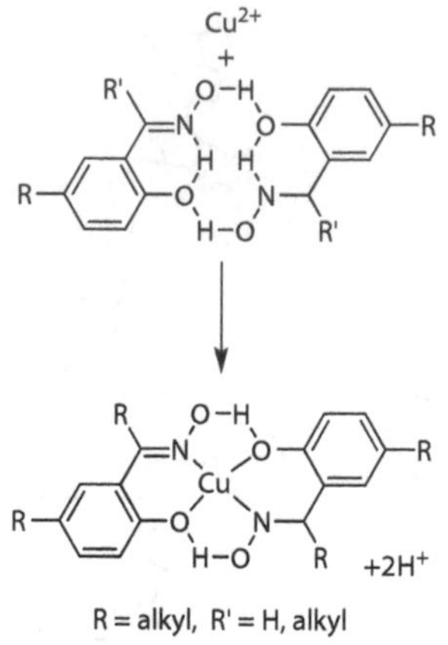

R = alkyl, R' = H, alkyl

FIGURE 1.15 Cu(II) complexation by the bisphenol-oxime shown involving loss of two protons.

Tasker et al. discussed the environmental advantages of employing hydrometallurgy for extracting base metals from their ores, especially for use with lower-grade ores. In the design of dual-site ligands of the above type for metal salt transport, the affinity of the anion binding site for uptake of the anion of interest was stressed, especially in relation to the binding of sulfate (reflecting the importance of sulfate in sulfide ore processing, especially for copper). In a conventional single-site cation extractant, the commercial phenol-oxime copper extractant shown in Figure 1.15, when used with a metal salt, results in loss of two protons and concomitant generation of acid in the aqueous phase. Namely, when a metal sulfate is involved, sulfuric acid is generated. The latter needs to be removed by either acid recovery or neutralization with base if the extraction system is to continue working efficiently. Alternatively, this difficulty can be addressed by employing a ditopic ligand, which simultaneously extracts both the metal cation and sulfate anion. For example, the use of a zwitterionic ligand that is capable of presenting positive and negatively charged binding sites allows uptake of both cation and anion to yield a charge-neutral species in the organic phase. At least in principle, stripping of these ions could then be carried out in two steps through pH control of the aqueous phase.

Zwitterionic extractants based on salen and related substituted salicylaldehyde-derived ligand derivatives (see, for example, Figure 1.16) were designed by the Tasker group for use in transporting Ni(II) or Cu(II) sulfate in two-phase systems.[80,164–168] The parent salen and its substituted derivatives have been long established to show strong binding for both these metal ions coupled with the loss of two protons. These extractants incorporate two tertiary amines that are capable of capturing the two protons released on metal binding at the O_2N_2 donor site, resulting in the formation of a zwitterion ligand system. In solution, the protonated amines are positioned to form a dicationic hydrogen bonding site for a sulfate ion (Figure 1.16), with the latter semiencapsulated by the surrounding organic structure as required for effective transport into the organic phase.

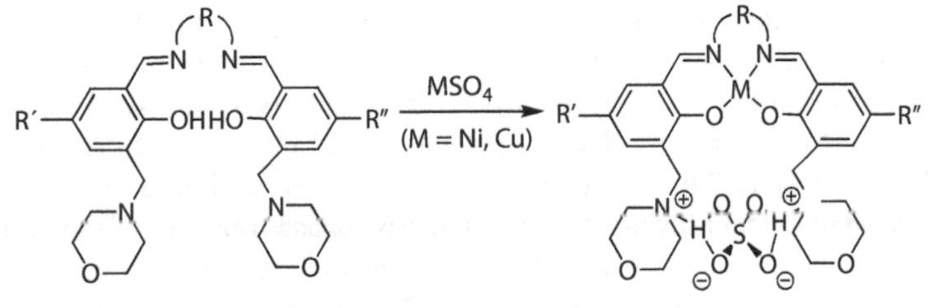

R' = -(CH$_2$)$_2$-, o-C$_6$H$_4$ or trans-1,2-cyclohexane
R" = tert-butyl branched nonyl

FIGURE 1.16 Ditopic extractants for the uptake of metal sulfates.

In an extension of the above studies, it was demonstrated that the attachment of dialkylaminomethyl to commercial phenol-oxime (salicyloxime) copper extractants (see Figure 1.15) gives rise to ligands showing enhanced transport of base metal salts through the formation of 1:1 or 1:2 complexes with zwitterionic forms of these substituted ligand derivatives.[169,170]

1.3.4 SMALL-MOLECULE BINDING AND EXTRACTION

Although the development of the field of solvent extraction has been motivated in large part by applications for metal ion separation, there is growing interest in the extraction of small molecules, which may include neutral species, zwitterions, or organic ions. In principle, recognition of such species by designed extractants may be expected to be governed by the same elements of complementarity and preorganization that governs the recognition of metal ions. However, coulombic interactions do not play a role in the case of neutral species, and there may be no requirement for accommodating counterions. Carbohydrate recognition and binding is an area of considerable current interest reflecting the biological importance of saccharides, especially with respect to areas such as carbohydrate metabolism, cell–cell recognition, pathogenic infection of cells, and the role of carbohydrates in immune responses.[171–176]

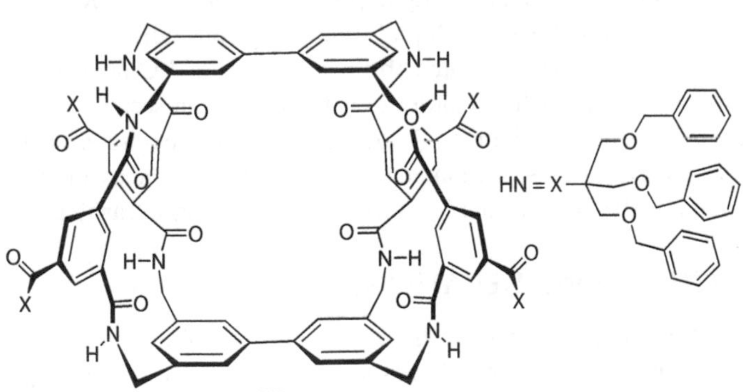

37

The solvent extraction of monosaccharides employing neutral lipophilic receptors has been reported by Davis et al.[177] In this study it was demonstrated that **37**, incorporating a lipophilic array of 12 benzyloxy groups, acts as an excellent extractant ($H_2O/CHCl_3$) for 3 hexoses (glucose, galactose, and mannose), 2 pentoses (ribose and xylose), as well as 2 methyl glucosides. For the hexoses, both strong affinity and selectivity were evident toward glucose. Clearly, phase transfers of the above type rely heavily on the presence of weak (noncovalent) supramolecular interactions between host and guest. It is interesting that **37** was shown to bind to *n*-octyl-β-D-glucoside in $CD_3OH/CDCl_3$ with a significant K value of 720 M^{-1}.

38

Based on the results of molecular modeling, the synthesis of the positively charged molecular receptor **38**, designed for recognition and binding of zwitterionic amino acids, was reported in 1996.[178] This receptor incorporates both a 4,10,16-triazacrown ether site for ammonium (RNH_3^+) binding and a bicyclic guanidinium site for carboxylate binding; both groups are connected by a flexible thioether-containing linking unit. As for the use of the guanidinium domain for carboxylate binding, the employment of a 4,10,16-triaza-18-crown-6 framework for primary ammonium group binding has long been established.[179] The bulky *tert*-butyl/silyl ether group was also included in the receptor structure to provide additional lipophilicity to enhance organic solvent solubility. Initial studies established that **38·**Cl in methanol acts as a solid-to-liquid phase-transfer system toward a number of amino acids. Following this study, liquid–liquid (aqueous buffer–dichloromethane) extraction experiments demonstrated the partitioning of ^{14}C-labeled amino acids between the two phases, with the order of extraction being serine << glycine < tryptophan < leucine < phenylalanine. Quantification of the extraction process using radiometry gave a clear 1:1 stoichiometry for the extracted species and also provided evidence that it is the zwitterionic form that undergoes phase transfer.

1.4 FURTHER CONSIDERATIONS

As discussed in previous sections, there are a number of key processes in liquid–liquid extraction that are intimately associated with supramolecular phenomena. Despite considerable progress in the solvent extraction field over more than a century, such aspects of the behavior of multicomponent two-phase extraction systems remain less than completely understood, largely due to the complicated nature of multicomponent two-phase extraction systems. In particular, interactions occurring at the liquid–liquid interface and association behavior in both phases, as well as at the interface, remain largely undefined in the majority of solvent extraction studies (for further discussion please see Chapters 2 and 5 in this volume). The role of solvation in extraction behavior is also in need of further study. Indeed, one can argue, albeit simplistically, that solvation is the essence of solvent extraction, which otherwise is simply coordination chemistry. Progress in the understanding of all of these areas is especially needed in order to improve the efficiency of extractive separation processes and for the development of new selective reagents.

In an extensive study, Schmidtchen et al.[16,17,136,181–185] have investigated the energetics (in terms of ΔG, ΔH, and $T\Delta S$) of anion binding in solution using isothermal calorimetry. This work was especially focused on the role of entropy in molecular association processes with respect to different environmental conditions. The weak supramolecular interactions in solution are the result of competition between solute–solute, solute–solvent, and solvent–solvent interactions, where each of these can dominate in a special case. The result is a sensitive and complicated interplay between synergic and antagonistic effects. The authors discussed how entropy can strongly influence both the binding efficiency and the selectivity for anions. An increase of the complexation entropy is also a major contributor to the well-known chelate effect in coordination chemistry largely based on solvation-desolvation processes. This effect plays an important role in the solvent extraction of metal cations with multifunctional ligand systems.

It is, of course, usually difficult to characterize the role of solvation in most solvent extraction systems. However, although caution invariably needs to be exercised when extrapolating from solid-state behavior to that occurring in solution, possible insight into solvation behavior can sometimes be gleaned from the inspection of solid-state structures of species postulated to also occur in a given solvent extraction system. An interesting example of this type was reported by Moyer et al.,[180] who determined the X-ray diffraction structure of [Cs(tetrabenzo-24-crown-8)(1,2-dichloroethane)$_2$](NO$_3$)·H$_2$O. This was shown to involve unprecedented bidentate coordination of each of two 1,2-dichloroethane solvent molecules to the Cs$^+$ center via this solvent's two chlorine atoms. The 24-membered crown ether, tetrabenzo-24-crown-8 (**39** in Figure 1.17), surrounds the metal center in a twisted configuration to give a cage-like structure in which all eight ether oxygen donors bind to the central Cs$^+$ ion (Figure 1.17, right-hand side). The dichloroethane molecules are bound between two clefts in this arrangement and lie between facing benzo groups such that both electronic and spatial complementarity occurs between each solvent guest and these arene residues. The anion is present in the lattice as a ligand-separated ion pair. The solid-state structure is in accord with the speciation behavior observed in the extraction (water–dichloromethane) of CsNO$_3$ (or CsClO$_4$) by the corresponding alkylated crown derivative 4,4″- or 4,5″-bis(*tert*-octylbenzo)dibenzo-24-crown-8 (the alkyl crown

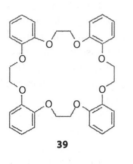 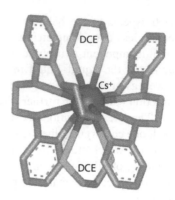

FIGURE 1.17 Structural formula of tetrabenzo-24-crown-8 (**39**) and X-ray structure of the corresponding complex cation in [Cs(**39**)(1,2-dichloroethane)$_2$]$^+$ showing the chelating 1,2-dichloroethane (DCE) molecules bound in clefts between pairs of arene rings. (From Levitskaia, T. G. et al., *J. Am. Chem. Soc.*, 122, 554–562, 2000.)

derivatives were chosen for the solution studies because of their increased solubilities in dichloromethane).

Modeling studies of the above liquid–liquid extraction system indicated 1:1 (M:L) complex formation with the nitrate (or perchlorate) anion being present as an ion pair in association with the cationic complex species. Of particular interest is the presence of weak electrostatic interactions between the chlorine atom donors of each 1,2-dichloroethane with the central metal while occupying the hydrophobic cleft within the coordination sphere of the metal ion. The host-guest complementarity of the 1,2-dichloroethane molecule for the respective clefts is reflected by the observation that five of the slightly electropositive hydrogen atoms of this guest are within 3.2 Å of an arene ring centroid, with the H⋯ring distances ranging from 2.64 to 2.94 Å. The absence of a contact ion-pair formation in the solid state appears surprising, since the anion is excluded from the coordination sphere in preference to the neutral 1,2-dichloroethane molecules. Finally, providing evidence for the favorable host properties of the hydrophobic clefts for this solvent, such behavior may also reflect the lower charge density of Cs$^+$ relative to the smaller alkali metal cations.

A further important aspect of extraction is the presence of a liquid–liquid interface where the hydrophobic extractant molecules interact with the hydrophilic species. This interfacial layer between the phases plays a key role in the aqueous-organic phase transfer of a solute. Reflecting the nanoscopic thickness of this layer, its properties are drastically different than those of the bulk phases. It has been found by surface tension measurements, surface spectroscopy, electrochemical studies, and especially computer simulations, that both the extractant and its complex are adsorbed at the interface.[20,186–188] Here possible association (aggregation) of the extractant and the complex can also take place, which may influence the whole extraction process. Again, this behavior is dominated by weak supramolecular interactions.[189,190]

The formation of such aggregates effectively corresponds to the generation of new extracting species of a supramolecular nature that will affect the extraction properties. One early example is the formation of macrocyclic phenol-oxime dimers based on intermolecular hydrogen bonding (see Figure 1.15) and their use for industrial Cu(II) extraction.[191] Another example is the aggregation of extractant molecules with amide functions, such as malonamides or diglycolamides, in nonpolar alkane diluents.[192] In these cases, pronounced hydrogen bonding interactions are observed, leading to the formation of reverse micelles, and ultimately to third-phase formation. Depending on the acid concentration, different aggregates may be formed, spanning dimeric to tetrameric species, with the latter showing high affinity for the trivalent lanthanide and actinide nitrates.[20,21,193–195] When the modifier n-octanol was added to the above extraction system employing malonamide as extractant, a distinctive hydrogen bond network was formed that prevents the formation of a third phase.[17]

While supramolecular aspects of solvent extraction behavior for different substrate types are discussed above, it needs to be noted that other aspects, such as favorable enhanced kinetics of metal ion extraction, have also been rationalized in terms of supramolecular influences in the case of particular solvent extraction systems.[23]

The application of supramolecular principles has opened up many opportunities for aiding the development of highly selective and efficient extraction processes, with interesting new possibilities for the winning and recovery of different ionic and molecular species being created. The understanding of the basic chemical and physical behavior of such processes has been enhanced considerably. A major advantage consists of the use of two basic processes of nature, molecular recognition and self-organization. Undoubtedly, the further development of supramolecular science will continue to make valuable contributions to both solvent extraction processes and sustainable development in general.

ACKNOWLEDGMENTS

We thank the German Research Foundation (DFG), the German Federal Ministry of Education and Research (BMBF project 02NUK014A), and the Australian Research Council (ARC) for support.

REFERENCES

1. Rydberg, J., Cox, M., Musikas, C., and Choppin, G. R. 2004. *Solvent extraction: principles and practice*. M. Dekker, New York.
2. Marcus, Y., and Kertes, A. S. 1969. *Ion exchange and solvent extraction of metal complexes*. Wiley-Interscience, London.
3. Sekine, T., and Hasegawa, Y. 1977. *Solvent extraction chemistry: fundamentals and applications*. M. Dekker, New York.
4. Lehn, J.-M. 1985. Supramolecular chemistry: receptors, catalysts, and carriers. *Science* 227: 849–856.
5. Steed, J. W., and Atwood, J. L. 2009. *Supramolecular chemistry*. Wiley, Chichester.

6. Swiegers, G., and Malefetse, T. 2001. Multiple-interaction self-assembly in coordination chemistry. *J. Incl. Phenom. Macrocycl. Chem.* 40: 253–264.
7. Steed, J. W., Turner, D. R., and Wallace, K. J. 2007. *Supramolecular chemistry and nanochemistry*. John Wiley & Sons, London.
8. Atwood, J. L., and Steed, J. W., Eds. 2004. *Encyclopedia of supramolecular chemistry*. M. Dekker, New York.
9. Pedersen, C. J. 1988. The discovery of crown ethers (Nobel lecture). *Angew. Chem. Int. Ed.* 27: 1021–1027.
10. Lehn, J.-M. 1988. Supramolecular chemistry—scope and perspectives molecules, supermolecules, and molecular devices (Nobel lecture). *Angew. Chem. Int. Ed.* 27: 89–112.
11. Gokel, G. W. 1991. *Crown ethers and cryptands*. Royal Society of Chemistry, Cambridge.
12. Vögtle, F., and Weber, E. 1979. Multidentate acyclic neutral ligands and their complexation. *Angew. Chem. Int. Ed.* 18: 753–776.
13. Vicens, J., and Böhmer, V. 1991. *Calixarenes: a versatile class of macrocyclic compounds*. Kluwer Academic, Dordrecht.
14. Böhmer, V. 1995. Calixarenes, macrocycles with (almost) unlimited possibilities. *Angew. Chem. Int. Ed.* 34: 713–745.
15. Gutsche, C. D. 1998. *Calixarenes revisited*. Royal Society of Chemistry, Cambridge.
16. Schmidtchen, F. P. 2008. The role of entropy in the design of molecular hosts. In *Proceedings of ISEC 2008*, Tucson, Moyer, B. A., Ed. Canadian Institute of Mining, Metallurgy and Petroleum, Montreal, pp. 59–81.
17. Schmidtchen, F. P. 2010. Hosting anions. The energetic perspective. *Chem. Soc. Rev.* 39: 3916–3935.
18. Nave, S., Mandin, C., Martinet, L., Berthon, L., Testard, F., Madic, C., and Zemb, T. 2004. Supramolecular organisation of tri-n-butyl phosphate in organic diluent on approaching third phase transition. *Phys. Chem. Chem. Phys.* 6: 799–808.
19. Abécassis, B., Testard, F., Zemb, T., Berthon, L., and Madic, C. 2003. Effect of n-octanol on the structure at the supramolecular scale of concentrated dimethyldioctylhexylethoxymalonamide extractant solutions. *Langmuir* 19: 6638–6644.
20. Jensen, M. P., Yaita, T., and Chiarizia, R. 2008. In *Proceedings of ISEC 2008*, Tucson, Moyer, B. A., Ed. Canadian Institute of Mining, Metallurgy and Petroleum, Montreal, pp. 1029–1034.
21. Ellis, R. J., and Antonio, M. R. 2012. Coordination structures and supramolecular architectures in a cerium(III)-malonamide solvent extraction system. *Langmuir* 28: 5987–5998.
22. Berny, F., Muzet, N., Troxler, L., and Wipff, G. 1999. In *Supramolecular science: where it is and where it is going*, Ungaro, R., and Dalcanale, E., Eds. Kluwer Academic, Dordrecht, pp. 95–124.
23. Cote, G. 2003. The supramolecular speciation: a key for improved understanding and modelling of chemical reactivity in complex systems. *Radiochim. Acta* 91: 639–644.
24. Moyer, B. A. 1996. Supramolecular reactivity and transport: extraction. In *Comprehensive supramolecular chemistry*, Vol. 1, Atwood, J. L., Lehn, J.-M., Davies, J. E. O., McNichol, D. D., and Vögtle, F., Eds. Pergamon Press, Oxford, pp. 377–416.
25. Beklemishev, M. K., Dmitrienko, S. G., and Isakova, N. V. 1997. In *Macrocyclic compounds in analytical chemistry*, Zolotov, Yu. A., Ed. Wiley, New York, pp. 63–208.
26. Bradshaw, J. S., and Izatt, R. M. 1997. Crown ethers: the search for selective ion ligating agents. *Acc. Chem. Res.* 30: 338–345.

27. Chartroux, C., Wichmann, K., Goretzki, G., Rambusch, T., Gloe, K., Müller, U., Müller, W., and Vögtle, F. 2000. Preorganized complexing agents as a tool for selective solvent extraction processes. *Ind. Eng. Chem. Res.* 39: 3616–3624.
28. Stephan, H., Kubeil, M., Gloe, K., and Gloe, K. 2012. Extraction methods. In *Analytical methods in supramolecular chemistry*, Schalley, Ch., Ed. Wiley-VCH, Weinheim, pp. 105–127.
29. Antonioli, B., Gloe, K., Gloe, K., and Lindoy, L. F. 2006. Anion extraction. In *Encyclopedia of supramolecular chemistry*, Atwood, J. L., and Steed, J. W., Eds. Taylor & Francis, online version, DOI: 10.1081/E-ESMC-120041521.
30. Moyer, B. A., Bonnesen, P. V., Custelcean, R., Delmau, L. H., and Hay, B. P. 2005. Strategies for using host-guest chemistry in the extractive separations of ionic guests. *Kem. Ind.* 54: 65–87.
31. Levitskaia, T. G., Bonnesen, P. V., Chambliss, C. K., and Moyer, B. A. 2003. Synergistic pseudo-hydroxide extraction: synergism and anion selectivity in sodium extraction using a crown ether and a series of weak lipophilic acids. *Anal. Chem.* 75: 405–412.
32. Moyer, B. A., McDowell, W. J., Ontko, R. J., Bryan, S. A., and Case, G. N. 1986. Complexation of strontium in the synergistic extraction system dicyclohexano-18-crown-6/versatic acid/carbon tetrachloride. *Solvent Extr. Ion Exch.* 4: 83–93.
33. Bryan, S. A., McDowell, W. J., Moyer, B. A., Baes, C. F., and Case, G. N. 1987. Spectral studies and equilibrium analysis of the didodecylnaphthalene sulfonic acid/dicyclohexano-18-crown-6/Sr^{2+} extraction system. *Solvent Extr. Ion Exch.* 5: 717–738.
34. Moyer, B. A., Delmau, L. H., Case, G. N., Bajo, S., and Baes, C. F. 1995. Comprehensive equilibrium analysis of the complexation of Cu(II) by tetrathia-14-crown-4 in a synergistic extraction system employing didodecylnaphthalene sulfonic acid. *Sep. Sci. Technol.* 30: 1047–1069.
35. Belova, V. V., Voshkin, A. A., Kholkin, A. I., and Payrtman, A. K. 2009. Solvent extraction of some lanthanides from chloride and nitrate solutions by binary extractants. *Hydrometallurgy* 97: 198–203.
36. Egorova, N. S., Belova, V. V., Voshkin, A. A., Kholkin, A. I., Pyartman, A. K., and Keskinov, V. A. 2008. Extraction of uranyl nitrate with a binary extractant based on di(2,4,4-trimethylpentyl)phosphinic acid. *Theor. Found. Chem. Eng.* 42: 708–713.
37. Belova, V. V., Zhidkova, T. I., Brenno, Y. Y., Zhidkov, L. L., and Kholkin, A. I. 2003. Extraction of platinum and palladium by diamine-organic acid binary extractants. *Russ. J. Inorg. Chem.* 48: 614–619.
38. Kholkin, A. I., Pashkov, G. L., and Belova, V. V. 2000. Binary extraction in hydrometallurgy. *Miner. Process. Extr. Metall. Rev.* 21: 89–141.
39. Belova, V. V., and Kholkin, A. I. 1998. Binary extraction of platinum metals. *Solvent Extr. Ion Exch.* 16: 1233–1255.
40. Kholkin, A. I., and Kusmin, V. I. 1984. *Isot. Environ. Health Stud.* 20: 339–344.
41. McDowell, W. J. 1988. Crown ethers as solvent extraction reagents: where do we stand? *Sep. Sci. Technol.* 23: 1251–1268.
42. Bond, A. H., Dietz, M. L., and Chiarizia, R. 2000. Incorporating size selectivity into synergistic solvent extraction: a review of crown ether-containing systems. *Ind. Eng. Chem. Res.* 39: 3442–3464.
43. Adam, K. R., Atkinson, I. M., Farquhar, S., Leong, A. J., Lindoy, L. F., Mahinay, M. S., Tasker, P. A., and Thorp, D. 1998. Tailoring molecular assemblies for metal ion binding. *Pure Appl. Chem.* 70: 2345–2350.
44. Adam, K. R., Antolovich, M., Atkinson, I. M., Leong, A. J., Lindoy, L. F., McCool, B. J., Davis, R. L., Kennard, C. H. L., and Tasker, P. A. 1994. On the nature of the host-guest interaction between cyclam and 4-tert-butylbenzoic acid—a system preassembled for metal complex formation. *J. Chem. Soc. Chem. Comm.* 1539–1540.

45. Byriel, K. A., Gasperov, V., Gloe, K., Kennard, C. H. L., Leong, A. J., Lindoy, L. F., Mahinay, M. S., Pham, H. T., Tasker, P. A., Thorp, D., and Turner, P. 2003. Host-guest assembly of ligand systems for metal ion complexation; synergistic solvent extraction of copper(II) ions by N_3O_2-donor macrocycles and carboxylic or phosphinic acids. *Dalton Trans.* 3034–3040.

46. Lindoy, L. F., Mahinay, M. S., Skelton, B. W., and White, A. H. 2003. Ligand assembly and metal ion complexation. Synthesis and X-ray structures of Ni(II) and Cu(II) benzoate and 4-*tert*-butylbenzoate complexes of cyclam. *J. Coord. Chem.* 56: 1203–1213.

47. Gasperov, V., Gloe, K., Lindoy, L. F., and Mahinay, M. S. 2004. Host-guest assembly of ligand systems for metal ion complexation. Synergistic solvent extraction of copper(II) and silver(I) by 1,4,8,11-tetrabenzyl-1,4,8,11-tetraazacyclodecane in combination with carboxylic acids. *Dalton Trans.* 3829–3834.

48. Gasperov, V., Gloe, K., Leong, A. J., Lindoy, L. F., Mahinay, M. S., Stephan, H., Tasker, P. A., and Wichmann, K. 2002. In *Proceedings of ISEC 2002*, Sole, K., Cole, P., Preston, J., and Robinson, D., Eds. South African Institute of Mining and Metallurgy, Capetown, pp. 353–359.

49. Kim, J.-H., Leong, A. J., Lindoy, L. F., Kim, J., Nachbaur, J., Nezhadali, A., Rounaghi, G., and Wei, G. 2000. Metal-ion recognition. Competitive bulk membrane transport of transition and post transition metal ions using oxygen-nitrogen donor macrocycles as ionophores. *J. Chem. Soc. Dalton Trans.* 3453–3450.

50. Shamsipur, M., and Akhoud, M. 1997. Highly efficient and selective membrane transport of silver(I) ion by a cooperative carrier composed of aza-18-crown-6 and palmitic acid. *Bull. Chem. Soc. Jpn.* 70: 339–343.

51. Akhond, M., and Shamsipur, M. 1997. Highly efficient cooperative membrane transport of lead(II) ions by aza-18-crown-6 and palmitic acid. *Sep. Sci. Technol.* 32: 1223–1232.

52. Masuda, Y., Zhang, Y., Yan, C., and Li, B. 1998. Studies on the extraction and separation of lanthanide ions with a synergistic extraction system combined with 1,4,10,13-tetrathia-7,16-diazacyclooctadecane and lauric acid. *Talanta* 46: 203–213.

53. Eyal, A. M., Bressler, E., Bloch, R., and Hazan, B. 1994. Extraction of metal salts by mixtures of water-immiscible amines and organic acids (acid-base couple extractants). 1. A review of distribution and spectroscopic data and of proposed extraction mechanisms. *Ind. Eng. Chem. Res.* 33: 1067–1075.

54. Eyal, A. M., Kogan, L., and Bressler, E. 1994. Extraction of metal salts by mixtures of water-immiscible amines and organic acids (acid-base couple extractants). 2. Theoretical treatment and analysis. *Ind. Eng. Chem. Res.* 33: 1076–1085.

55. Kholkin, A. I., Belova, V. V., Pashkov, G. L., Fleitlikh, I. Y., and Sergeev, V. V. 1999. Solvent binary extraction. *J. Mol. Liq.* 82: 131–146.

56. Aggett, J., and Woollard, G. A. 1981. Solvent extraction of nickel(II) by combinations of aliphatic diamines and carboxylic acids. *J. Inorg. Nucl. Chem.* 43: 2959–2964.

57. Alivertis, D., Paraskevopoulos, G., Theodorou, V., and Skobridis, K. 2009. Dendritic effects of crown ether-functionalized dendrimers on the solvent extraction of metal ions. *Tetrahedron Lett.* 50: 6019–6021.

58. Gale, P. A. 2010. Anion receptor chemistry: highlights from 2008 and 2009. *Chem. Soc. Rev.* 39: 3746–3771.

59. Steed, J. W. 2009. Coordination and organometallic compounds as anion receptors and sensors. *Chem. Soc. Rev.* 38: 506–519.

60. Savoia, D., and Gualandi, A. 2009. Chiral perazamacrocycles: synthesis and applications. Part 2. *Curr. Org. Synth.* 6: 119–142.

61. Gong, W. T., Hiratani, K., and Lee, S. S. 2008. Macrocyclic bis(amidonaphthol)s for anion sensing: tunable selectivity by ring size in proton transfer process. *Tetrahedron* 64: 11007–11011.

62. Custelcean, R., and Moyer, B. A. 2007. Anion separation with metal–organic frameworks. *Eur. J. Inorg. Chem.* 1321–1340.
63. Nguyen, B. T., and Anslyn, E. V. 2006. Indicator–displacement assays. *Coord. Chem. Rev.* 250: 3118–3127.
64. Bayly, S., and Beer, P. 2008. Metal-based anion receptor systems. In *Recognition of anions*, Vol. 129, Vilar, R., Ed. Springer, Berlin, pp. 45–94.
65. Bowman-James, K. 2005. Alfred Werner revisited: the coordination chemistry of anions. *Acc. Chem. Res.* 38: 671–678.
66. Sessler, J. L., Gale, P. A., and Cho, W.-S. 2006. *Anion receptor chemistry*. RSC, Cambridge.
67. Gale, P. A. 2011. Anion receptor chemistry. *Chem. Comm.* 47: 82–86.
68. Wenzel, M., Hiscock, J. R., and Gale, P. A. 2012. Anion receptor chemistry: highlights from 2010. *Chem. Soc. Rev.* 41: 480–520.
69. Zaccheddu, M., Filippi, C., and Buda, F. 2008. Anion-π and π-π cooperative interactions regulating the self-assembly of nitrate–triazine complexes. *J. Phys. Chem. A* 112: 1627–1632.
70. Rotger, C., Soberats, B., Quiñonero, D., Frontera, A., Ballester, P., Benet-Buchholz, J., Deyà, P. M., and Costa, A. 2008. Crystallographic and theoretical evidence of anion-π and hydrogen-bonding interactions in a squaramide–nitrate salt. *Eur. J. Org. Chem.* 1864–1868.
71. Götz, R. J., Robertazzi, A., Mutikainen, I., Turpeinen, U., Gamez, P., and Reedijk, J. 2008. Concurrent anion-π interactions between a perchlorate ion and two π-acidic aromatic rings, namely pentafluorophenol and 1,3,5-triazine. *Chem. Comm.* 3384–3386.
72. Mooibroek, T. J., Black, C. A., Gamez, P., and Reedijk, J. 2008. What's new in the realm of anion-π binding interactions? Putting the anion-π interaction in perspective. *Cryst. Growth Des.* 8: 1082–1093.
73. Yorita, H., Otomo, K., Hiramatsu, H., Toyama, A., Miura, T., and Takeuchi, H. 2008. Evidence for the cation-π interaction between Cu^{2+} and tryptophan. *J. Am. Chem. Soc.* 130: 15266–15267.
74. Valencia, L., Pérez-Lourido, P., Bastida, R., and Macías, A. 2008. Dinuclear Zn(II) polymer consisting of channels formed by π, π-stacking interactions with a flow of nitrate anions through the channels. *Cryst. Growth Des.* 8: 2080–2082.
75. Gural'skiy, I. A., Escudero, D., Frontera, A., Solntsev, P. V., Rusanov, E. B., Chernega, A. N., Krautscheid, H., and Domasevitch, K. V. 2009. 1,2,4,5-Tetrazine: an unprecedented mu(4)-coordination that enhances ability for anion-π interactions. *Dalton Trans.* 2856–2864.
76. Antonioli, B., Büchner, B., Clegg, J. K., Gloe, K., Gloe, K., Götzke, L., Heine, A., Jäger, A., Jolliffe, K. A., Kataeva, O., Kataev, V., Klingeler, R., Krause, T., Lindoy, L. F., Popa, A., Seichter, W., and Wenzel, M. 2009. Interaction of an extended series of N-substituted di(2-picolyl)amine derivatives with copper(II). Synthetic, structural, magnetic and solution studies. *Dalton Trans.* 4795–4805.
77. Hollis, C. A., Hanton, L. R., Morris, J. C., and Sumby, C. J. 2009. 2-D coordination polymers of hexa(4-cyanophenyl)[3]-radialene and silver(I): anion-π interactions and radialene CH-anion hydrogen bonds in the solid-state interactions of hexaaryl[3]-radialenes with anions. *Cryst. Growth Des.* 9: 2911–2916.
78. Colacio, E., Aouryaghal, H., Mota, A. J., Cano, J., Sillanpää, R., and Rodriguez-Diéguez, A. 2009. Anion encapsulation promoted by anion-π interactions in rationally designed hexanuclear antiferromagnetic wheels: synthesis, structure and magnetic properties. *CrystEngComm* 11: 2054–2064.
79. Hay, B. P., and Custelcean, R. 2009. Anion-π interactions in crystal structures: common place or extraordinary? *Cryst. Growth Des.* 9: 2539–2545.

80. Wichmann, K., Antonioli, B., Söhnel, T., Wenzel, M., Gloe, K., Gloe, K., Price, J. R., Lindoy, L. F., Blake, A. J., and Schröder, M. 2006. Polyamine-based anion receptors: extraction and structural studies. *Coord. Chem. Rev.* 250: 2987–3003.

81. Dalla Cort, A., De Bernardin, P., Forte, G., and Mihan, F. Y. 2010. Metal-salophen-based receptors for anions. *Chem. Soc. Rev.* 39: 3863–3874.

82. Mercer, D. J., and Loeb, S. J. 2010. Metal-based anion receptors: an application of second-sphere coordination. *Chem. Soc. Rev.* 39: 3612–3620.

83. Mullen, K. M., and Beer, P. D. 2009. Sulfate anion templation of macrocycles, capsules, interpenetrated and interlocked structures. *Chem. Soc. Rev.* 38: 1701–1713.

84. Glasson, C. R. K., Meehan, G. V., Clegg, J. K., Lindoy, L. F., Turner, P., Duriska, M. B., and Willis, R. 2008. New anion-encapsulating tetrahedral M_4L_6 cages based on Fe^{II} and a linear quaterpyridine. *Chem. Comm.* 1190–1192.

85. Ward, M. D. 2009. Polynuclear coordination cages. *Chem. Comm.* 4487–4499.

86. Tidmarsh, I. S., Taylor, B. F., Hardie, M. J., Russo, L., Clegg, W., and Ward, M. D. 2009. Further investigations into tetrahedral M_4L_6 cage complexes containing guest anions: new structures and NMR spectroscopic studies. *New J. Chem.* 33: 366–375.

87. Custelcean, R., Bosano, J., Bonnesen, P. F., Kertesz, V., and Hay, B. P. 2009. Computer-aided design of a sulfate-encapsulating receptor. *Angew. Chem. Int. Ed.* 48: 4025–4029.

88. Riddell, I. A., Smulders, M. M., Clegg, J. K., and Nitschke, J. R. 2011. Encapsulation, storage and controlled release of sulfur hexafluoride from a metal-organic capsule. *Chem. Comm.* 47: 457–459.

89. Glasson, C. R. K., Clegg, J. K., McMurtrie, J. C., Meehan, G. V., Lindoy, L. F., Motti, C. A., Moubaraki, B., Murray, K. S., and Cashion, J. D. 2011. Unprecedented encapsulation of a [Fe^{III}Cl$_4$]$^-$ anion in a cationic [Fe$^{II}_4L_6$]$^{8+}$ tetrahedral cage derived from 5,5'''-dimethyl-2,2':5',5'':2'',2'''-quaterpyridine. *Chem. Sci.* 2: 540–543.

90. Glasson, C. R. K., Meehan, G. V., Motti, C. A., Clegg, J. K., Turner, P., Jensen, P., and Lindoy, L. F. 2011. New nickel(II) and iron(II) helicates and tetrahedra derived from expanded quaterpyridines. *Dalton Trans.* 40: 10481–10490.

91. Gloe, K., Stephan, H., and Grotjahn, M. 2003. Where is the anion extraction going? *Chem. Eng. Technol.* 26: 1107–1117.

92. Stephan, H., Gloe, K., Kraus, W., Spies, H., Johannsen, B., Wichmann, K., Reck, G., Chand, D. K., Bharadwaj, P. K., Müller, T., Müller, W. M., and Vögtle, F. 2004. Binding and extraction of pertechnetate and perrhenate by azacages. In *Fundamentals and applications of anion separations*, Moyer, B. A., and Singh, R. P., Eds. Kluwer Academic/Plenum, New York, pp. 151–168.

93. Moyer, B. A., Delmau, L. H., Fowler, C. J., Ruas, A., Bostick, D. A., Sessler, J. L., Katayev, E., Pantos, G. D., Llinares, J. M., Hossain, A., Kang, S. O., and Bowman-James, K. 2007. Supramolecular chemistry of environmentally relevant anions. *Adv. Inorg. Chem.* 59: 175–204.

94. Katayev, E. A., Kolesnikov, G. V., and Sessler, J. L. 2009. Molecular recognition of pertechnetate and perrhenate. *Chem. Soc. Rev.* 38: 1572–1586.

95. Kim, S. K., and Sessler, J. L. 2010. Ion pair receptors. *Chem. Soc. Rev.* 39: 3784–3809.

96. Kubik, S., Kirchner, R., Nolting, D., and Seidel, J. 2002. A molecular oyster: a neutral anion receptor containing two cyclopeptide subunits with a remarkable sulfate affinity in aqueous solution. *J. Am. Chem. Soc.* 124: 12752–12760.

97. Katayev, E. A., Boev, N. V., Khrustalev, V. N., Ustynyuk, Y. A., Tananaev, I. G., and Sessler, J. L. 2007. Bipyrrole- and dipyrromethane-based amido-imine hybrid macrocycles. New receptors for oxoanions. *J. Org. Chem.* 72: 2886–2896.

98. Katayev, E. A., Sessler, J. L., Khrustalev, V. N., and Ustynyuk, Y. A. 2007. Synthetic model of the phosphate binding protein: solid-state structure and solution-phase anion binding properties of a large oligopyrrolic macrocycle. *J. Org. Chem.* 72: 7244–7252.

99. Caltagirone, C., Hiscock, J. R., Hursthouse, M. B., Light, M. E., and Gale, P. A. 2008. 1,3-Diindolylureas and 1,3-diindolylthioureas: anion complexation studies in solution and the solid state. *Chem. Eur. J.* 14: 10236–10243.

100. Wu, B., Liang, J., Yang, J., Jia, C., Yang, X.-J., Zhang, H., Tang, N., and Janiak, C. 2008. Sulfate ion encapsulation in caged supramolecular structures assembled by second-sphere coordination. *Chem. Comm.* 1762–1764.

101. Custelcean, R., Remy, P., Bonnesen, P. V., Jiang, D., and Moyer, B. A. 2008. Sulfate recognition by persistent crystalline capsules with rigidified hydrogen-bonding cavities. *Angew. Chem. Int. Ed.* 47: 1866–1870.

102. Custelcean, R., Moyer, B. A., and Hay, B. P. 2005. A coordinatively saturated sulfate encapsulated in a metal-organic framework functionalized with urea hydrogen-bonding groups. *Chem. Comm.* 5971–5973.

103. Li, Y., Mullen, K. M., Claridge, T. D. W., Costa, P. J., Felix, V., and Beer, P. D. 2009. Sulfate anion templated synthesis of a triply interlocked capsule. *Chem. Comm.* 7134–7136.

104. Ravikumar, I., Lakshminarayanan, P. S., Arunachalam, M., Suresh, E., and Ghosh, P. 2009. Anion complexation of a pentafluorophenyl-substituted tripodal urea receptor in solution and the solid state: selectivity toward phosphate. *Dalton Trans.* 4160–4168.

105. Gasperov, V., Galbraith, S. G., Lindoy, L. F., Rumbel, B. R., Skelton, B. W., Tasker, P. A., and White, A. H. 2005. A study of the complexation and extraction of Cu(II) sulfate and Ni(II) sulfate by N_3O_2-donor macrocycles. *Dalton Trans.* 139–145.

106. Jia, C. D., Wu, B. A., Li, S. G., Huang, X. J., Zhao, Q. L., Li, Q. S., and Yang, X. J. 2011. Highly efficient extraction of sulfate ions with a tripodal hexaurea receptor. *Angew. Chem. Int. Ed.* 50: 486–490.

107. Katayev, E. A., Ustynyuk, Y. A., and Sessler, J. L. 2006. Receptors for tetrahedral oxyanions. *Coord. Chem. Rev.* 250: 3004–3037.

108. Kang, S. O., Begum, R. A., and Bowman-James, K. 2006. Amide-based ligands for anion coordination. *Angew. Chem. Int. Ed.* 45: 7882–7894.

109. Caltagirone, C., and Gale, P. A. 2009. Anion receptor chemistry: highlights from 2007. *Chem. Soc. Rev.* 38: 520–563.

110. Gale, P. A., García-Garrido, S. E., and Garric, J. 2008. Anion receptors based on organic frameworks: highlights from 2005 and 2006. *Chem. Soc. Rev.* 37: 151–190.

111. Amendola, V., and Fabbrizzi, L. 2009. Anion receptors that contain metals as structural units. *Chem. Comm.* 513–531.

112. Sessler, J. L., Katayev, E. A., Pantos, G. D., and Ustynyuk, Y. A. 2004. Synthesis and study of a new diamidodipyrromethane macrocycle. An anion receptor with a high sulfate-to-nitrate binding selectivity. *Chem. Comm.* 1276–1277.

113. Fowler, C. J., Haverlock, T. J., Moyer, B. A., Shriver, J. A., Gross, D. E., Marquez, M., Sessler, J. L., Hossain, M. A., and Bowman-James, K. 2008. Enhanced anion exchange for selective sulfate extraction: overcoming the Hofmeister bias. *J. Am. Chem. Soc.* 130: 14386–14387.

114. Yang, Z., Wu, B., Huang, X., Liu, Y., Li, S., Xia, Y., Jia, C., and Yang, X.-J. 2011. Sulfate encapsulation in a metal-assisted capsule based on a mono-pyridylurea ligand. *Chem. Comm.* 47: 2880–2882.

115. Marcus, Y. 1991. Thermodynamics of solvation of ions. Part 5. Gibbs free energy of hydration at 298.15 K. *J. Chem. Soc. Faraday Trans.* 87: 2995–2999.

116. Hofmeister, F. 1888. Zur lehre von der wirkung der Salze (About the science of the effects of salts). *Arch. Exp. Pathol. Pharmakol.* 24: 247–260.
117. Wegmann, D., Weiss, H., Ammann, D., Morf, W. E., Pretsch, E., Sugahara, K., and Simon, W. 1984. Anion-selective liquid membrane electrodes based on lipophilic quaternary ammonium compounds. *Microchim. Acta III* 84: 1–16.
118. Cacace, M. G., Landau, E. M., and Ramsden, J. J. 1997. The Hofmeister series: salt and solvent effects on interfacial phenomena. *Quart. Rev. Biophysics* 30: 241–277.
119. Stephan, H., Gloe, K., Schiessl, P., and Schmidtchen, F. P. 1995. Lipophilic ditopic guanidinium receptors: selective extractants for tetrahedral oxoanions. *Supramol. Chem.* 5: 273–280.
120. Gloe, K., Stephan, H., Krüger, T., Czekalla, M., and Schmidtchen, F. P. 1996. In *Proceedings of ISEC'96*, Vol. 1, Shallcross, D. C., Paimin, R. L., and Prvcic, M., Eds. University of Melbourne, Melbourne, pp. 287–292.
121. Eller, L. R., Stepień, M., Fowler, C. J., Lee, J. T., Sessler, J. L., and Moyer, B. A. 2007. Octamethyl-octaundecylcyclo[8]pyrrole: a promising sulfate anion extractant. *J. Am. Chem. Soc.* 129: 11020–11021.
122. Ritcey, G. M. 2006. *Solvent extraction, principles and applications to process metallurgy*, Vols. 1 and 2. G. M. Ritcey & Associates, Ottawa.
123. Lumetta, G. J. 2004. In *Fundamentals and applications of anion separations*, Moyer, B. A., and Singh, R. P., Eds. Kluwer Academic/Plenum, New York, pp. 107–114.
124. Bianchi, A., Bowman-James, K., and Garcia-Espana, E., Eds. 1997. *Supramolecular chemistry of anions*. Wiley-VCH, New York.
125. Reetz, M. T. 1996. In *Comprehensive supramolecular chemistry*, Vol. 1, Lehn, J.-M., Atwood, J. L., Davies, J. E. O., McNichol, D. D., and Vögtle, F., Eds. Pergamon Press, Oxford, pp. 553–562.
126. Moyer, B. A., Sloop, F. V., Fowler, C. J., Haverlock, T. J., Kang, H. A., Delmau, L. H., Bau, D. M., Hossain, M. A., Bowman-James, K., Shriver, J. A., Bill, N. L., Gross, D. E., Marquez, M., Lynch, V. M., and Sessler, J. L. 2010. Enhanced liquid–liquid anion exchange using macrocylic anion receptors: effect of receptor structure on sulphate-nitrate exchange selectivity. *Supramol. Chem.* 22: 653–671.
127. Borman, C. J., Custelcean, R., Hay, B. P., Bill, N. L., Sessler, J. L., and Moyer, B. A. 2011. Supramolecular organization of calix[4]pyrrole with a methyl-trialkylammonium anion exchanger leads to remarkable reversal of selectivity for sulfate extraction vs. nitrate. *Chem. Comm.* 47: 7611–7613.
128. Custelcean, R., Delmau, L. H., Moyer, B. A., Sessler, J. L., Cho, W.-S., Gross, D., Bates, G. W., Brooks, S. J., Light, M. E., and Gale, P. A. 2005. Calix[4]pyrrole: an old yet new ion-pair receptor. *Angew. Chem. Int. Ed.* 44: 2537–2542.
129. Sessler, J. L., Gross, D. E., Cho, W.-S., Lynch, V. M., Schmidtchen, F. P., Bates, G. W., Light, M. E., and Gale, P. A. 2006. Calix[4]pyrrole as a chloride anion receptor: solvent and countercation effects. *J. Am. Chem. Soc.* 128: 12281–12288.
130. Gross, D. E., Schmidtchen, F. P., Antonius, W., Gale, P. A., Lynch, V. M., and Sessler, J. L. 2008. Cooperative binding of calix[4]pyrrole–anion complexes and alkylammonium cations in halogenated solvents. *Chem. Eur. J.* 14: 7822–7827.
131. Sessler, J. L., Cho, W.-S., Gross, D. E., Shriver, J. A., Lynch, V. M., and Marquez, M. 2005. Anion binding studies of fluorinated expanded calixpyrroles. *J. Org. Chem.* 70: 5982–5986.
132. Anzenbacher, P., Try, A. C., Miyaji, H., Jursíková, K., Lynch, V. M., Marquez, M., and Sessler, J. L. 2000. Fluorinated calix[4]pyrrole and dipyrrolylquinoxaline: neutral anion receptors with augmented affinities and enhanced selectivities. *J. Am. Chem. Soc.* 122: 10268–10272.

133. Clare, J. P., Ayling, A. J., Joos, J.-B., Sisson, A. L., Magro, G., Pérez-Payán, M. N., Lambert, T. N., Shukla, R., Smith, B. D., and Davis, A. P. 2005. Substrate discrimination by cholapod anion receptors: geometric effects and the "affinity–selectivity principle." *J. Am. Chem. Soc.* 127: 10739–10746.

134. Gloe, K., Antonioli, B., Gloe, K., and Stephan, H. 2005. Dendrimers in separation processes. In *Green separation processes*, Afonso, C. A. M., and Crespo, J. G., Eds. Wiley-VCH, Weinheim, pp. 304–322.

135. Stephan, H., Spies, H., Johannsen, B., Klein, L., and Vögtle, F. 1999. Lipophilic urea-functionalized dendrimers as efficient carriers for oxyanions, *Chem. Comm.* 1875–1876.

136. Schmidtchen, F. P., and Berger, M. 1997. Artificial organic host molecules for anions. *Chem. Rev.* 97: 1609–1646.

137. Beer, P. D., and Gale, P. A. 2001. Anion recognition and sensing: the state of the art and future perspectives. *Angew. Chem. Int. Ed.* 40: 486–516.

138. Stephan, H., Spies, H., Johannsen, B., Gloe, K., Gorka M., and Vögtle, F. 2001. Synthesis and host-guest properties of multi-crown dendrimers towards sodium pertechnetate and mercury(II) chloride. *Eur. J. Inorg. Chem.* 2957–2963.

139. Gasa, T. B., Valente, C., and Stoddart, J. F. 2011. Solution-phase counterion effects in supramolecular and mehanostereochemical systems. *Chem. Soc. Rev.* 40: 57–78.

140. Kim, S. K., Sessler, J. L., Gross, D. E., Lee, C. H., Kim, J. S., Lynch, V. M., Delmau, L. H., and Hay, B. P. 2010. A calix[4]arene strapped calix[4]pyrrole: an ion-pair receptor displaying three different cesium cation recognition modes. *J. Am. Chem. Soc.* 132: 5827–5836.

141. Howard, S. T., Hibbs, D. E., Amoroso, A. J., and Platts, J. A. 2006. Quantum-chemical design of cryptand-like ditopic salt binders. *J. Chem. Theory Comput.* 2: 354–363.

142. Deetz, M. J., Shang, M., and Smith, B. D. 2000. A macrobicyclic receptor with versatile recognition properties: simultaneous binding of an ion pair and selective complexation of dimethylsulfoxide. *J. Am. Chem. Soc.* 122: 6201–6207.

143. Mahoney, J. M., Beatty, A. M., and Smith, B. D. 2001. Selective recognition of an alkali halide contact ion-pair. *J. Am. Chem. Soc.* 123: 5847–5858.

144. Mahoney, J. M., Nawaratna, G. U., Beatty, A. M., Duggan, P. J., and Smith, B. D. 2004. Transport of alkali halides through a liquid organic membrane obtaining a ditopic salt-binding receptor. *Inorg. Chem.* 43: 5902–5907.

145. Mahoney, J. M., Beatty, A. M., and Smith, B. D. 2004. Selective solid-liquid extraction of lithium halide salts using a ditopic macrobicyclic receptor. *Inorg. Chem.* 43: 7617–7621.

146. Mahoney, J. M., Stucker, K. A., Jiang, H., Carmichael, I., Brinkmann, N. R., Beatty, A. M., Noll, B. C., and Smith, B. D. 2005. Molecular recognition of trigonal oxyanions using a ditopic salt receptor: evidence for anisotropic shielding surface around nitrate anion. *J. Am. Chem. Soc.* 127: 2922–2928.

147. Liu, H., Shao, X.-B., Jia, M.-X., Jiang, X.-K., Li, Z.-T., and Chen, G.-J. 2005. Selective recognition of sodium cyanide and potassium cyanide by diaza-crown ether-capped Zn-porphyrin receptors in polar solvents. *Tetrahedron* 61: 8095–8100.

148. Piotrowski, H., and Severin, K. 2002. A self-assembled, redox-responsive receptor for the selective extraction of LiCl from water. *Proc. Natl. Acad. Sci.* 99: 4997–5000.

149. Marcus, Y. 1994. A simple empirical model describing the thermodynamics of hydration of ions of widely varying charges, sizes, and shapes. *Biophys. Chem.* 51: 111–127.

150. Kotch, F. W., Sidorov, V., Lam, Y. F., Kayser, K. J., Li, H., Kaucher, M. S., and Davis, J. T. 2003. Water-mediated association provides an ion pair receptor. *J. Am. Chem. Soc.* 125: 15140–15150.

151. Bonar-Law, R. P., and Sanders, J. K. M. 1995. Polyol recognition by a steroid-capped porphyrin. Enhancement and modulation of misfit guest binding by added water or methanol. *J. Am. Chem. Soc.* 117: 259–271.

152. Arena, G., Casnati, A., Mirone, L., Sciotto, D., and Ungaro, R. 1997. A new water-soluble calix[4]arene ditopic receptor rigidified by microsolvation: acid-base and inclusion properties. *Tetrahedron Lett.* 38: 1999–2002.

153. Mansikkamaki, H., Nissinen, M., and Rissanen, K. 2002. Encapsulation of diquats by resorcinarenes: a novel staggered anion-solvent mediated hydrogen bonded capsule. *Chem. Comm.* 1902–1903.

154. Shivanyuk, A., Friese, J. C., Döring, S., and Rebek, J. 2003. Solvent-stabilized molecular capsules. *J. Org. Chem.* 68: 6489–6496.

155. Avram, L., and Cohen, Y. 2002. Spontaneous formation of hexameric resorcinarene capsule in chloroform solution as detected by diffusion NMR. *J. Am. Chem. Soc.* 124: 15148–15149.

156. Avram, L., and Cohen, Y. 2002. The role of water molecules in a resorcinarene capsule as probed by NMR diffusion measurements. *Org. Lett.* 4: 4365–4368.

157. Avram, L., and Cohen, Y. 2003. Effect of a cationic guest on the characteristics of the molecular capsule of resorcinarene: a diffusion NMR study. *Org. Lett.* 5: 1099–1102.

158. Shivanyuk, A., and Rebek, J. 2003. Assembly of resorcinarene capsules in wet solvents. *J. Am. Chem. Soc. 125*: 3432–3233.

159. Sidorov, V., Kotch, F. W., Abdrakhmanova, G., Mizani, R., Fettinger, J. C., and Davis, J. T. 2002. Ion channel formation from a calix[4]arene amide that binds HCl. *J. Am. Chem. Soc.* 124: 2267–2278.

160. Guschlbauer, W., Chantot, J.-F., and Thiele, D. 1990. Four-stranded nucleic acid structures 25 years later: from guanosine gels to telomer DNA. *J. Biomol. Struct. Dyn.* 8: 491–511.

161. Davis, J. T., and Spanda, G. P. 2007. Supramolecular architectures generated by self-assembly of guanosine derivatives. *Chem. Soc. Rev.* 36: 296–313.

162. Galbraith, S. G., and Tasker, P. A. 2005. The design of ligands for the transport of metal salts in extractive metallurgy. *Supramol. Chem.* 17: 191–207.

163. Tasker, P. A., Tong, C. C., and Westra, A. N. 2007. Co-extraction of cations and anions in base metal recovery. *Coord. Chem. Rev.* 251: 1868–1877.

164. Miller, H. A., Laing, N., Parsons, S., Parkin, A., Tasker, P. A., and White, D. J. 2000. Solvent extraction of metal sulfates by zwitterionic forms of ditopic ligands. *J. Chem. Soc. Dalton Trans.* 3773–3782.

165. Coxall, R. A., Lindoy, L. F., Miller, H. A., Parkin, A., Parsons, S., Tasker, P. A., and White, D. J. 2003. Solvent extraction of metal sulfates by zwitterionic forms of ditopic ligands. *Dalton Trans.* 55–64.

166. Galbraith, S. G., Lindoy, L. F., Tasker, P. A., and Plieger, P. G. 2006. Simple procedures for assessing and exploiting the selectivity of anion extraction and transport. *Dalton Trans.* 1134–1136.

167. Galbraith, S. G., Wang, Q., Li, L., Blake, A. J., Wilson, C., Collinson, S. R., Lindoy, L. F., Plieger, P. G., Schröder, M., and Tasker, P. A. 2007. Anion selectivity in zwitterionic amide-functionalised metal salt extractants. *Chem. Eur. J.* 13: 6091–6107.

168. Lin, T., Gasperov, V., Smith, K. J., Tong, C. C., and Tasker, P. A. 2010. Double loading of $ZnCl_2$ by polytopic ligands which co-extract Zn^{2+} and tetrachloridozincate. *Dalton Trans.* 39: 9760–9762.

169. Forgan, R. S., Davidson, J. E., Galbraith, S. G., Henderson, D. K., Parsons, S., Tasker, P. A., and White, F. J. 2008. Transport of metal salts by zwitterionic ligands; simple but highly efficient salicylaldoxime extractants. *Chem. Comm.* 4049–4051.

170. Forgan, R. S., Davidson, J. E., Fabbiani, F. P. A., Galbraith, S. G., Henderson, D. K., Moggach, S. A., Parsons, S., Tasker, P. A., and White, F. J. 2010. Cation and anion selectivity of zwitterionic salicylaldoxime metal salt extractants. *Dalton Trans.* 39: 1763–1770.
171. Bertozzi, C. R., and Kiessling, L. L. 2001. Chemical glycobiology. *Science* 291: 2357–2364.
172. Williams, S. J., and Davies, G. J. 2001. Protein carbohydrate interactions: learning lessons from nature. *Trends Biotechnol.* 19: 356–362.
173. Feizi, T., and Mulloy, B. 2001. Carbohydrates and glycoconjugates. Progress at the frontiers of structural glycobiology. *Curr. Opin. Struct. Biol.* 11: 585–586.
174. Roseman, S. 2001. Reflections on glycobiology. *J. Biol. Chem.* 276: 41527–41542.
175. Aoyama, Y., Tanaka, Y., and Sugahara, S. 1989. Molecular recognition. 5. Molecular recognition of sugars via hydrogen-bonding interaction with a synthetic polyhydroxy macrocycle. *J. Am. Chem. Soc.* 111: 5397–5404.
176. Kobayashi, K., Ikeuchi, F., Inaba, S., and Aoyama, Y. 1992. Molecular recognition. 19. Accommodation of polar guests in unimolecular polyamine-polyhydroxy cores: solubilization of sugars in apolar organic media via intramolecular polar microsolvation. *J. Am. Chem. Soc.* 114: 1105–1107.
177. Ryan, T. J., Lecollinet, G., Velasco, T., and Davis, A. P. 2002. Phase transfer of monosaccharides through noncovalent interactions: selective extraction of glucose by a lipophilic cage receptor. *Proc. Natl. Acad. Sci.* 99: 4863–4866.
178. Metzger, A., Gloe, K., Stephan, H., and Schmidtchen, F. P. 1996. Molecular recognition and phase transfer of underivatized amino acids by a foldable artificial host. *J. Org. Chem.* 61: 2051–2055.
179. Lindoy, L. F. 1989. *The chemistry of macrocyclic ligand complexes.* CUP, Cambridge.
180. Levitskaia, T. G., Bryan, J. C., Sachleben, R. A., Lamb, J. D., and Moyer, B. A. 2000. A surprising host-guest relationship between 1,2-dichloroethane and the cesium complex of tetrabenzo-24-crown-8. *J. Am. Chem. Soc.* 122: 554–562.
181. Ursu, A., and Schmidtchen, F. P. 2012. Selective host–guest binding of anions without auxiliary hydrogen bonds: entropy as an aid to design. *Angew. Chem. Int. Ed.* 51: 242–246.
182. Walter, S. M., Kniep, F., Rout, L., Schmidtchen, F. P., Herdtweck, E., and Huber, S. M. 2012. Isothermal calorimetric titrations on charge-assisted halogen bonds: role of entropy, counterions, solvent, and temperature. *J. Am. Chem. Soc.* 134: 8507–8512.
183. Schmidtchen, F. P. 2012. Isothermal titration calorimetry in supramolecular chemistry. In *Supramolecular chemistry: from molecules to nanomaterials*, Vol. 2, Steed, J. W., and Gale, P. A., Eds. Wiley, New York, pp. 275–296.
184. Schmidtchen, F. P. 2006. Reflections on the construction of anion receptors: is there a sign to resign from design? *Coord. Chem. Rev.* 250: 2918–2928.
185. Schmidtchen, F. P. 2005. Artificial host molecules for the sensing of anions. In Anion sensing, Stibor, I., Ed., *Top. Curr. Chem.* 255: 1–29.
186. Watarai, H. 2008. In *Proceedings of ISEC 2008*, Tucson, Moyer, B. A., Ed. Canadian Institute of Mining, Metallurgy and Petroleum, Montreal, pp. 1029–1034.
187. Chaumont, A., Schurhammer, R., Vayssiere, P., and Wipff, G. 2005. In *Macrocyclic chemistry-current trends and future perspectives*, Gloe, K., Ed. Springer, Dordrecht, pp. 327–348.
188. Chaumont, A., Chevrot, G., Galand, N., Sieffert, N., Schurhammer, R., and Wipff, G. 2008. In *Proceedings of ISEC 2008*, Tucson, Moyer, B. A., Ed. Canadian Institute of Mining, Metallurgy and Petroleum, Montreal, pp. 43–58.
189. Testard, F. Berthon, L., and Zemb, T. 2007. Liquid–liquid extraction: an adsorption isotherm at divided interface? *C. R. Chimie* 10: 1034–1041.

190. Chandler, D. 2005. Interfaces and the driving force of hydrophobic assembly. *Nature* 437: 640–647.
191. Smith, A. G., Tasker, P. A., and White, D. J. 2003. The structures of phenolic oximes and their complexes. *Coord. Chem. Rev.* 241: 61–85.
192. Bauduin, P., Testard, F., Berthon, L., and Zemb, T. 2007. Relation between the hydrophile/hydrophobe ratio of malonamide extractants and the stability of the organic phase: investigation at high extractant concentrations. *Phys. Chem. Chem. Phys.* 9: 3776–3785.
193. Testard, F., Bauduin, P., Martinet, L., Abecassis, B., Berthon, L., Madic, C., and Zemb, Th. 2008. Self-assembling properties of malonamide extractants used in separation processes. *Radiochim. Acta* 96: 265–272.
194. Meridiano, Y., Berthon, L., Crozes, X., Sorel, C., Dannus, P., Antonio, M. R., Chiarizia, R., and Zemb, T. 2009. Aggregation in organic solutions of malonamides: consequences for water extraction. *Solvent Extr. Ion Exch.* 27: 607–637.
195. Berthon, L., Testard, F., Martinet, L., Zemb, T., and Madic, C. 2010. Influence of the extracted solute on the aggregation of malonamide extractant in organic phases: consequences for phase stability. *C. R. Chimie* 13: 1326–1334.

2 Supramolecular Interactions in the Outer Coordination Spheres of Extracted Metal Ions

Peter A. Tasker
School of Chemistry, University of
Edinburgh, Edinburgh, UK

Benjamin D. Roach
Chemical Sciences Division, Oak Ridge National
Laboratory, Oak Ridge, Tennessee

CONTENTS

2.1   INTRODUCTION

Using solvent extraction to effect the separation and concentration of metal values in extractive metallurgy in flowsheets of the type shown in Figure 2.1 is now well established.[1-4] Processes can be operated continuously and on large scales using technology first established in the Manhattan Project in the 1940s.[2,5]

For most large-scale operations, relatively high-boiling hydrocarbons are the preferred water-immiscible solvents on the grounds of cost and safety. Also, their very low solubility in water reduces the risk of pollution and promotes phase disengagement. The very low polarity of these solvents favors the formation of secondary bonds, particularly hydrogen bonds (H-bonds) between the ligands

49

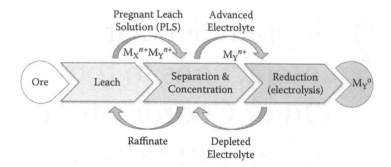

FIGURE 2.1   The key steps in a hydrometallurgical flowsheet.

SELECTIVITY
TRENGTH
SPEED
SEPARATION
SOLUBILITY
STABILITY
SAFETY
SYNTHESIS
SYSTEM

FIGURE 2.2   The S-criteria needed for an efficient solvent extraction process in extractive hydrometallurgy. (From Tasker, P. A. et al., Eds., *Comprehensive Coordination Chemistry II*, Elsevier Ltd., Oxford, 2004.)

used as metal extractants. While it has been recognized for some time[6,7] that the formation of supramolecular assemblies arising from such secondary bonding is very important in defining the modes of action of extractants, only relatively recently have attempts been made at the design stage to exploit such interactions in the outer spheres of metal complexes to control the strength and selectivity of new reagents. Supramolecular chemistry is involved in most aspects of the formulation of extractant mixtures to achieve the criteria needed for an efficient process. These criteria are listed in Figure 2.2, in which "separation" refers to phase disengagement and "system" to ensuring that all components of the pregnant leach solution are accounted for.

This chapter focuses on the better-understood aspects of supramolecular chemistry that influence the thermodynamic stability of metal complexes in hydrocarbon solvents. The effects of secondary bonding on the formation of assemblies at liquid–liquid interfaces that have a major effect on phase-transfer processes and phase disengagement (*speed* and *separation* in Figure 2.2) have very important practical consequences,[8] but are not considered in any detail because the structure and bonding in the assemblies are generally poorly defined. However, some of the methodologies, particularly computational techniques, needed to understand the supramolecular chemistry of interfacial phenomena (see Chapter 5) are similar to those discussed in this review for assemblies formed in the bulk solution.

In the sections that follow, we consider how supramolecular chemistry influences the *strength* and *selectivity* of extraction for the transport of either metal cations, metalate anions, or metal salts into water-immiscible solvents.

2.2 EXTRACTION OF METAL CATIONS

The transport of metal cations into a water-immiscible solvent without anion co-extraction requires exchange with another cation to ensure charge neutrality of the system. Most frequently, the exchanged cations are protons, and organic acids, LH, are used as extractants in equilibria of the type shown in Equation (2.1), where metal loading and stripping are pH dependent.[1]

$$n\mathrm{LH}_{org} + M^{n+} \rightleftharpoons [ML_n]_{org} + nH^+ \qquad (2.1)$$

The presence of highly polarized X–H groups in the acidic extractants favors the formation of H-bonded assemblies in hydrocarbon solvents. The propensity for hydrophobic carboxylic acids and phosphorus(V) acids to form stable H-bonded assemblies provides an explanation for the unusual stoichiometries observed when they are used in solvent extraction processes. Commonly, two or more times the number of moles of acid that are required to give a neutral complex as in Equation (2.1) are involved in forming assemblies in water-immiscible solvents.[6] Carboxylic acids can show particularly complicated speciation on metal complex formation, both in the solid state and in solution.[6,7,9] The motif involved in the classic carboxylic acid dimer, $[RCO_2H]_2$, shown in Figure 2.3a, is particularly stable in hydrocarbon solvents and can be preserved on the formation of mono- and dinuclear complexes (Figure 2.3b and c). This, and the observation that water can be retained in the inner coordination sphere of the metal, is of particular relevance to metal extraction processes, leading to the generic equation in Figure 2.3. With copper, for example, the mononuclear $[Cu(RCO_2)_2]$, $[Cu(RCO_2)_2 \cdot (RCO_2H)]$, $[Cu(RCO_2)_2 \cdot 2(RCO_2H)]$, $[Cu(RCO_2)_2 \cdot 4(RCO_2H)]$, and the dinuclear $[Cu(RCO_2)_2]_2$, $[Cu(RCO_2)_2 \cdot (RCO_2H)]_2$, and $[Cu(RCO_2)_2 \cdot 2(RCO_2H)]_2$ species have all been suggested.[10]

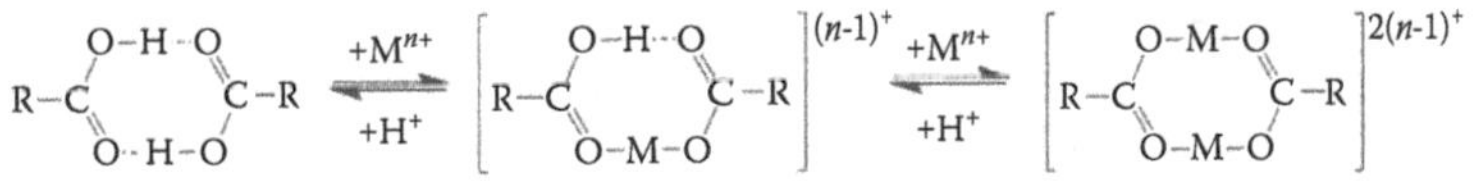

$$M^{n+} + (^n/_2 + {}^m/_2)[RCO_2H]_2 + pH_2O \geq 1/x[M(RCO_2)_n \cdot m(RCO_2H) \cdot p(H_2O)]_x + nH^+$$

FIGURE 2.3 Motifs involved in the association of carboxylic acids in metal-free and mono- and dinuclear complexes.

In principle, very large polynuclear carboxylate assemblies could be formed in solvent extraction experiments, similar to those that have been crystallized from a single liquid phase and characterized by X-ray structure determination.[11–14] These most commonly involve metals in >+2 oxidation states and separate most readily at elevated pH when they often also contain bridging hydroxide or alkoxide groups, or when the solution contains other small bridging anions such as fluoride.[15] As the formation of such assemblies is usually slow, and elevated temperatures and pressure are often required,[16] they are unlikely to be observed frequently in two-phase systems. However, in the context of supramolecular chemistry of the type considered in this chapter, the fact that alkylammonium and other cations are frequently used[17] to template the assembly of polynuclear carboxylate complexes indicates the importance of secondary bonding in the outer coordination spheres in determining the outcome of complex formation.

Assemblies with similar motifs can be formed[6,18] by the phosphorus(V) acid represented generically in Figure 2.4. These usually have lower pK_a values and are stronger extractants than hydrophobic carboxylic acids.[7] Interligand H-bonding of the type shown in Figure 2.4 makes it possible to use ligand design features operating in the *outer* coordination sphere[19] to influence the properties of the *inner* coordination sphere of complexed metal ions.

Extraction of divalent base metal cations by the commercial reagent di-(2-ethylhexyl)phosphoric acid (D2EHPA = LH) usually gives 4:1 complexes, $[ML_2(LH)_2]$.[6] The eight-membered pseudochelate rings in these complexes can subtend O–M–O angles significantly greater than 90°, and consequently, D2EHPA shows a marked selectivity for base metals that favor tetrahedral coordination geometry. The ease of extraction of the divalent metals of the first transition series reflects this and follows[6] the sequence Zn > Cr > Mn > Cu > Fe > Co > Ni $\simeq$ V, which is not the Irving–Williams order[20] of stability of first-row transition M^{2+} complexes.[7] The high affinity for zinc that results from

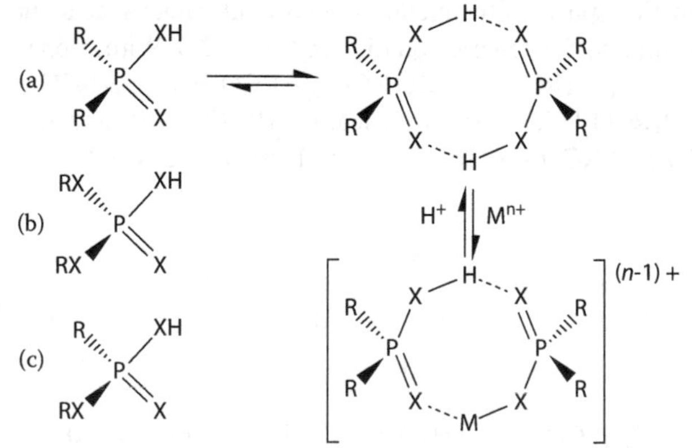

FIGURE 2.4 Formation of eight-membered pseudochelate rings by dialkylphosphinic acid extractants (a) X = O or S. Similar structures can be formed by diesters of phosphoric acid (b) Y = O, or monoesters of phosphonic acids (c) Y = O.

the supramolecular chemistry of D2EHPA forms the basis for its application in a recently commissioned plant for the hydrometallurgical recovery of zinc on a 150,000 tonne p.a. scale.[21,22]

The selective recovery of cobalt(II) from nickel(II)-containing solutions is an important issue in the development of circuits to process lateritic ores[23] and can be also be approached by exploiting the former's greater preference for tetrahedral coordination geometry and the propensity for phosphorus(V) extractants to favor this geometry. The use of D2EHPA (LH) to effect this separation has been the subject of many fundamental investigations.[6] It is a poor extractant for nickel, as this metal shows a preference for a pseudo-octahedral structure in which two axial sites are occupied by either neutral extractant molecules, LH, extractant dimers [(LH)$_2$], or water molecules, depending on extractant concentration. The four equatorial sites are occupied by the two monoanionic chelate units, LH·L$^-$, with interligand H-bonding as shown in Figure 2.5. For cobalt it is proposed that both tetrahedral and octahedral forms exist in equilibrium. The entropy increase associated with the dissociation of the octahedral cobalt complex to give the tetrahedral form has been exploited to enhance the separation of cobalt from nickel by performing the separation at elevated temperatures.[24] As the theoretical maximum loading of cobalt is approached, based on formation of the neutral 1:2 complexes [CoL$_2$], there are no protons left to provide H-bonds between extractant molecules, and it has been proposed that oligomeric complexes, [CoL$_2$]$_n$, form in which the (RO)$_2$PO$_2^-$ units act as a bridging ligand. The stepwise change in viscosity observed upon high metal loading in working systems employing D2EHPA (and other phosphorus acid reagents) is attributed to the formation of such high-molecular-weight complexes.[6]

Subtle differences in the structure of the alkyl substituents in the organophosphorus acids shown in Figure 2.4 influence the stabilities of the interligand H-bonded eight-membered chelate rings in complexes, and thus have an important role in tuning the strength and selectivity of extractants. This has been most extensively investigated for the separation of cobalt and nickel, and cobalt from acidic media.[25-33] Cyanex® 272 and bis(1,3,3-trimethylbutyl)phosphoric acid (Figure 2.6)

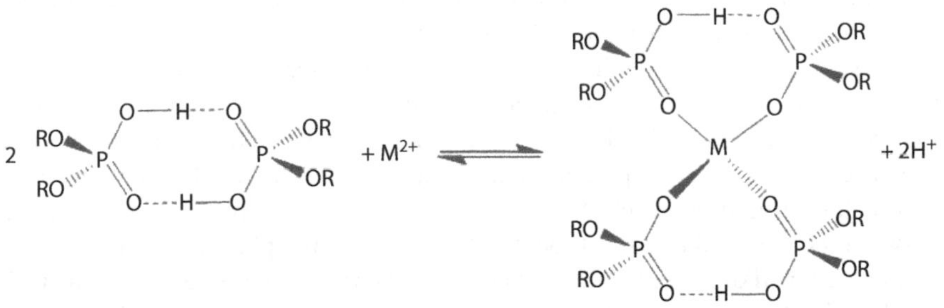

FIGURE 2.5 The 4:1 stoichiometry that is commonly involved in extraction of divalent metal cations by phosphoric acid diesters, LH, such as D2EHPA in pH-swing extractions of the type shown in Equation (2.1).

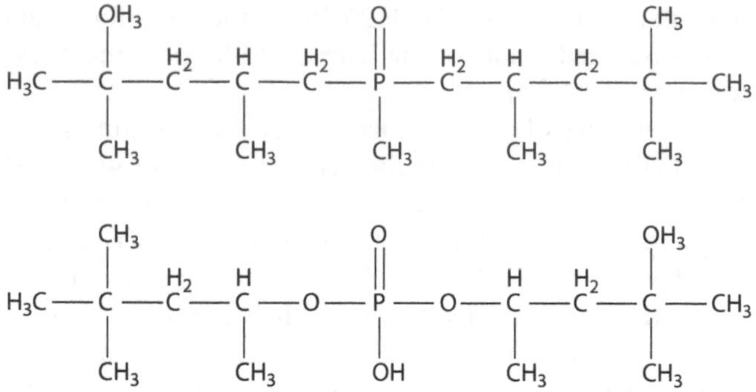

FIGURE 2.6 Structures of the phosphinic and phosphoric acids Cyanex 272 (top) and bis(1,3,3-trimethylbutyl)phosphoric acid (bottom).

TABLE 2.1

Calculated R–P–R' Angles and Dihedral Angles between the RPR' and OPO' Planes in D2EHPA, PC-88A, and Cyanex 272

	Substituent			Dihedral
Extractant	R	R'	R–P–R' Angle/°	Angle/°
D2EHPA	Oxy-2-ethylhexyl	Oxy-2-ethylhexyl	104.7	90.2
PC-88A	Oxy-2-ethylhexyl	2-Ethylhexyl	115.7	91.2
Cyanex 272	2,4,4-Trimethylpentyl	2,4,4-Trimethylpentyl	120.1	89.5

Source: Zhu, T., in *International Solvent Extraction Conference*, Cape Town, South Africa, March 17–21, 2002, pp. 203–207.

---

have separation factors $\alpha_{ex(Co/Ni)}$ of 3.8 × 10^2 and 3.7, respectively.[34] The steric bulk of the alkoxy groups on the phosphorus atom in the latter is presumed to be similar to that of the alkyl group in Cyanex 272 because, although the oxygen atom carries no hydrogen substituents, the P–O bond length (on average 162 pm) is significantly shorter than the P–C bond (on average 185 pm).

These observation led Zhu to investigate[35] the steric and electronic effects arising from variation in substituents in the commercial reagents D2EHPA, PC-88A, and Cyanex 272 (Table 2.1) using molecular mechanics and molecular orbital calculations. The predominant structural parameter defining the preference for formation of tetrahedral over octahedral complexes was the angle between the alkyl-alkoxy substituents, R and R', and the phosphorus atom.

Replacing alkoxy with alkyl groups results in an increase in the R–P–R' bond angle. The larger bond angle in the phosphinic acid (Cyanex 272) leads to greater steric hindrance between the equatorial and axial ligands in an octahedral complex. This is consistent with the observed[35] selectivity for cobalt over nickel, increasing in the order: D2EHPA < PC-88A < Cyanex 272. The effect

TABLE 2.2
Values of pK_as, Extraction Constants, Separation Factors, and Calculated Point Charges in D2EHPA, PC-88A, and Cyanex 272

Extractant	pK_a	$K_{ex(Co)}$	$K_{ex(Ni)}$	$\alpha_{ex(Co/Ni)}$	Point Charge (q) P	O	O'
D2EHPA	3.57	5.50×10^{-6}	1.10×10^{-6}	5	0.161	−0.345	−0.772
PC-88A	4.10	1.07×10^{-6}	1.51×10^{-8}	71	0.028	−0.367	−0.794
Cyanex 272	5.05	1.66×10^{-7}	2.00×10^{-10}	830	0.092	−0.393	−0.814

Source: Zhu, T., in *International Solvent Extraction Conference*, Cape Town, South Africa, March 17–21, 2002, pp. 203–207.

complements that of the interligand H-bonding, which also favors tetrahedral geometry (see above).

The electron density of the oxygen atoms of the reagents also plays a role in determining their strength/selectivity (Table 2.2).[35] The calculated point charges on the oxygen donor atoms follow the order D2EHPA < PC-88A < Cyanex 272, consistent with the increasing pK_a along the series and decreased extraction constants for both metals. While increasing the acidity clearly favors metal extraction equilibria such as those shown in Figure 2.5, there will be a concomitant decrease in the basicity of the anionic donor atoms that will lower the stability of the resulting metal complexes. In practice, the strength and selectivity of phosphorus acid extractants depend on a number of factors, but importantly for this review, these include the supramolecular effects in which ligand–ligand interactions in the outer coordination spheres of complexed metal ions have a significant influence on the strength and geometry of bonds formed in the inner coordination spheres.

Hydrophobic carboxylic acids[6,7,19] are generally relatively weak and nonselective extractants, but their propensity to form supramolecular assemblies has been exploited in enhancing their performance by using them in combination with synergists. Versatic® 10, a multibranched decanoic acid, has been used to recover nickel from sulfate feeds after iron has been removed by raising pH, but as it is a weak extractant, several extraction stages and interstage neutralization are needed to achieve effective recovery.[36] The addition of the pyridine-3,5-dicarboxylic acid ester, CLX50® (Figure 2.7), enhances its strength and selectivity over alkaline earths.[37,38] The origin of this synergism in terms of ligand–ligand secondary bonding is not well understood. More information on the mode of action of the synergism of nickel extraction is available[39,40] on the use of carboxylic acids such as Versatic 10 in combination with α-hydroxyoximes (e.g., LIX® 63, shown in Figure 2.7). H-bonding between ligands in the outer coordination sphere of a complexed Ni(II) cation is thought to involve the deprotonation of the Versatic acid, rather than the hydroxyoxime, as in Equation (2.2).

$$Ni^{2+} + 2RCOOH_{org} + 2LIX63_{org} \rightleftharpoons [Ni(RCOO \cdot LIX63)_2] + 2H^+ \quad (2.2)$$

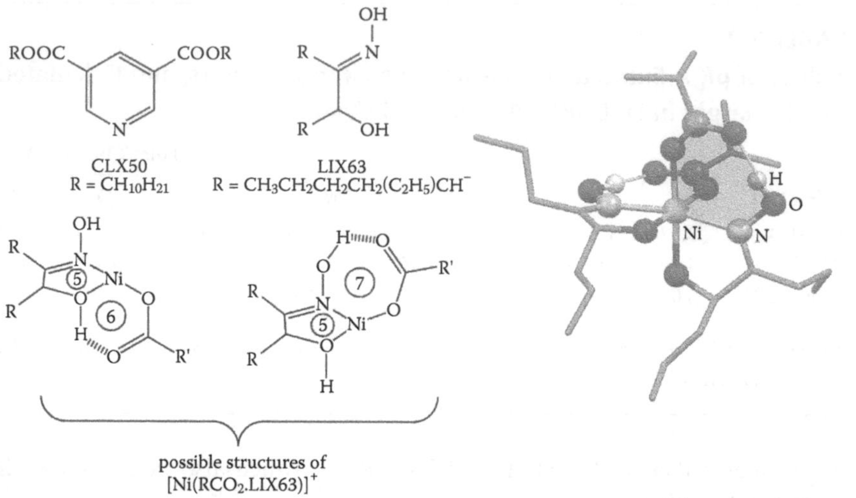

FIGURE 2.7 CLX50 and LIX63 synergists for Versatic acid 10 and the outer-sphere H-bonding found in the crystal structure of a model system using a di-*n*-propyl analogue of LIX63 (R = CH$_3$CH$_2$CH$_2$–) and isobutyric acid (R′ = (CH$_3$)$_2$CH–) to replace Versatic acid. (From Barnard, K. R. et al., *Solvent Extr. Ion Exch.*, 28, 2010, 778–792.)

X-ray structure determinations of model systems (see Figure 2.7)[41] have defined the likely modes of ligand–ligand H-bonding in the outer coordination sphere, confirming that it is the carboxylic acid rather than the oxime that has been deprotonated in forming the complex. These arrangements effectively provide the Ni(II) ion with two planar pseudotridentate [NO$_2$]⁻ donor sets. A sequence of 5/7-membered chelate rings results from the OH group of the oxime unit acting as the H-bond donor to the carboxylate rather than the α-hydroxyl group, which would give the 5/6-membered chelate ring sequence shown on the left in Figure 2.7.

Striking examples of synergism for base metal extraction that are likely to depend on similar supramolecular chemistry have been known for some time.[42–46] One of the earliest[42] is the combination of LIX63 and D2EHPA, which prompted Flett et al. to investigate mixtures of LIX 63 with carboxylic acids.[43–45] Synergistic systems involving combinations of neutral and acidic extractants for the separation of trivalent lanthanides and transuranic elements have been reviewed recently.[47] Three very different types of neutral extractants have been studied (diamides, carbamoylmethylphosphine oxides, and multidentate heterocyclic nitrogen compounds). These were shown to have great potential in achieving a single-step separation of trivalent actinides from acidic wastes,[47] but the "fundamental chemistry underlying these combined systems is not yet well understood." It is clearly timely to investigate the supramolecular chemistry involved in these and earlier recorded examples of synergism.

One of the best examples of ligand-to-ligand H-bonding enhancing the performance of cation extractants is provided by the use of phenolic oximes[48] in the recovery of copper from sulfate streams. These reagents currently account for approximately 25% of worldwide copper production.[49,50] Their strength and selectivity as copper extractants is associated with the formation of the 14-membered pseudomacrocyclic H-bonded assembly (Figure 2.8), which provides a cavity of nearly ideal size for the Cu(II) ion.[51] This motif is also observed in the preorganized ligand dimers (Figure 2.8) that have *intra*molecular phenolic-to-imine and *inter*molecular oxime-to-phenol H-bonds.[51,52] The same H-bonding sequences can also generate ribbon structures in the solid state.[51,52]

The importance of the H-bonding between extractant molecules has been probed by investigating the effect of introducing H-bond acceptor groups (X in Figure 2.9) into the 3-positions of salicylaldoximes.[52,53]

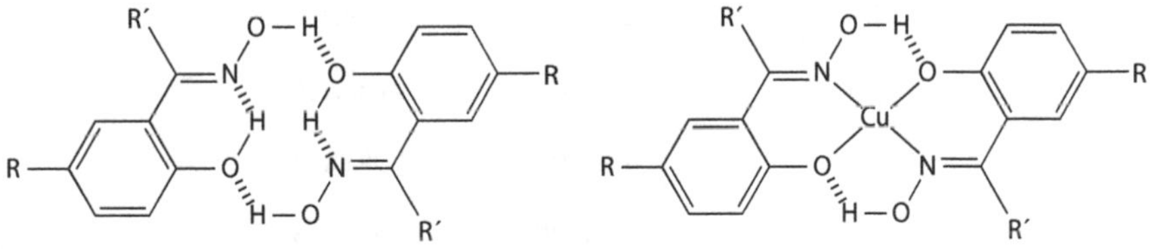

FIGURE 2.8 The 14-membered pseudomacrocyclic structures formed by the phenolic oxime extractants. (From Smith, A. G. et al., *Coord. Chem. Rev.*, 241, 2003, 61–85.)

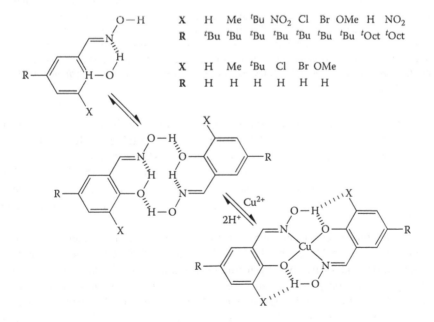

X	H	Me	^{t}Bu	NO$_2$	Cl	Br	OMe	H	NO$_2$
R	^{t}Bu	^{t}Bu	^{t}Bu	^{t}Bu	^{t}Bu	^{t}Bu	^{t}Bu	^{t}Oct	^{t}Oct

X	H	Me	^{t}Bu	Cl	Br	OMe
R	H	H	H	H	H	H

FIGURE 2.9 Possible buttressing of interligand H-bonding by 3-X substituents in the formation of dimers and pseudomacrocyclic Cu^{II} complexes of salicylaldoximes.

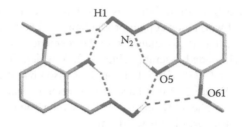

3-X-group	Hole Size/Å[a]	
	5-t-Bu	5-H
H	_[b]	2.0048(15)
Me	2.003(2)	2.0237(18)
t-Bu	2.025(1)	2.0367(19)
Cl	1.973(8)	1.9837(12)
Br	1.968(8)	1.9726(53)
OMe	_[b]	1.9492(19)

FIGURE 2.10   The structure of the 3-methoxysalicylaldoxime dimer in the solid state showing the interaction (H1⋯O61 = 2.96(3) Å) between the 3-OMe group and the oximic OH group and a comparison of the size of the cavity defined by the N_2O_2 donor set (1.9492(19) Å) with those in related pseudomacrocyclic dimers. The mean distance of the oximic nitrogen and phenolic oxygen atoms from their centroid; these compounds form assemblies of linear ribbons rather than pseudomacrocyclic dimers. (From Smith, A. G. et al., *Coord. Chem. Rev.*, 241, 2003, 61–85; Forgan, R. S. et al., *Inorg. Chem.*, 50, 2011, 4515–4522.)

Salicylaldoximes that contain a *t*-butyl group or no hydrocarbon substituent in the 5-position (Figure 2.10) form crystals suitable for X-ray structure determination. The "cavity sizes," the average distance of the O and N donor atoms from their centroid, are significantly smaller when the 3-X substituent is a H-bond acceptor. The gas phase energies of dimerization calculated by density function theory (DFT) methods correlate with the cavity sizes (Table 2.3).[51,52]

Interaction energies (Table 2.3) within the salicylaldoxime dimers in the solid state have been analyzed[53] by the PIXEL method,[54,55] which models Coulombic polarization and dispersive repulsion contributions. *Differences* between the energy contributions, $E_{Coulombic}$, etc., are presented in Table 2.3. These "normalized" values demonstrate that the ligands with H-bond acceptor substituents (OMe, Br, Cl) show the most favorable ligand–ligand attraction energies. The $E_{Coulombic}$ term is most favorable for the methoxy-substituted compound, which accords with the good H-bond acceptor properties of the OMe group,[56] despite the much longer

TABLE 2.3

Interaction Energies between the Halves of Each of the 3-X-Substituted Salicylaldoxime Dimers Relative to Their Unsubstituted Analogue (X = H)

3-X Substituent	OMe	Br	Cl	H	Me	^{t}Bu
$E_{Coulombic}$/kJ mol^{-1}	−4.8	−2.2	−0.5	0.0	+0.2	+2.1
$E_{repulsion}$/kJ mol^{-1}	−0.4	+0.7	0.0	0.0	+0.6	+6.2
$E_{polarization}$/kJ mol^{-1}	−1.0	−1.4	−1.0	0.0	−0.8	−2.7
$E_{dispersion}$/kJ mol^{-1}	−0.8	−1.3	−1.5	0.0	−1.5	−5.8
E_{TOTAL}/kJ mol^{-1}	**−7.0**	**−4.2**	**−3.0**	**0.0**	**−1.5**	**−0.2**
Cavity radius/Å (XRD)	1.949(2)	1.973(5)	1.982(1)	2.005(1)	2.024(2)	2.037(2)
Cavity radius/Å (DFT)	1.972	1.966	1.988	2.005	2.007	2.045
$\Delta H_{dimerization}$/kJ mol^{-1} (DFT)	**−50.2**	**−45.2**	**−45.7**	**−40.7**	**−39.3**	**−29.1**

Source: Forgan, R. S. et al., *Inorg. Chem.*, 50, 2011, 4515–4522; Forgan, R. S. et al., *Chem. Commun.*, 2007, 4940–4942.

than ideal distance between the methoxy and phenolic oxygen atoms for a strong H-bond (see Figure 2.10). The large repulsion term (E_{rep}) for the 3-t-butyl substituted compound is consistent with the bulk of this substituent and with this dimer having the largest cavity.

The influence of electronegative 3-X substituents in buttressing the H-bonding between salicylaldoxime molecules on their strength as metal extractants is of considerable practical importance. In the early stages of development of the phenolic oxime reagents, it was reported that 3-nitro- and 3-chloro-substituted ligands were stronger extractants than unsubstituted derivatives.[57,58] This feature was ascribed to the increased acidity of the phenol group.[59] This is an important factor, but an analysis of the strengths of a coherent series of 5-t-butyl-salicylaldoximes as copper extractants (Figure 2.11), coupled with DFT calculations, suggests that buttressing is the dominant factor.[52] 3-Substitution clearly has a major influence on extractant strength, which is found to follow the order Br > NO$_2$ > Cl > OMe > Me ≥ H > ^{t}Bu. The difference in the pH$_{0.5}$ values for copper loading by the 3-t-butyl- and 3-bromo-substituted extractants represents a difference in the distribution coefficient for copper extraction by two orders of magnitude.

The size of the H-bond acceptor atom X appears to be more important than its electronegativity, based on the 3-Br reagent being a stronger extractant than its 3-Cl analogue. Formation enthalpies for the Cu complexes obtained using a high level of theory DFT calculations [B3LYP/6–31+G(d,p)] on the H-, MeO-, Cl-, and Br-substituted extractants correlate remarkably well (Figure 2.12) with the observed strengths as copper extractants.[52] As the incorporation of a methoxy substituent increases extractant strength, despite lowering the acidity of the phenol, it appears that H-bond buttressing is the dominant factor in determining the performance of these extractants.

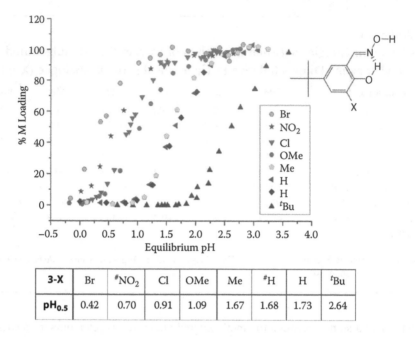

3-X	Br	#NO$_2$	Cl	OMe	Me	#H	H	^{t}Bu
pH$_{0.5}$	0.42	0.70	0.91	1.09	1.67	1.68	1.73	2.64

FIGURE 2.11   pH profiles and pH$_{0.5}$ values for Cu(II) loading by 0.01 M CHCl$_3$ solutions of the 3-X-5-^{t}Bu-salicylaldoxime ligands from equal volumes of 0.01 M aqueous CuSO$_4$. #: extractants carrying a 5-nonyl substituent instead of 5-^{t}Bu.

As mentioned in the introduction to this chapter, most commercial operations that involve the solvent extraction of metal cations usually have formulations of modifiers and phase disengagement additives in the water-immiscible phase.[59,60] Particularly complicated, multicomponent formulations are used in the nuclear industry.[61,62] Almost invariably, the modifiers used to tune the strength and selectivity of metal recovery are H-bond donors or acceptors, and their modes of action involve supramolecular chemistry to form assemblies with both complexed and metal-free extractant molecules. Studies have been carried out to understand the stoichiometries of the resultant metal-containing assemblies, including the use of computational methods to define the intermolecular bonding involved.[63] In many such systems, the extraction process can also involve the transport of a metalate anion from the aqueous phase, which is considered in the next section.

2.3   EXTRACTION OF METALATE ANIONS

The solvent extraction of metalate anions, MX$_y{}^{n-}$, from an aqueous solution has commonly been ascribed to the formation of ion pairs in the water-immiscible phase.[64,65] In practice, these will involve assemblies in which a *combination* of secondary bonds is responsible for the formation of an assembly between the metalate anions and cationic extractant molecules. When the latter are generated

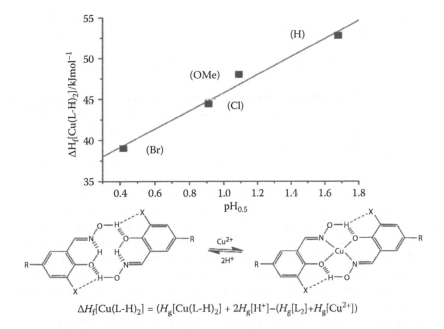

$$\Delta H_f[Cu(L\text{-}H)_2] = (H_g[Cu(L\text{-}H)_2] + 2H_g[H^+] - (H_g[L_2] + H_g[Cu^{2+}]))$$

FIGURE 2.12 The correlation between calculated (B3LYP/6–31+G(d,p)) enthalpies of formation (ΔH_f) of the copper(II) complexes, $[Cu(L\text{-}H)_2]$, in the gas phase with the measured extraction strength ($pH_{0.5}$) of the corresponding 3-substituted salicylaldoximes, showing the equilibrium and equation used to calculate the ΔH_f values. (From Forgan, R. S. et al., *Inorg. Chem.*, 50, 2011, 4515–4522.)

by protonation in a "pH swing" process as in Equation (2.3), it is possible for the added proton to be involved in H-bond formation to the metalate.

$$nH^+ + nL_{org} + MX_y^{n-} \rightleftharpoons [(LH)_nMX_y]_{org} \tag{2.3}$$

When the extractant carries a permanent positive charge and an "anion swing" process is used to control the loading and stripping of the metalate (Equation (2.4)), other, often weaker, types of H-bonds will complement the electrostatic forces in the assembly with the metalate anion.

$$n[R^+Y^-]_{org} + MX_y^{n-} \rightleftharpoons [R_nMX_y]_{org} + nY^- \tag{2.4}$$

Equations (2.3) and (2.4) describe the simplest types of supramolecular assemblies that can be formed in metalate extraction. Hydration of the cationic extractant, for example, by using hydroxonium ions instead of protons to generate the cationic component of the assembly, and the incorporation of "solvating" neutral extractant molecules, as shown in Equation (2.5), will lead to more complex structures. So too will the introduction of neutral modifier molecules, such as phosphine oxides, which are often used to enhance the solubility of assemblies and to improve phase disengagement in practicable operations.

$$n\text{H}_3\text{O}^+ + m\text{L}_{org} + \text{MX}_y^{n-} \rightleftharpoons [(\text{L})_{m-n}(\text{L} \cdot \text{H}_3\text{O})_n\text{MX}_y]_{org} \qquad (2.5)$$

Trialkylamines, trialkyphosphine oxides, and amides are most commonly used in the "proton swing" processes represented by Equations (2.3) and (2.5).[6,7,66–70] Quaternary ammonium salts are the most commonly used reagents for extractions represented by Equation (2.4).[7,71,72]

With the emergence of the branch of supramolecular chemistry that deals with anion recognition and the design of selective complexing agents,[73–83] it should be possible to design extractants that show the strength and selectivity needed to achieve the *concentration* and *separation* operations in particular hydrometallurgical circuits. Software, such as Host Designer,[78] which has been developed to define receptors that will be a good fit for various anions, can be used to assist in this endeavor, which, until recently, has been undertaken on an empirical basis, largely using existing reagents, often in blends. In the examples of metalate extractants described below, we focus on systems in which the bonding in the host-guest assemblies has been investigated.

Reagents that extract chloridometalates are of particular interest, because chloride-leaching processes have been developed for most base, precious, and f-block metals,[69,70,84,85] and as the pregnant leach solutions contain high concentrations of chloride, the formation of chloridometalate species, $[\text{MCl}_x]^{y-}$, is favored.

As the precious metals in their higher oxidation states form particularly kinetically inert $[\text{MCl}_x]^{y-}$ complexes,[86] they are well suited to studies to define the design criteria for good extractants because the speciation in aqueous solutions is often well defined and species do not interconvert on the timescales of extraction experiments. Computational and NMR studies on the solvation of $[\text{PdCl}_6]^{2-}$ and $[\text{PtCl}_6]^{2-}$, and an analysis of the solid-state structures of their salts by X-ray crystallography, suggest that the most favorable locations for H-bond donors to interact with the chlorine atoms are the centers of faces and edges of the octahedral.[87–92] On this basis, tripodal reagents (L) with a protonatable bridgehead nitrogen atom and H-bond donors in the three arms should be able to address the outer coordination sphere of $[\text{PtCl}_6]^{2-}$ in the manner shown in Figure 2.13 (formation of an extractable, charge-neutral assembly $[(\text{LH})_2\text{PtCl}_6]$ requires a second LH^+ unit centered over the opposite face).

Figure 2.14 provides examples of tripodal extractants containing either amide or urea groups in each arm and compares their $[\text{PtCl}_6]^{2-}$ loading efficiency with that of tri-*n*-octylamine (TOA), which is a model for the commercial trialkylamines.[93,94] The incorporation of the additional H-bond donors clearly enhances extractant strength and selectivity over chloride, which is present in 60-fold excess in the aqueous feed solutions, but no evidence was available from X-ray structure determinations of $[(\text{LH})_2\text{PtCl}_6]$ complexes that could be crystallized for structures having the threefold symmetry shown in Figure 2.13. In the solid-state structures, the extractants do use their H-bonding groups to address the chlorine atoms

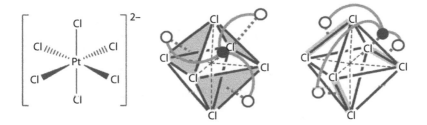

FIGURE 2.13 Tripodal reagents with H-bond donor groups (hollow spheres) to address the faces or edges of a $PtCl_6^{2-}$ octahedron and a bridgehead ammonium NH^+ group (black sphere) positioned centrally over a triangular face of the octahedron. (From Bell, K. J. et al., *Angew. Chem. Int. Ed.*, 47, 2008, 1745–1748.)

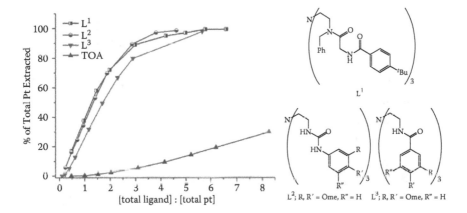

FIGURE 2.14 Platinum recovery from a 0.001 M solution of H_2PtCl_6 in 0.6 M HCl into equal volumes of $CHCl_3$ solutions of the tripodal amides, L, as a function of [L]:[Pt]. (From Bell, K. J. et al., *Angew. Chem. Int. Ed.*, 47, 2008, 1745–1748; Bell, K. J. et al., *Solvent Extr.: Fundam. Ind. Appl., Proc. ISEC 2008, Int. Solvent Extr. Conf.*, 2008, 1457–1462.)

in the chloridoplatinate, but form bridges between neighboring $[PtCl_6]^{2-}$ units to give polymeric structures.[95]

When using solvent extraction to recover chloridometalates via a pH-swing process of the type shown in Equation (2.3), attention needs to be paid to the possibility that reagents could enter the inner coordination sphere of the metal when the pH is raised to strip the metal from the loaded organic phase as in Equation (2.6). This could make metal recovery difficult and is more likely to become an issue when kinetically labile base metal ions are involved, as these exchange chloride ligands rapidly.[86,96]

$$[(LH)_nMCl_y]_{org} + nOH^- \rightleftharpoons [(L)_nMCl_{y-n}]_{org} + nH_2O + nCl^- \qquad (2.6)$$

One approach to suppressing inner-sphere complex formation is to increase the steric hindrance around the protonatable atom to the point where it effectively

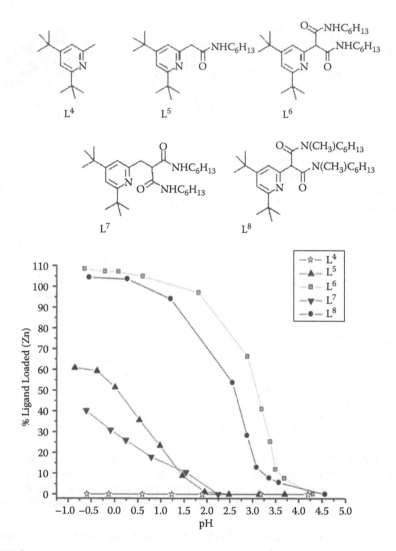

FIGURE 2.15 Amidopyridyl reagents with sterically hindered pyridine N atoms and their $ZnCl_4^{2-}$ uptake in 0.01 M CHCl$_3$ solutions from aqueous ZnCl$_2$ (0.1 M, total HCl/LiCl 6.0 M). (From Ellis, R. J. et al., *Chem. Eur. J.*, 18, 2012, 7715–7728.)

becomes a proton-specific Lewis base. The 6-*t*-butyl-substituted pyridines shown in Figure 2.15 incorporate this design feature.[97,98]

The presence of amido substituents greatly enhances chloridozincate loading, as the "control" reagent, L[4], shows very little zinc uptake. The strength of $ZnCl_4^{2-}$ extraction and selectivity over chloride arise from a combination of supramolecular interactions. These are best illustrated by the strongest extractant, the malonamide L[6]. On protonation, the pyridinium NH group forms intramolecular H-bonds to the two amido oxygen atoms (Figure 2.16), setting up an array of two

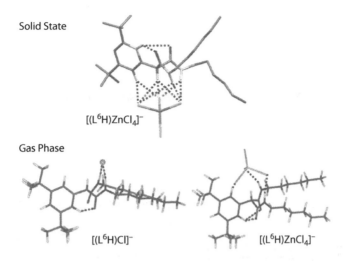

Solid State

$[(L^6H)ZnCl_4]^-$

Gas Phase

$[(L^6H)Cl]^-$

$[(L^6H)ZnCl_4]^-$

FIGURE 2.16 Part of the X-ray crystal structure of $[(L^6)_2ZnCl_4]$ (at the top) in which one similarly bonded L^6H^+ unit has been omitted for clarity to allow comparison with DFT energy-minimized structures (below) of $[(L^6H)Cl]$ and $[(L^6H)ZnCl_4]^-$. Only three of the nine close contacts to the $ZnCl_4^{2-}$ ion are shown in the latter for clarity. (From Ellis, R. J. et al., *Chem. Eur. J.*, 18, 2012, 7715–7728.)

N–H and two C–H groups to interact with a face of the $ZnCl_4^{2-}$ and its edges,[97] DFT calculations show that this array of "soft" H-bond donors in L^6H^+ results in a higher binding energy to the charge diffuse $ZnCl_4^{2-}$ anion than to a "hard" chloride ion. In the energy-minimized gas-phase structure of the $[(L^6H)Cl]$ ion pair, the chloride ion only makes short contacts with the amido N–H groups, whereas in that of the chloridozincate, complex $[(L^6H)ZnCl_4]^-$, the bonding is very similar to that observed in the solid state of $[(L^6H)_2ZnCl_4]$, with bonds also to the C–H groups of the 3-pyridino and the central malonamide atoms.

There is no evidence for the pyridinium N–H unit in ligands L^5H^+ to L^8H^+ forming H-bonds to $ZnCl_4^{2-}$ or $CoCl_4^{2-}$, or for the cationic nitrogen atom making close contact with these chloridometalates, as would be expected in an electrostatic (ion pairing) model for bonding in the neutral assemblies, $[(LH)_2MCl_4]$.[98–100] This arises from the structural feature whereby the added proton is chelated by the reagent, as shown in Figure 2.17 for L^5H^+ and for two series of reagents that contain the atom sequences R_2N–CH_2CH_2–CO–NR'_2 and R_2N–CH_2–NR'–CO–R''. These sequences also favor the formation of a six-membered proton chelate ring.[97,101,102]

A striking feature of these reagents is their strength as zinc extractants (Figure 2.18), which is dependent on them showing high selectivity for $[ZnCl_4]^{2-}$ over Cl^-. For L^{12}, the selectivity K_s, as calculated[101] from Equation (2.7), is 11×10^5. It is this remarkable selectivity that distinguishes them from the commercial trialkylamine (Alamine®) reagents for which TEHA (Figure 2.18) is a model.

$$K_s = [(LH)_2ZnCl_4][Cl^-]^2/[(LH)Cl]^2[ZnCl_4^{2-}] \qquad (2.7)$$

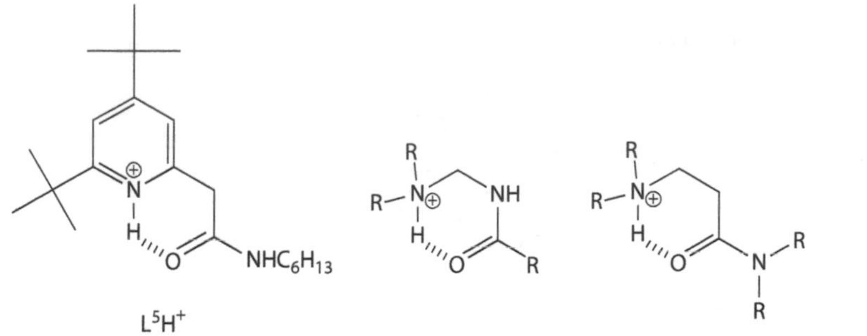

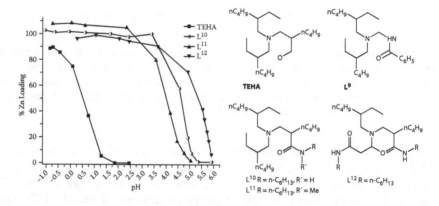

FIGURE 2.17 The six-membered proton chelate structures formed by the pyridinoamides L^5, L^6, L^8 and by the amide-substituted tertiary amine extractants shown in Figure 2.18.

FIGURE 2.18 Examples of extractants having the atom sequences in Figure 2.17, showing the pH dependence of loading of $[ZnCl_4]^{2-}$ by the mono- (L^{10} and L^{11}) and di- (L^{12}) amido derivatives of tris(2-ethylhexyl-amine) (TEHA). (From Ellis, R. J. et al., *Solvent Extr. Ion Exch.*, 29, 2011, 657–672.)

Like the sterically hindered pyridines (L^5–L^8), the tertiary amine-containing reagents in Figure 2.18 show no tendency to enter the inner coordination spheres of metals when stripping. Their modes of binding to the outer coordination spheres of chloridometalates are not as well established, but again appear to involve the cationic reagent providing an array of soft N–H and C–H hydrogen-bond donors, as judged by an analysis of proton NMR data.[101] An example of the $ZnCl_4^{2-}$ showing a preference for softer H-bond donors is provided by a comparison of its uptake by the extractants L^{10} and L^{11}. The latter, which has no amido N–H group, is a slightly stronger extractant (Figure 2.18).

Supramolecular interactions are key to understanding the strength and selectivity shown by these amidoamine extractants, and gas-phase DFT calculations have provided a very useful insight into how these interactions contribute to the properties observed in solution. For a series of extractants having the structures $R_{3-x}(N-CH_2-NR'-CO-R'')_x$, of which L^9 is an example, the strength as

$ZnCl_4^{2-}$ extractants varies in the order monoamides > diamides > triamides. At first sight, this is counterintuitive, as the binding energies to the chloridometalate might be expected to *increase* with the number of hydrogen bond donor (amide) groups in the reagent. DFT calculations accord with this assumption, but also reveal that other important factors favor reagents with smaller numbers of amide groups. The lowest-energy forms of the LH^+ cations all contain the six-membered proton-chelate rings shown in the middle of Figure 2.17, but the ease of protonation is dependent on how many intramolecular amido N–H to amido C=O H-bonds have to be sacrificed to generate such structures.[101] This results in the monoamide L^9 having the most favorable proton affinity, as defined by Equation (2.8). Once formed, the LH^+ ligands have the possibility of forming complexes with either $ZnCl_4^{2-}$ or Cl^-, and as the latter is usually present in large excess, strong Zn extractants must show a higher affinity for the chloridometalate than chloride. DFT calculations reveal that the more amide N–H groups present, the higher the binding energy to the hard chloride ion.[101]

$$L + H_3O^+ \rightleftharpoons [(LH)Cl] + H_2O \qquad (2.8)$$

A striking and potentially useful feature of many of the amidoamine extractants with the generic structures shown in Figure 2.17 is their unprecedented selectivity for $[ZnCl_4]^{2-}$ over $[FeCl_4]^-$, making it possible to effect a clean separation of these metals (Figure 2.19). Such a separation is much more difficult to achieve with the commercial trialkylamine (Alamine) reagents for which tris(2-ethylhexyl)amine (TEHA in Figure 2.19) is a model, and is unexpected because the Hofmeister bias[103,104] predicts that the less hydrated monoanion, $[FeCl_4]^-$, will be more readily dehydrated, and thus more easily transported into the water-immiscible phase.

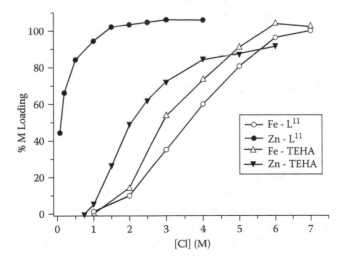

FIGURE 2.19 Loadings of $[ZnCl_4]^{2-}$ and $[FeCl_4]^-$ by L^{11} and TEHA as a function of the chloride concentration in the aqueous feed solution.

Information on the origins of the selectivity shown by L^9 has been obtained from DFT calculations of the binding energies calculated[100,101] for the gas phase reactions shown in Equation (2.9).

$$xL + xH_3O^+ + MCl_4^{x-} \rightleftharpoons [(LH)_xMCl_4] + xH_2O \qquad (2.9)$$

The binding of a single cationic ligand L^9H^+ to the outer sphere of $[ZnCl_4]^{2-}$ is very much more favorable[101] than that to the outer sphere of $[FeCl_4]^-$. This arises principally from the higher negative charge on $[ZnCl_4]^{2-}$, and is reflected by the considerably stronger interactions revealed in natural bond-order calculations.[105] The formation of $[(L^9H)_2ZnCl_4]$ over $[(L^9H)FeCl_4]$ is also favored because it involves the binding of *two* L^9H^+ ligands to form charge-neutral extractable species. The much higher thermodynamic stability of $[(L^9H)_2ZnCl_4]$ must compensate for the less favorable dehydration energy for $[ZnCl_4]^{2-}$ that is involved in its formation in solvent extraction experiments.

For the reagents having the structural motifs shown in Figure 2.17, DFT calculations of the gas-phase energies of formation of the ligands, LH^+, and of the assemblies, $[(LH)_2ZnCl_4]$, $[(LH)FeCl_4]$, and $[(LH)Cl]$, have provided explanations for the extraction strengths and selectivities shown in *solution*, and how these vary with the structures of the reagents.[98,100,101] At first sight, this is quite remarkable and probably arises in part from the unusual simplicity of speciation in the water-immiscible phase shown by these reagents. Based on nuclear magnetic resonance (NMR) spectra, there is no evidence for the inclusion of water in the assemblies formed. This is most likely due to the reagents having an built-in ability to chelate the proton needed to generate the cation (see Figure 2.17), thus facilitating its dehydration on transfer to the water-immiscible phase. In contrast, in many other systems the assemblies formed in the organic phase contain more than two components because modifiers are often required to achieve sufficient solubility and effective phase disengagement.[61,62,106]

2.4 METAL SALT EXTRACTION

Understanding the nature of the supramolecular interactions in the extraction of metal salts (MX_y) is complicated by structural issues such as whether the attendant anion (e.g., X^-) is present in the inner or outer coordination sphere, or both, in the formation of water-immiscible complexes. In metal salt extractions (Equation (2.10)), because the reagents, L, appear merely to make a target metal salt soluble in a water-immiscible liquid, they have been frequently referred to by extractive metallurgists as "solvating extractants."[7,107,108] With the benefit of hindsight, this terminology is often misleading, as the reagent is frequently present as a ligand in the inner coordination sphere of the metal cation, as in the very important process[6] to concentrate and purify uranium(VI) as its uranyl nitrate salt. Tri-*n*-butylphosphate (TBP) acts as both an inner-sphere ligand and the diluent in the "nitrate swing" process shown in Equation (2.11). Clearly any interactions between the inner- and outer-sphere components of the assembly will contribute to determining its stability.

$$n L_{org} + MX_y \rightleftharpoons [L_n MX_y]_{org} \qquad (2.10)$$

$$UO_2^{2+} + 2NO_3^- + 2TBP_{(org)} \rightleftharpoons [UO_2(NO_3)_2(TBP)_2]_{(org)} \qquad (2.11)$$

One of the major achievements of supramolecular chemistry in the last two decades has been the development of highly selective anion receptors.[73-80,109-114] Often, successful systems exploit ion pairing in which an electrostatic interaction between the anion and a positively charged metal in the anion receptor are used to enhance the strength and selectivity of binding. Systems in which the anion can be considered to be present in the outer coordination sphere of the metal can also be used as anion sensors because the presence of X^- often significantly changes the spectroscopic or electrochemical properties of the metal center. Most of the research in this area has involved single-phase solvent systems, and the development of polytopic solvent extractants that contain sites to bind *both* the cations and anions of a metal salt is much less common.[115] However, as the design features needed to achieve strong and selective binding of anions are now quite well understood, it is timely to consider this approach, especially for metal salts containing anions that are weak inner-sphere ligands, because these are much less likely to generate stable neutral complexes of the type shown in Equation (2.10). Sulfate is an example of a weak inner-sphere ligand.[116] It is the counteranion in feeds from sulfuric acid leaching, which are less corrosive than chloride streams. Ditopic metal sulfate extractants could potentially open up new flowsheets for metal recovery. Prototypes[117] for this class of reagent are the *salen*-based systems shown in Figure 2.20.

These ditopic extractants function in a zwitterionic form; incorporation of a M^{2+} ion into the salen donor set releases two protons, which are transferred to two pendant amine groups to form the anion-binding site that has been preorganized (templated) by the formation of the metal salen unit.[118] This is revealed by the X-ray crystal structure[119] of the charge-neutral complex $[NiL^{13}(SO_4)_2]$ shown in Figure 2.21. The ligand L^{13} (Figure 2.22) is present in a zwitterionic form, providing two morpholinium groups that form bifurcated H-bonds to two edges of the tetrahedral sulfate ion.

The uranyl dication templates the anion-binding site in related *salphen* complexes (Figure 2.23) that contain pendant amide groups, and its electronic properties contribute to the selectivity and sensing of bound anions.[121]

The ditopic ligands in these uranyl complexes, $[L(UO_2)]$ ($L = L^{16}-L^{19}$), do not exist in a zwitterionic form, and consequently, the loading and stripping of a metal salt cannot be achieved by variation of pH in the same way as for the systems shown in Figure 2.20. For these, metal loading is favored by raising the pH (equilibrium A, Figure 2.20), while sulfate loading (equilibrium B) is favored by lowering the pH.[119] For practicable systems, there must be a range of pH values at which *both* the metal cation and its attendant sulfate anion are loaded. The reagent L^{14}

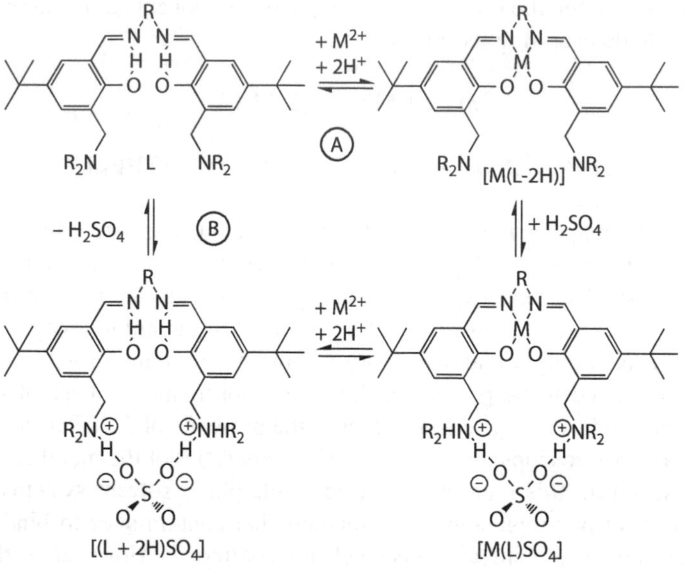

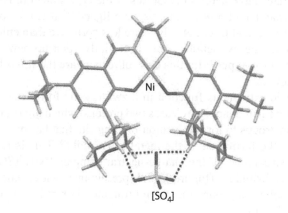

FIGURE 2.20 Equilibria describing M^{2+} and SO_4^{2-} loading of a ditopic Schiff base ligand, which is in a zwitterionic form in the metal sulfate complex.

FIGURE 2.21 The X-ray crystal structure structure of a nickel(II) sulfate complex of L^{13} that has morpholinium arms in the sulfate-binding site. (From Miller, H. A. et al., *J. Chem. Soc., Dalton Trans.*, 2000, 3773–3782.)

provides an example (Figure 2.24) in which Cu^{2+} and SO_4^{2-} are fully loaded in the pH range 2 to 4; below pH 2 sulfate loadings >100% are observed because two HSO_4^- ions can be incorporated in the assembly.

A challenge in the development of ditopic reagents of this type that depend on subtle features of supramolecular chemistry is to achieve *selectivity* of anion transport into the water-immiscible phase. For sulfate-based flowsheets,

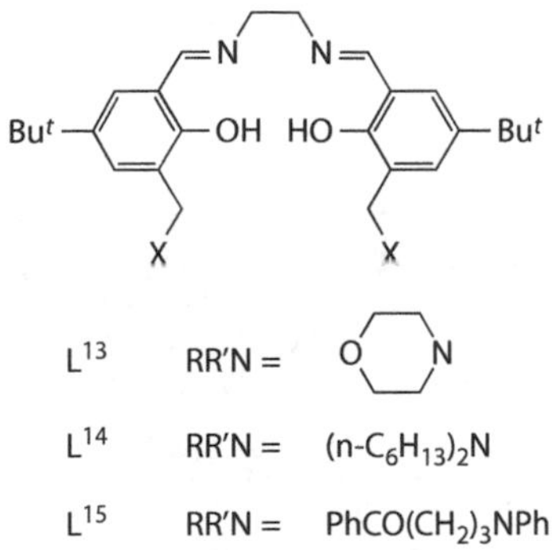
N
N
Bu^t
OH
HO
Bu^t
X
X
L^{13} RR'N = O N
L^{14} RR'N = $(n\text{-}C_6H_{13})_2N$
L^{15} RR'N = $PhCO(CH_2)_3NPh$

FIGURE 2.22 Examples of ditopic salen-based metal salt extractants. (From Galbraith, S. G. et al., *Chem. Commun.*, 2002, 2662–2663.)

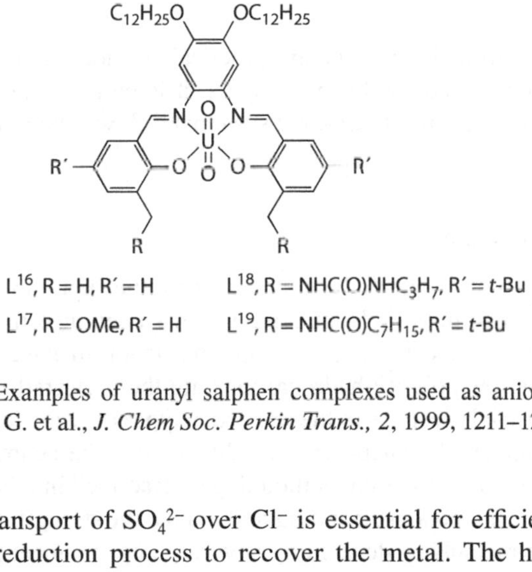
$C_{12}H_{25}O$
$OC_{12}H_{25}$
N
O
N
U
R′
O
O
O
R′
R
R
L^{16}, R = H, R′ = H
L^{17}, R = OMe, R′ = H
L^{18}, R = $NHC(O)NHC_3H_7$, R′ = *t*-Bu
L^{19}, R = $NHC(O)C_7H_{15}$, R′ = *t*-Bu

FIGURE 2.23 Examples of uranyl salphen complexes used as anion sensors. (From Antonisse, M. M. G. et al., *J. Chem Soc. Perkin Trans., 2*, 1999, 1211–1218.)

selectivity of transport of SO_4^{2-} over Cl^- is essential for efficient operation of an electroytic reduction process to recover the metal. The higher hydration energy of sulfate makes it intrinsically more difficult to transport into a water-immiscible phase.[104] The ditopic extractant, L^{13} (Figure 2.22), contains additional H-bond donor groups in the anion-binding site to help compensate for the loss of water from the sulfate ion.[103,120] Studies on the pH dependence of anion binding indicate that in a single phase (95% methanol-water), this does indeed show a preference for sulfate binding over chloride, but this is still not sufficient to overcome the Hofmeister bias[103,104] for its selective extraction into chloroform.[103,120]

While the ditopic systems discussed above provide good examples of the potential for exploiting supramolecular chemistry in extraction processes, the structures of these reagents are complicated, and their molecular weights/sizes are high. These factors will mitigate against their being used in large-scale

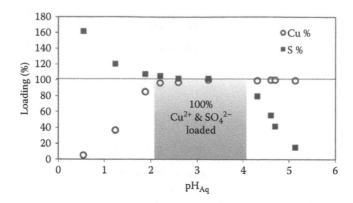

FIGURE 2.24 Copper and sulfate loading of a chloroform solution of L^{14} as a function of pH; 100% loading corresponds to formation of the 1:1:1 complex $[Cu(L^{14})SO_4]$. (From Miller, H. A. et al., *J. Chem. Soc., Dalton Trans.*, 2000, 3773–3782.)

processes on the grounds of costs and practical considerations, such as limited solubility and high viscosity. In this respect, it is more likely that mixtures of cation and anion extractants (*dual-host* systems)[122,123] will be more commercially viable.

2.5 CONCLUSIONS

In this chapter, we have selected examples where the supramolecular chemistry of assemblies formed in the solvent extraction has been investigated to the point where the structures and the types of bonding involved have been fairly well defined. Because low-polarity hydrocarbons are the preferred water-immiscible solvents for industrial use, most of the important types of secondary bonding interactions in supramolecular chemistry are able to stabilize neutral, hydrocarbon-soluble assemblies. The research methodologies described in this review could be used to exploit such secondary bonding and to tune the strength and selectivity of metal solvent extractants by developing new reagents that meet the requirements of particular hydrometallurgical flowsheets. Determining the levels of hydration in the assemblies and the extent to which they are solvated by excess reagent molecules presents problems in defining the speciation of metal assemblies. So too, in particular, does the addition of modifiers, most of which are capable of forming strong H-bonds. However, this review presents examples where it has proved possible to *design* reagents that overcome these problems, and consequently, it should be increasingly possible to use an understanding of supramolecular chemistry to develop new reagents more effectively than the largely empirical approaches that are used at present. This will be particularly challenging when considering the supramolecular chemistry at liquid interfaces, which has a major role in phase disengagement and the kinetics of phase transport of metal ions, and at solid-liquid interfaces, which is relevant to crud formation and control.

ACKNOWLEDGMENTS

We thank Professor Mike Cox for very helpful information on the background to the development of the synergistic systems described in Section 2.2 and Matt Wilson, Jennifer Turkington, and James Roebuck for stimulating discussions and their help in preparing figures. The contribution of BDR was sponsored by the Division of Chemical Sciences, Geosciences, and Biosciences, Office of Basic Energy Sciences, U.S. Department of Energy.

REFERENCES

1. Habashi, F., *A Textbook of Hydrometallurgy*, 3rd ed., Metallurgie Extractive Quebec, Quebec, 1993.
2. Habashi, F., A Short History of Hydrometallurgy, *Hydrometallurgy*, 79, (2005), 15–22.
3. Flett, D. S., The Role of Hydrometallurgy in Extractive Metallurgy, *Chemistry and Industry*, (1981), 427–431.
4. Flett, D. S., Solvent Extraction in Hydrometallurgy, *Journ. Metals*, 35, (1982), A42–A42.
5. Irish, E. R., Reas, W. H. *The Purex Process—A Solvent Extraction Reprocessing Method for Irradiated Uranium*, Hanford Laboratories, Richmond, Washington, 1957.
6. Nicol, M. J., Fleming, C. A., Preston, J. S. *Comprehensive Coordination Chemistry*, Vol. 6, L Pergamon, Oxford, 1987.
7. Tasker, P. A., Plieger, P. G., West, L. C., *Comprehensive Coordination Chemistry II*, Elsevier Ltd., Oxford, 2004.
8. Perera, J. M., Stevens, G. W., The Role of Additives in Metal Extraction in Oil/Water Systems, *Solvent Extr. Ion Exch.*, 29, (2011), 363–383.
9. Yamada, H., Tanaka, M., Solvent Extraction of Metal Carboxylates, *Adv. Inorg. Chem. Radiochem.*, 29, (1985), 143–168.
10. Yamada, H., Tanaka, M., Solvent Extraction of Metal Carboxylates, *Adv. Inorg. Chem.*, 29, (1985), 143–168.
11. Aromi, G., Brechin, E. K., Structure and Bonding in *Single-Molecule Magnets and Related Phenomena*, Winpenny, R., Ed., Vol. 122, Springer, Berlin, 2006, pp. 1–67.
12. Gatteschi, D., Sessoli, R., Cornia, A., Single-Molecule Magnets Based on Iron(III) Oxo Clusters, *Chem. Commun.*, (2000), 725–732.
13. Aromi, G., Aubin, S. M. J., Bolcar, M. A., Christou, G., Eppley, H. J., Folting, K., Hendrickson, D. N., Huffman, J. C., Squire, R. C., Tsai, H. L., Wang, S., Wemple, M. W., Manganese Carboxylate Clusters: From Structural Aesthetics to Single-Molecule Magnets, *Polyhedron*, 17, (1998), 3005–3020.
14. Wemple, M. W., Coggin, D. K., Vincent, J. B., McCusker, J. K., Streib, W. E., Huffman, J. C., Hendrickson, D. N., Christou, G., $[M_4(\mu_3-O)_2]^{8+}$ (M = Mn^{III} or Fe^{III}) cores: crystal structures and properties of $[Mn_4O_2Cl_2(O_2CC_6H_3F_2-3,5)_6(py)_4]$, $[Fe_4O_2Cl_2(O_2CMe)_6(bpy)_2]$ and $[NBu^n_4][Fe_4O_2(O_2CMe)_7(pic)_2]$†, *J. Chem. Soc., Dalton Trans.*, (1998), 719–725.
15. Piligkos, S., Weihe, H., Bill, E., Neese, F., El Mkami, H., Smith, G. M., Collison, D., Rajaraman, G., Timco, G. A., Winpenny, R. E. P., McInnes, E. J. L., EPR Spectroscopy of a Family of $Cr^{III}_7M^{II}$ "Wheels": Studies of Isostructural Compounds with Different Spin Ground States, *Chem. Eur. J.*, 15, (2009), 3152–3167.

16. Laye, R. H., McInnes, E. J. L., Solvothermal Synthesis of Paramagnetic Molecular Clusters, *Eur. J. Inorg. Chem.*, (2004), 2811–2818.
17. Timco, G. A., McInnes, E. J. L., Winpenny, R. E. P., Physical Studies of Heterometallic Rings: An Ideal System for Studying Magnetically-Coupled Systems, *Chem. Soc. Rev.*, 42, (2013), 1796–1806.
18. Danesi, P. R., Reichley-Yinger, L., Mason, G., Kaplan, L., Horwltz, E. P., Diamond, H., Selectivity-Structure Trends in the Extraction of Co(II) and Ni(II) by Dialkyl Phosphoric, Alkyl Alkylphosphonic Acids, *Solvent Extr. Ion Exch.*, 3, (1985), 435–452.
19. Turkington, J. R., Bailey, P. J., Love, J. B., Wilson, A. M., Tasker, P. A., Exploiting Outer-Sphere Interactions to Enhance Metal Recovery by Solvent Extraction, *Chem. Commun.*, (2013), 1891–1899.
20. Irving, H., Williams, R. J. P., The Stability of Transition-Metal Complexes, *J. Chem. Soc.*, (1953), 3192–3210.
21. Cole, P. M., Sole, K. C., Feather, A. M., Solvent Extraction Developments in Southern Africa, *Tsinghua Sci. Technol.*, 11, (2006).
22. Sole, K. C., Feather, A. M., Cole, P. M., Solvent Extraction in Southern Africa: An Update of Some Recent Hydrometallurgical Developments, *Hydrometallurgy*, 78, (2005), 52–78.
23. King, M. G., Nickel Laterite Technology—Finally a New Dawn? *JOM*, 57, (2005), 35–39.
24. Preston, J. S., Solvent Extraction of Cobalt and Nickel by Organophosphorus Acids. I. Comparison of Phosphoric, Phosphonic, and Phosphinic Acid Systems, *Hydrometallurgy*, 9, (1982), 115–133.
25. Cheng, C. Y., Purification of Synthetic Laterite Leach Solution by Solvent Extraction Using D2EHPA, *Hydrometallurgy*, 56, (2000), 369–386.
26. Dreisinger, D. B., Cooper, W. C., The Kinetics of Zinc, Cobalt and Nickel Extraction in the D2EHPA-Heptane-HCRO$_4$ System Using the Rotating Diffusion Cell Technique, *Solvent Extr. Ion Exch.*, 7, (1989), 335–360.
27. Devi, N. B., Nathsarma, K. C., Chakravortty, V., Sodium-Salts of D2EHPA, Pc-88a and Cyanex-272 and Their Mixtures as Extractants for Nickel(II), *Scand. J. Metall.*, 23, (1994), 194–200.
28. Devi, N. B., Nathsarma, K. C., Chakravortty, V., Sodium-Salts of D2EHPA, PC-88A and Cyanex-272 and Their Mixtures as Extractants for Cobalt(II), *Hydrometallurgy*, 34, (1994), 331–342.
29. Golding, J. A., Xun, F., Zhao, S. Z., Hu, Z. S., Sui, S. P., Hao, J. M., Extraction of Nickel from Aqueous Sulfate-Solution into Bis(2,2,4-trimethylpentyl) phosphinic Acid, Cyanex 272™ Equilibrium and Kinetic-Studies, *Solvent Extr. Ion Exch.*, 11, (1993), 91–118.
30. Lindell, E., Jaaskelainen, E., Paatero, E., Nyman, B., Effect of Reversed Micelles in Co/Ni Separation by Cyanex 272, *Hydrometallurgy*, 56, (2000), 337–357.
31. Sarangi, K., Reddy, B. R., Das, R. P., Extraction Studies of Cobalt(II) and Nickel(II) from Chloride Solutions Using Na-Cyanex 272. Separation of Co(II)/Ni(II) by the Sodium Salts of D2EHPA, PC-88A and Cyanex 272 and Their Mixtures, *Hydrometallurgy*, 52, (1999), 253–265.
32. Reddy, B. R., Sarma, P., Separation and Recovery of Cobalt and Nickel from Sulfate Solutions of Indian Ocean Nodules Using Cyanex 272, *Miner. Metall. Proc.*, 18, (2001), 172–176.
33. Hubicki, Z., Hubicka, H., Studies on the Extraction Process of Nickel(II) Sulphate Purification Using Cyanex 272, *Hydrometallurgy*, 40, (1996), 65–76.

34. Danesi, P. R., Reichley-Yinger, L., Mason, G., Kaplan, L., Horwitz, E. P., Diamond, H., Selectivity-Structure Trends in the Extraction of Cobalt(II) and Nickel(II) by Dialkyl Phosphoric, Alkyl Alkylphosphonic, and Dialkylphosphinic Acids, *Solvent Extr. Ion Exch.*, 3, (1985), 435–452.

35. Zhu, T., A Structure Parameter Characterizing the Steric Effect of Organophosphorus Acid Extractants, in *ISEC*, Cape Town, South Africa, March 17–21, 2002, pp. 203–207.

36. Donegan, S., Direct Solvent Extraction of Nickel at Bulong Operations, *Miner. Eng.*, 19, (2006), 1234–1245.

37. du Preez, A. C., Preston, J. S., Separation of Nickel and Cobalt from Calcium, Magnesium and Manganese by Solvent Extraction with Synergistic Mixtures of Carboxylic Acids, *J. S. Afr. Inst. Min. Metall.*, 104, (2004), 333–338.

38. Cheng, C. Y., Urbani, M., Houchin, M., Synergistic Solvent Extraction and Its Potential Application to Nickel and Cobalt Recovery, In *Hydrometallurgy*, Young, C., Alfantazi, A., Anderson, C., James, A., Dreisinger, D., Harris, B, Eds., Vancouver, Canada, 2003, pp. 787–800.

39. Barnard, K. R., Nealon, G. L., Ogden, M. I., Skelton, B. W., Crystallographic Determination of Three Ni-α-Hydroxyoxime-Carboxylic Acid Synergist Complexes, *Solvent Extr. Ion Exch.*, 28, 778–792.

40. Barnard, K. R., McIldowie, M., Nealon, G. L., Ogden, M. I., Skelton, B. W., in *19th ISEC*, L. Valenzuela, F. V., Moyer, B. A., Eds., Santiago, Chile, 2011.

41. Barnard, K. R., Nealon, G. L., Ogden, M. I., Skelton, B. W., Crystallographic Determination of Three Ni-Alpha-Hydroxyoxime-Carboxylic Acid Synergist Complexes, *Solvent Extr. Ion Exch.*, 28, (2010), 778–792.

42. Joe, E. G., Ritcey, G. M., Ashbrook, A. W., Uranium and Copper Extraction by Liquid Ion Exchange, *JOM*, 18, (1966), 18.

43. Flett, D. S., Titmuss, S., Synergistic Effect of Lix63 on Extraction of Copper and Cobalt by Napthenic Acid, *J. Inorg. Nucl. Chem.*, 31, (1969), 2612.

44. Flett, D. S., Cox, M., Heels, J. D., Extraction of Cobalt with a Propriety Alkylated 8-Quinolinol, *J. Inorg. Nucl. Chem.*, 37, (1975), 2197–2201.

45. Flett, D. S., West, D. W., in *ISEC*, 1971, p. 214.

46. Cox, M., Flett, D. S., in *ISEC*, 1971, p. 204.

47. Lumetta, G. J., Gelis, A. V., Vandegrift, G. F., Review: Solvent Systems Combining Neutral and Acidic Extractants for Separating Trivalent Lanthanides from the Transuranic Elements, *Solvent Extr. Ion Exch.*, 28, (2010), 287–312.

48. Szymanowski, J., *Hydroxyoximes and Copper Hydrometallurgy*, CRC Press, Boca Raton, FL, 1993.

49. Mackey, P. J., in *CIM Magazine*, 2, (2007), 35.

50. Kordosky, G. A., in *ISEC*, Cape Town, 2002, p. 853.

51. Smith, A. G., Tasker, P. A., White, D. J., The Structures of Phenolic Oximes and Their Complexes, *Coord. Chem. Rev.*, 241, (2003), 61–85.

52. Forgan, R. S., Roach, B. D., Wood, P. A., White, F. J., Campbell, J., Henderson, D. K., Kamenetzky, E., McAllister, F. E., Parsons, S., Pidcock, E., Richardson, P., Swart, R. M., Tasker, P. A., Using the Outer Coordination Sphere to Tune the Strength of Metal Extractants, *Inorg. Chem.*, 50, (2011), 4515–4522.

53. Forgan, R. S., Wood, P. A., Campbell, J., Henderson, D. K., McAllister, F. E., Parsons, S., Pidcock, E., Swart, R. M., Tasker, P. A., Supramolecular Chemistry in Metal Recovery; H-Bond Buttressing to Tune Extractant Strength, *Chem. Commun.*, (2007), 4940–4942.

54. Dunitz, J. D., Gavezzotti, A., Molecular Recognition in Organic Crystals: Directed Intermolecular Bonds or Nonlocalized Bonding? *Angew. Chem. Int. Ed.* 44, (2005), 1766–1787.
55. Gavezzotti, A., Calculation of Lattice Energies of Organic Crystals: The Pixel Integration Method in Comparison with More Traditional Methods, *Z. Kristallogr.*, 220, (2005), 499–510.
56. Palusiak, M., Grabowski, S. J., Methoxy Group as an Acceptor of Proton in Hydrogen Bonds, *J. Mol. Struct.*, 642, (2002), 97–104.
57. Parrish, J. R., Selective Liquid Ion-Exchangers. 2. Derivatives of Salicylaldoxime, *Journal of the South African Chemical Institute*, 23, (1970), 129.
58. Lakshmanan, V. I., Lawson, G. J., Extraction of Copper from Aqueous Chloride Solutions with LIX-70 in Kerosine, *J. Inorg. Nucl. Chem.*, 37, (1975), 207–209.
59. Szymanowski, J., Borowiakresterna, A., Chemistry and Analytical Characterization of the Effect of Hydroxyoxime Structure Upon Metal-Complexing and Extraction Properties, *Crit. Rev. Anal. Chem.*, 22, (1991), 65–112.
60. Rydberg, J., Cox, M., Musikas, C., Choppin, G. R., Eds., *Solvent Extraction Principles and Practice*, 2nd ed., Marcel Dekker, New York, 1992.
61. Mathur, J. N., Murali, M. S., Nash, K. L., Actinide Partitioning—A Review, *Solvent Extr. Ion Exch.*, 19, (2001), 357–390.
62. Nilsson, M., Nash, K. L., Review Article: A Review of the Development and Operational Characteristics of the Talspeak Process, *Solvent Extr. Ion Exch.*, 25, (2007), 665–701.
63. Braley, J. C., Carter, J. C., Sinkov, S. I., Nash, K. L., Lumetta, G. J., The Role of Carboxylic Acids in Talsqueak Separations, *J. Coord. Chem.*, 65, (2012), 2862–2876.
64. McNaught, A. D., Wilkinson, A., *IUPAC Compendium of Chemical Terminology,* 2nd ed., Blackwell Scientific Publications, Oxford, 1997.
65. Nic, M., Jirat, J., Kosata, B., Enkins, A., *IUPAC Compendium of Chemical Terminology*, 2nd ed., XML online corrected version, 2006–.
66. Malik, P., Paiva, A. P., Solvent Extraction Studies for Platinum Recovery from Chloride Media by a N,N'-Tetrasubstituted Malonamide Derivative, *Solvent Extr. Ion Exch.*, 27, (2009), 36–49.
67. Grant, R. A., in *Precious Metals Recovery and Refining Seminar*, Historical Publications, Scottsdale, AZ, 1989, pp. 7–39.
68. Grant, R. A., Burnham, R. F., Collard, S., *The High Efficiency Separation of Iridium from Rhodium by Solvent Extraction Using a Mono-N-Substituted Amide*, International Solvent Extraction Conference, Japan, 1990, 961–966.
69. Belair, S. C., Breeze, B. A., Grant, R. A., O'Shaughnessy, P. N., Schofield, E. R., Woollam, S. F., in 19th *ISEC*, L. Valenzuela, F., Moyer, B. A., Eds., Santiago, Chile, 2011.
70. Ribeiro, L. C., Paiva, A. P., in 19th *ISEC*, L. Valenzuela, F., Moyer, B. A., Eds., Santiago, Chile, 2011.
71. McDonald, R. G., Whittington, B. I., Atmospheric Acid Leaching of Nickel Laterites Review. Part II. Chloride and Bio-Technologies, *Hydrometallurgy*, 91, (2008), 56–69.
72. Flett, D. S., in *Chloride Metallurgy, 32nd Annual Hydrometallurgy Meeting*, Peek, E., Van Weert, G., Eds., Canadian Institute of Mining, Metallurgy, and Petroleum, Montreal, Canada, Vol. 2, 2002, p. 255.
73. Kim, S. K., Sessler, J. L., Ion Pair Receptors, *Chem. Soc. Rev.*, 39, (2010), 3784–3809.
74. Kang, S. O., Hossain, M. A., Bowman-James, K., Influence of Dimensionality and Charge on Anion Binding in Amide-Based Macrocyclic Receptors, *Coord. Chem. Rev.*, 250, (2006), 3038–3052.

75. Kang, S. O., Day, V. W., Bowman-James, K., Geometrical Highlights of Anion Recognition and Beyond, *Abst. Papers Am. Chem. Soc.*, 239, (2010).

76. Kang, S. O., Day, V. W., Bowman-James, K., Tricyclic Host for Linear Anions, *Inorg. Chem.*, 49, (2010), 8629–8636.

77. Hay, B. P., Firman, T. K., Moyer, B. A., Structural Design Criteria for Anion Hosts: Strategies for Achieving Anion Shape Recognition through the Complementary Placement of Urea Donor Groups, *J. Am. Chem. Soc.*, 127, (2005), 1810–1819.

78. Hay, B. P., De Novo Structure-Based Design of Anion Receptors, *Chemical Society Reviews*, 39, (2010), 3700–3708.

79. Gale, P. A., Gunnlaugsson, T., Preface: Supramolecular Chemistry of Anionic Species Themed Issue, *Chem. Soc. Rev.*, 39, (2010), 3595–3596.

80. Gale, P. A., Anion Receptor Chemistry: Highlights from 2008 and 2009, *Chem. Soc. Rev.*, 39, (2010), 3746–3771.

81. Custelcean, R., Bock, A., Moyer, B. A., Selectivity Principles in Anion Separation by Crystallization of Hydrogen-Bonding Capsules, *J. Am. Chem. Soc.*, 132, (2010), 7177–7185.

82. Busschaert, N., Wenzel, M., Light, M. E., Iglesias-Hernandez, P., Perez-Tomas, R., Gale, P. A., Structure-Activity Relationships in Tripodal Transmembrane Anion Transporters: The Effect of Fluorination, *J. Am. Chem. Soc.*, 133, (2011), 14136–14148.

83. Bianchi, A., Bowman-James, K., Garcia-Espana, E., Eds., *Supramolecular Chemistry of Anions*, Wiley VCH, New York, 1997.

84. Collins, D. N., Flett, D. S., *Role of Chloride Hydrometallurgy in Processing of Complex (Massive) Sulphide Ores.*, in Sulphide Deposits—Their Origin and Processing, Gray, P. M. J. et al., Eds, The Institution of Mining and Metallurgy, 1990, p. 233.

85. Winand, R., Chloride Hydrometallurgy, *Hydrometallurgy*, 27, (1991), 285–316.

86. Reedijk, J., Metal-Ligand Exchange Kinetics in Platinum and Ruthenium Complexes, *Plat. Met. Rev.*, 52, (2008), 2–11.

87. Naidoo, K. J., Lopis, A. S., Westra, A. N., Robinson, D. J., Koch, K. R., Contact Ion Pair between Na$^+$ and PtCl$_6^{2-}$ Favored in Methanol, *J. Am. Chem. Soc.*, 125, (2003), 13330–13331.

88. Naidoo, K. J., Klatt, G., Koch, K. R., Robinson, D. J., Geometric Hydration Shells for Anionic Platinum Group Metal Chloro Complexes, *Inorg. Chem.*, 41, (2002), 1845–1849.

89. Brammer, L., Swearingen, J. K., Bruton, E. A., Sherwood, P., Hydrogen Bonding and Perhalometal Late Ions: A Supramolecular Synthetic Strategy for New Inorganic Materials, *Proc. Natl. Acad. Sci. USA*, 99, (2002), 4956–4961.

90. Brammer, L., Bruton, E. A., Sherwood, P., Understanding the Behavior of Halogens as Hydrogen Bond Acceptors, *Cryst. Growth Des.*, 1, (2001), 277–290.

91. Felloni, M., Hubberstey, P., Wilson, C., Schroder, M., Conserved Hydrogen-Bonded Supramolecular Synthons in Pyridinium Tetrachlorometallates, *Cryst. Eng. Comm.*, 6, (2004), 87–95.

92. Dolling, B., Gillon, A. L., Orpen, A. G., Starbuck, J., Wang, X. M., Homologous Families of Chloride-Rich 4,4'-Bipyridinium Salt Structures, *Chem. Commun.*, (2001), 567–568.

93. Bell, K. J., Westra, A. N., Warr, R. J., Chartres, J., Ellis, R., Tong, C. C., Blake, A. J., Tasker, P. A., Schroder, M., Outer-Sphere Coordination Chemistry: Selective Extraction and Transport of the [PtCl$_6$]$^{2-}$ Anion, *Angew. Chem. Int. Ed.*, 47, (2008), 1745–1748.

94. Bell, K. J., Chartres, J., Ellis, R. J., Robinson, D. J., Tong, C. C., Warr, R. J., Westra, A. N., Schroder, M., Tasker, P. A., Anion Recognition: The Selective Extraction of [PtCl$_6$]$^{2-}$, *Solvent Extr.: Fundam. Ind. Appl., Proc. ISEC 2008 Int. Solvent Extr. Conf.*, 2, (2008), 1457–1462.

95. Warr, R. J., Westra, A. N., Bell, K. J., Chartres, J., Ellis, R., Tong, C., Simmance, T. G., Gadzhieva, A., Blake, A. J., Tasker, P. A., Schroder, M., Selective Extraction and Transport of the [PtCl$_6$]$^{2-}$ Anion through Outer-Sphere Coordination Chemistry, *Chem. Eur. J.,* 15, (2009), 4836–4850.

96. Purcell, K. F., Kotz, J. C., *Inorganic Chemistry*, Holt-Saunders, London, 1977.

97. Ellis, R. J., Chartres, J., Sole, K. C., Simmance, T. G., Tong, C. C., White, F. J., Schroder, M., Tasker, P. A., Outer-Sphere Amidopyridyl Extractants for Zinc(II) and Cobalt(II) Chlorometallates, *Chem. Commun.*, (2009), 583–585.

98. Ellis, R. J., Chartres, J., Henderson, D. K., Cabot, R., Richardson, P. R., White, F. J., Schroder, M., Turkington, J. R., Tasker, P. A., Sole, K. C., Design and Function of Pre-Organised Outer-Sphere Amidopyridyl Extractants for Zinc(II) and Cobalt(II) Chlorometallates: The Role of C-H Hydrogen Bonds, *Chem. Eur. J.*, 18, (2012), 7715–7728.

99. Ellis, R. J., Chartres, J., Tasker, P. A., Sole, K. C., Amide Functionalised Aliphatic Amine Extractants for Co(II) and Zn(II) Recovery from Acidic Chloride Media, *Solvent Extr. Ion Exch.*, 29, (2011), 657–672.

100. Turkington, J. R., Bailey, P. J., Chartres, J., Ellis, R. J., Henderson, D. K., Tasker, P. A., Kamenetzky, E., Sassi, T., Sole, K. C., in *19th ISEC*, L. Valenzuela, F., Moyer, B. A., Eds., Santiago, Chile, 2011.

101. Turkington, J. R., Cocalia, V., Kendall, K., Morrison, C. A., Richardson, P., Sassi, T., Tasker, P. A., Bailey, P. J., Sole, K. C., Outer Sphere Coordination Chemistry: Amido-Ammonium Ligands as Highly Selective Tetrachloridozinc(II)ate Extractants, *Inorg. Chem.*, 52 (2012), 12805–12814.

102. Ellis, R., Chartres, J., Tasker, P. A., Sole, K. C., Amide Functionalised Aliphatic Amine Extractants for Co(II) and Zn(II) Recovery from Acidic Chloride Media, *Solvent Extr. Ion Exch.* (2011), 657–672.

103. Moyer, B. A., Bonnesen, P. V., Custelcean, R., Delmau, L. H., Hay, B. P., Strategies for Using Host-Guest Chemistry in the Extractive Separations of Ionic Guests, *Kem. Ind.*, 54, (2005), 65–87.

104. Hofmeister, F., On Regularities in the Albumin Precipitation Reactions with Salts and Their Relationship to Physiological Behaviour, *Archiv. Exptl. Pathol. Pharmakol.*, 24, (1888).

105. Glendening, E. D., Reed, A. E., Carpenter, J. E., Weinhold, F., NBO version 3.1.

106. Delmau, L. H., Birdwell, J. F., McFarlane, J., Moyer, B. A., Robustness of the Cssx Process to Feed Variation: Efficient Cesium Removal from the High Potassium Wastes at Hanford, *Solvent Extr. Ion Exch.*, 28, 19–48.

107. Muller, P., Glossary of Terms Used in Physical Organic Chemistry, *Pure Appl. Chem.*, 66, (1994), 1077–1184.

108. Manchanda, V. K., Pathak, P. N., Mohapatra, P. K., in *Solvent Extr. Ion Exch.*, Moyer, B. A., Ed., CRC Press, Boca Raton, FL, 2010, p. 71.

109. Antonisse, M. M. G., Reinhoudt, D. N., Neutral Anion Receptors: Design and Application, *Chem. Commun.*, (1998), 443–448.

110. Beer, P. D., Schmitt, P., Molecular Recognition of Anions by Synthetic Receptors, *Curr. Opin. Chem. Biol.*, 1, (1997), 475–482.

111. Gale, P. A., Garcia-Garrido, S. E., Garric, J., Anion Receptors Based on Organic Frameworks: Highlights from 2005 and 2006, *Chem. Soc. Rev.*, 37, (2008), 151–190.

112. Goetz, S., Kruger, P. E., A New Twist in Anion Binding: Metallo-Helicate Hosts for Anionic Guests, *Dalton Trans.*, (2006), 1277–1284.
113. Caballero, A., White, N. G., Beer, P. D., A Bidentate Halogen-Bonding Bromoimidazoliophane Receptor for Bromide Ion Recognition in Aqueous Media, *Angew. Chem. Int. Ed.*, 50, (2011), 1845–1848.
114. Kilah, N. L., Wise, M. D., Serpell, C. J., Thompson, A. L., White, N. G., Christensen, K. E., Beer, P. D., Enhancement of Anion Recognition Exhibited by a Halogen-Bonding Rotaxane Host System, *J. Am. Chem. Soc.*, 132, (2010), 11893–11895.
115. Akkus, N., Campbell, J. C., Davidson, J., Henderson, D. K., Miller, H. A., Parkin, A., Parsons, S., Plieger, P. G., Swart, R. M., Tasker, P. A., West, L. C., Exploiting Supramolecular Chemistry in Metal Recovery: Novel Zwitterionic Extractants for Nickel(II) Salts, *Dalton Trans.*, (2003), 1932–1940.
116. Martell, A. E., Smith, R. M. *Critical Stability Constants*, Vol. 4, Plenum Press, New York, 1976.
117. White, D. J., Laing, N., Miller, H., Parsons, S., Coles, S., Tasker, P. A., Ditopic Ligands for the Simultaneous Solvent Extraction of Cations and Anions, *Chem. Commun.*, (1999), 2077–2078.
118. Plieger, P. G., Tasker, P. A., Galbraith, S. G., Zwitterionic Macrocyclic Metal Sulfate Extractants Containing 3-Dialkylaminomethylsalicylaldimine Units, *Dalton Trans.*, (2004), 313–318.
119. Miller, H. A., Laing, N., Parsons, S., Parkin, A., Tasker, P. A., White, D. J., Supramolecular Assemblies from Ditopic Ligands and Transition Metal Salts, *J. Chem Soc., Dalton Trans.*, (2000), 3773–3782.
120. Galbraith, S. G., Plieger, P. G., Tasker, P. A., Cooperative Sulfate Binding by Metal Salt Extractants Containing 3-Dialkylaminomethylsalicylaldimine Units, *Chem. Commun.*, (2002), 2662–2663.
121. Antonisse, M. M. G., Snellink-Ruel, B. H. M., Ion, A. C., Engbersen, J. F. J., Reinhoudt, D. N., Synthesis of Novel Uranyl Salophene Derivatives and Evaluation as Sensing Molecules in Chemically Modified Field Effect Transistors (Chemfets), *J. Chem. Soc. Perkin Trans.*, 2, (1999), 1211–1218.
122. Steed, J. W., Atwood, J. L. *Supramolecular Chemistry*, Wiley, New York, 2009.
123. Moyer, B. A., Singh, R. P., Meeting, A. C. S. *Fundamentals and Applications of Anion Separations*, Kluwer Academic, Dordrecht, 2004.

3 Molecular Design and Metal Extraction Behavior of Calixarene Compounds as Host Extractants

Keisuke Ohto
Department of Chemistry and Applied Chemistry, Faculty of Science and Engineering, Saga University, Saga, Japan

CONTENTS

3.1 INTRODUCTION

In hydrometallurgical processing, the role of solvent extraction reagents is predominant among many factors to improve extraction and separation efficiencies. Ten requirements have been identified for industrial use of commercially available extractants,[1–3] as follows:

1. High selectivity for the target ion
2. Direct extraction from leachate without pH adjustment
3. Sufficiently fast extraction rate
4. Reversible extraction and easy stripping
5. High solubility of the extractant and its metal complex in suitable diluents (organic solvents)
6. Chemically long-term stability for use (stability to radiation in nuclear waste solution is also required)

   7. Low loss of the extractant to the aqueous phase
   8. Clean and quick phase separation with minimal entrained carryover
   9. Low price for commercial use
  10. Safe for operation (nontoxicity, low volatility, nonflammability)

Although various kinds of extractants have been prepared and examined and even brought into industrial use, no extractant has been found that satisfies all of the above requirements. Thus, in actual practice, compromises must be accepted, leaving considerable opportunity for improvement through dedicated research. The most important factor among the 10 requirements is the first: high selectivity for a given target ion. Conventional preparation of the extractants has been carried out by introducing functional groups for extraction of a target ion, together with adequately long alkyl chains and phenyl groups for high lipophilicity and minimal solubility loss to the aqueous phase.

Crown ethers, first reported by C. J. Pedersen in 1967, introduced the era of host-guest chemistry with distinctly different possibilities for developing improved separations as compared with conventional extractants.[4] While many synthetic variants of crown ethers emerged, driven by applications in industrial and analytical chemistry, an even more versatile synthetic chemistry arose about a decade later with the advancement of calixarenes. In the present review, the focus will be on molecular design of calixarene compounds as hosts for the extraction of guest metal cations.

3.2 MACROCYCLIC LIGANDS

3.2.1 STRUCTURES AND NAMES OF THE COMPOUNDS

Since many kinds of macrocyclic polyether ligands called crown ethers were reported by C. J. Pedersen in 1967,[4,5] the crown ethers were considered in one line of research as enzyme-mimic materials. These compounds are multidentate ligands providing three-dimensionally arranged cavities for coordination and have been called "host molecules"[6,7] or, when complexed, "supramolecules."[8] Consequently, they have stimulated research on "molecular recognition" and "ion discrimination," and similar type compounds have been prepared one after another, such as cyclic coronand (i.e., crown ether), cryptand, spherand, cavitand, carcerand, hemispherand, and linear podand, derived from their shapes.[9] It is important to distinguish the true appearance of these compounds as three-dimensional steric structures from drawn two-dimensional flat structures. Thus, macrocyclic ligands exhibit structural specificity, such as ligand topology including connectivity, cyclic order and dimensionality, cavity diameter for size fitting, kind of coordinating elements related to hard and soft acids and bases (HSAB) theory,[10] cyclic order and macrocyclic effect related to dehydration, functional group effect related to coordination ability and acid dissociation of ion-exchangeable group, conformation related to geometry of coordination space, lipophilicity related to the polarity of the ligand itself and its complex, distribution, molecular volume, hydrophile–lipophile balance (HLB) parameter, etc., for crown ethers.[11,12] Based on these considerations, macrocyclic host compounds with high separation efficiency can be designed. Although ionophores are defined as intracavital ion carriers in a more limited meaning, such enzyme-mimic host compounds are also classified as ionophores in the broad sense. With the elucidation of their complex structures with guest molecules or ions and specific properties, research has spread to applications in organic syntheses, metal uptake and mutual separation, spectroscopic analyses and ion-selective electrodes, optical resolution, and so on. Most of them, however, are not available as practical extractants due to the limited requirements to be used as described above.

From the viewpoint of solvent extraction, the molecular design and metal extraction of calixarene compounds are described in the present review. An excellent review by Izatt et al., including a large number of the above-mentioned ligands,[13] summarizes early developments.

3.2.2 CHARACTERISTICS OF CALIXARENES

Calixarenes, classified as metacyclophanes and prepared by the condensation reaction of p-alkylphenol with formaldehyde, are also macrocyclic oligomers. They are different in functionality from macrocyclic polysaccharides (known as cyclodextrins) and crown ethers and are called "the third host molecules."[14] They consist of units of phenols and cross-linked methylenes, the cavity size being different by unit number (n), as shown in Figure 3.1.

Calixarene chemistry started from phenol resin in the latter half of the nineteenth century. Zinke and Ziegler in the early 1940s suggested that the prepared compounds possess cyclic tetrameric structure, although they still could not confirm it due to poor analytical tools for characterization.[15] Multistep synthesis of calix[4]arene was exploited in the 1970s by Kaemmerer et al.,[16] and the structure was finally identified in 1978 by Gutsche et al.[17] Calixarenes attracted much attention from many chemists after a single-step large-scale synthesis was established in 1981 by Gutsche et al.[18] Gutsche suggested the name *calix* for such cyclic compounds, which was derived from the shape of a "Greek vase or chalice,"[19] and *arene*, indicating the presence of aryl residues in the macrocyclic array. Calixarenes as macrocyclic compounds provide cavities with variable size and offer preorganized constrained structures for guests. There are some advantages and disadvantages compared with cyclodextrins and crown ethers.[14] Systematic

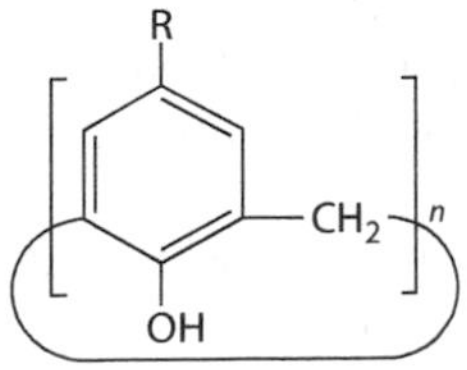

FIGURE 3.1   Structures of calix[n]arene.

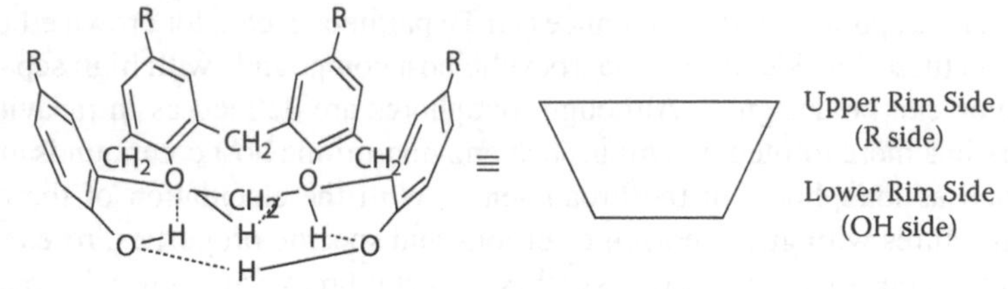

FIGURE 3.2   Cyclic structure of calix[4]arene and its rims.

change of ring size even in large-scale preparation is a common advantage for the three families of compounds. Unique advantages of calixarenes include their easy conversion to various derivatives that offer a variety of functionality as ionophores and cavity-shaped host molecules. Disadvantages possibly lie in their transparency for spectroscopic analyses, neutrality under working conditions, and optical activity.

In order to improve the functionality of the extractants, chemical modifications are necessary. In general, there are two major features of calixarenes that readily permit modifications:[20]

1. The parent calixarene ring possesses phenolic hydroxyl groups, which permit easy introduction of functional groups.
2. Aromatic substitution reactions are facilitated by dealkylation at the *p*-position.

Consequently, both chemical modifications at the lower (OH side) and upper (alkyl branch side) rims of calixarene are possible (see Figure 3.2).

The calixarene framework as a "platform" provides chemically tailored space to accept metal ions. The remarkable features of discrimination for appropriately modified calixarenes are additionally listed as follows:

3. They possess different-sized cavities consisting of phenol and methylene units.
4. They provide two frames of different sizes at their upper and lower rims.
5. They can be fixed at different conformations by introducing functional groups of suitable length.
6. They show symmetry.
7. They provide convergence of multiple functionalities.
8. They can provide complementary binding of guest species.
9. They can provide allosteric effects for sensing and switching, for example.

By using some of these features, novel molecules with specific functionality can be designed and synthesized. An additional dimension of structural control of calixarenes arises in the conformational variability within the calixarene ring itself. Original calixarenes are in the cone conformation due to their strong

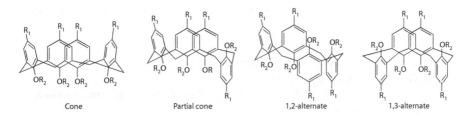

FIGURE 3.3   Structures of four conformers of the modified calix[4]arene ($R_2 \geq n$-propyl group).

intramolecular hydrogen bonding. In general, after modification on hydroxyl oxygen atoms by longer groups than propyl ones, calix[4]arene derivatives are locked into four conformers. The three-dimensional cyclic structure of unmodified p-alkylcalix[4]arene in its cone conformation and the conformations of the modified calix[4]arene are shown in Figures 3.2 and 3.3.

3.3 MOLECULAR DESIGN OF CALIXARENE COMPOUNDS FOR METAL EXTRACTION

Up to the previous section, preliminarily, macrocyclic host molecules built using the calixarene platform were described to design solvent extraction reagents. Since the design of the extractants is important, as described above, various calixarene types have been prepared by many chemists to investigate and control extraction behavior. Although there are some other important factors from the industrial point of view, such as effect of diluent, operating temperature, and type of extraction apparatus, this review will focus primarily on the effect of calixarene structure.

First, fundamental concepts for molecular design of common extractants are described. Even before that, targets should be well known, and much care should be exercised in understanding and defining the separation problem to be solved. For example, in the case of metal ions as targets, understanding of their physical properties and parameters is very important, as follows:

A. Charge
B. Coordination number
C. Coordination geometry-related coordination number
D. Hardness and softness based on HSAB theory
E. Ion radius
F. Dehydration energy
G. Dehydration rate based on Eigen mechanism[21]
H. Stability constants with anions, etc.
I. Solubility products with anions such as hydroxyl ion
J. Redox states and potentials
K. Radioactivity
L. Lability to substitution
M. Other specific property applied to the extraction

Furthermore, understanding of the extraction system is also important, as follows:

a. Individual or competitive extraction system
b. Group or mutual separation
c. Extraction purpose, such as removal (e.g., decontamination or purification), recovery, and separation
d. pH region and extraction medium, such as coexisting anions
e. Special limitation during the extraction

Then, the structures of the conventional extractants will be determined by such information, together with the required terms described in section 3.1, as follows:

I. Determination of extraction mechanism such as ion-exchange, solvation, and chelate extraction
II. Determination of functional groups interacting with metal ions
III. Determination of lipophilic groups such as aliphatic and aromatic groups, carbon number, number of branches, position on the alkyl chain, and so on
IV. Determination of the extractant structure

Next, the molecular design of the extractants containing calixarenes is explained. The features of calixarenes for modification and their discriminating properties are cited in numbers 1, 2, and 3–9, as described previously. The characteristics of calixarene compounds required for their use as the extractants are cited as follows:

i. Size-discriminating effect (remarkable features in numbers 3 and 4)
ii. Conformation effect (number 5)
iii. Converging effect of multifunctional group (number 7; 1 and 2)
iv. Complementarity effect (numbers 7 and 8)
v. Allosteric effect (e.g., numbers 5 and 9)

The size-discriminating effect is one of the most understandable structural effects based on the increased and effective dehydration of water molecules surrounding metal ions. Size effects are available by changing the number of phenol and cross-linked methylene units as in calix[6]arene and calix[4]arene, and by using the corresponding broader upper and narrower lower rims. In the case of calix[4]arene, the coordination site at the lower rim is fixed to the sodium ion (approx. 0.2 nm) and at the upper rim is fixed to toluene (approx. 0.5–0.6 nm). In the case of metal ions with different sizes, as shown in Figure 3.4, the size (structural) effect based on the rim and functional group effects on metal ion affinity should act synergistically. In general, since the van der Waals force for hydrophobic interaction and the electrostatic interaction based on localized electric charge

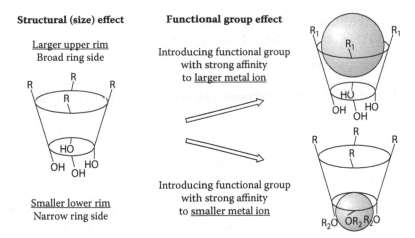

Structural (size) effect

Larger upper rim
Broad ring side

Smaller lower rim
Narrow ring side

Functional group effect

Introducing functional group
with strong affinity
to larger metal ion

Introducing functional group
with strong affinity
to smaller metal ion

FIGURE 3.4 Example of multiplier effect of structural effect and functional group effect.

density are energetically much lower than ionic bonding, the functional group effect in the extraction by the ion-exchange mechanism is much higher than the structural effect and other effects. This means that molecular design should not be too much dependent on the structural effect of calixarenes. If both effects of size and functional group do not act synergistically, the separation efficiency for host type of the extractants may be depressed compared with that for the conventional extractants.

The conformational effect is based on the structures of the conformational isomers. The conformation may be related to the stoichiometry of the metal complex. The converging effect of multifunctional groups is related to the entropy effect together with the size-discriminating and the complementarity effects. Since calixarenes provide adequate space for metal inclusion, the complexation ability should relatively increase due to the compensation effect of entropy based on the preorganized host structure compared with the corresponding linear compounds. When a unit structure of calixarenes forms five- or six-membered ring chelates, the extraction ability will be significantly enhanced not only by the converging effect of the functional groups, but also by a multiple chelate effect. Homogeneous functional groups contribute to enhance the extraction ability as a complementarity effect. The allosteric effect in host–guest chemistry means that the host molecule acquires an affinity for a second guest species after the first guest is bound. If calixarene compounds possess two different binding sites and a metal ion as an effecter is bound by one of them (allosteric site), allosteric effects, such as changes of complexation ability and structure of the binding site, would then take place.

A change or improvement of extraction behavior is possible by using the structural properties of calixarenes. In the next section, various research on metal extraction with calixarene compounds is described.

3.4 RESEARCH ON METAL CATION EXTRACTION WITH CALIXARENE DERIVATIVES

The first group reporting on metal extraction with calixarene compounds was Izatt et al.[22] They did not carry out batchwise extraction, but rather transported alkali metal ions through liquid–membranes with unmodified p-t-butylcalixarenes, 1–3. Due to the low acid dissociation constants of the calixarenes, they showed poor transporting ability, and no size-discriminating effect was observed, although basic solutions prepared by dissolving metal hydroxides were employed. Afterward, many papers have reported on the size effect. For example, Chang and Cho reported on a series of alkali and alkaline earth metal ions with esters 5a and 6a and diethers 8 and 9, derivatives of p-t-butylcalixarenes.[23] In their subsequent paper, they investigated the extraction with calix[4]arene derivatives and referred to the size effect by the unit structure number of the calixarene.[24] The extraction of metal picrates by calixarene derivatives is listed in Table 3.1. Ester derivatives 4a, 5a, and 6a showed high extractability compared with diethers 7–9 and O-methylated derivatives 10–12. The 4a derivative exhibited Na$^+$ selectivity. The cations Cs$^+$ and K$^+$ were selectively extracted rather than divalent cations due to an ion-pair extraction mechanism. In particular, 5a exhibited high extractability based on the size-discriminating effect. Since they reported that calix[8]arene derivatives are too large to discriminate among metal ions, the work was limited to calix[8]arene compounds from then on. In this paper, although they successfully prepared the tetracarboxylic acid of calix[4]arene, 13, they did not investigate metal extraction due to its limited solubility in common organic solvents.

Ungaro et al. reported the extraction of a series of alkaline earth metal ions, the uranyl ion, and the divalent lead ion with unmodified 1, tetraacetic acid 13, diacetic acid 16, and diacetic crown-5-bridged p-t-butylcalix[4]arene 17.[25] The fractions of cations extracted are listed in Table 3.2. Three derivatives showed

TABLE 3.1
Percent Extraction of Metal Picrates by Calixarene Derivatives

Extractant	Li$^+$	Na$^+$	K$^+$	Rb$^+$	Cs$^+$	Ca^{2+}	Ba^{2+}
1, 2, 3, 11	<1	<1	<1	<1	<1	<1	<1
4a	48.9	87.7	51.2	41	52.8	45.7	47.3
5a	6.7	15.6	66.2	60.5	88.9	5.3	8.2
6a	<1	4.5	21.5	16.4	17.0	6.4	17.9
10	<1	6.2	3.8	<1	<1	<1	<1
12	<1	<1	<1	<1	2.9	<1	1.7
7	<1	3.8	<1	<1	<1	3.2	2.3
8	<1	2.9	~0	~0	2.6	2.0	2.4
9	2.3	1.3	1.7	1.8	2.7	2.0	3.0

Source: Chang, S.-K., and Cho, I. J., *Chem. Soc. Perkin Trans. 1*, 1986, 211–214.

TABLE 3.2
Percent Extraction of Cations

Extractant	Type	Mg^{2+}	Ca^{2+}	Sr^{2+}	Ba^{2+}	UO_2^{2+}	Pb^{2+}
1	Unmodified	0	0	0	0	0	10
13	Tetraacetic acid	5	81	79	63	83	98
16	Diacetic acid	10	61	42	32	69	95
17	Monocrown-5	0	83	33	37	16	87

Source: Ungaro, R., and Pochini, A., *J. Inclusion Phenom.*, 2, 1984, 199–206.

high extractability, while **1** hardly extracted any cations at all. Among alkaline earths, the tetraacetic acid derivative showed high selectivity for Ca^{2+} and Sr^{2+}, the diacetic acid derivative showed Ca^{2+} selectivity, and the diacetic crown-5-bridged compound also showed high Ca^{2+} selectivity due to its obvious size effect and the converging effect of the functional groups. They also reported on stoichiometry and structural change caused by Ca^{2+} from ^{1}H-NMR (nuclear magnetic resonance) spectra, which gave evidence for the inclusion of the calcium cation inside the polyethereal cavity as a 1:1 complex.

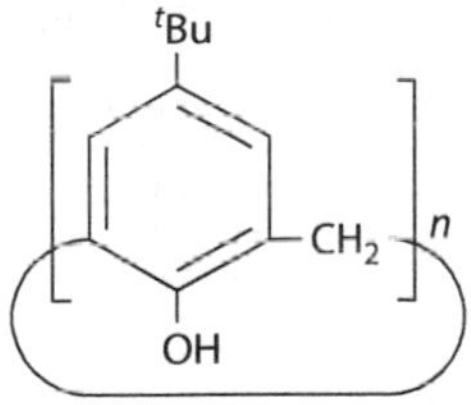

$n = 4$ **1**
$n = 6$ **2**
$n = 8$ **3**

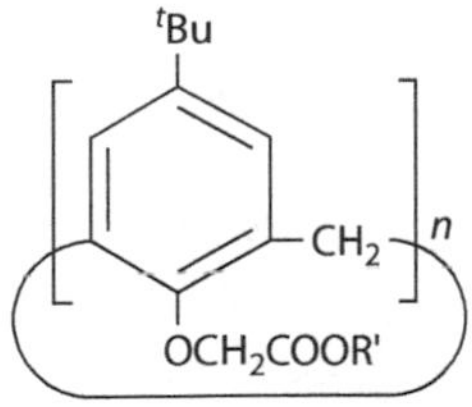

$n = 4$ R' = Et **4a**
R' = Me **4b**
R' = ^{t}Bu **4c**
$n = 6$ R' = Et **5a**
R' = Me **5b**
$n = 8$ R' = Et **6a**
R' = Me **6b**

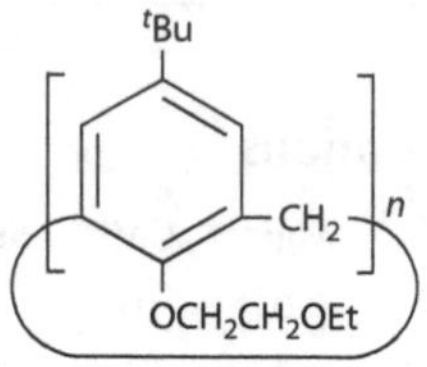

$n = 4$ **7**
$n = 6$ **8**
$n = 8$ **9**

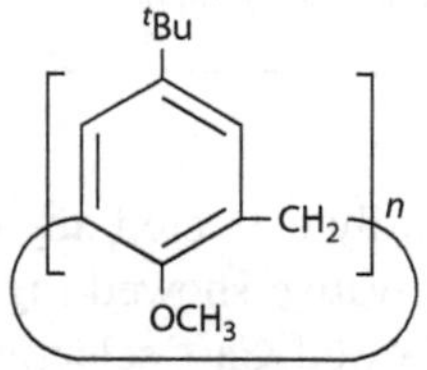

$n = 4$ **10**
$n = 6$ **11**
$n = 8$ **12**

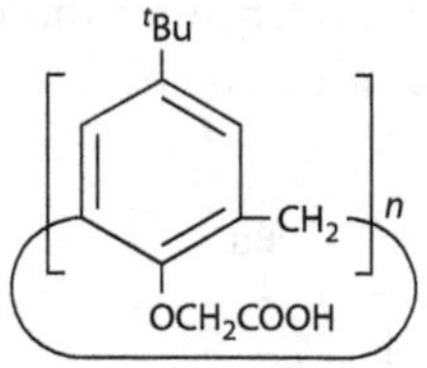

$n = 4$ **13**
$n = 6$ **14**
$n = 8$ **15**

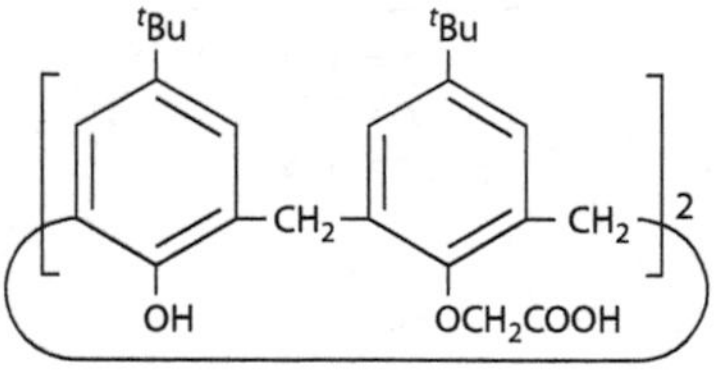

16

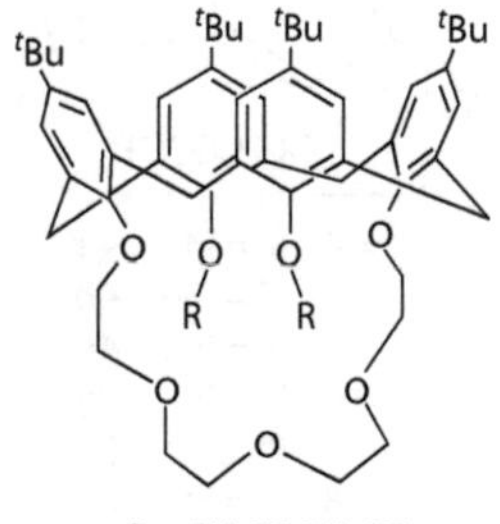

R = CH_2COOH **17**

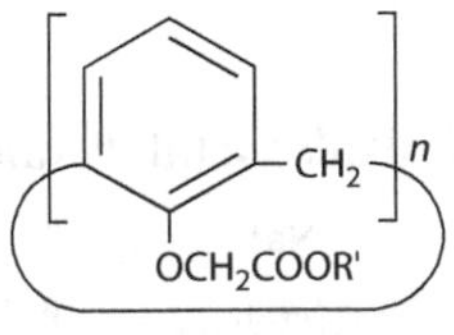

$n = 4$ R' = Et **18a**
    R' = Me **18b**
$n = 6$ R' = Et **19a**
    R' = Me **19b**
$n = 8$ R' = Et **20a**
    R' = Me **20b**

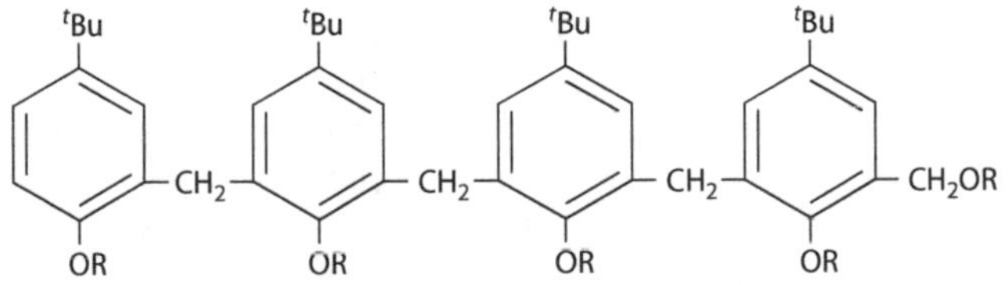

R : CH$_2$COOEt   **21**

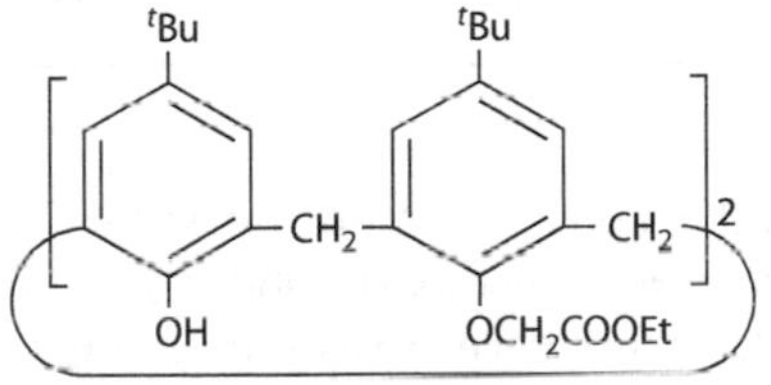

22

McKervey et al. reported on the extraction of alkali metals with 16 extractants, including calix[4]arene, calix[6]arene, calix[8]arene, **4a**, **4b**, **18a**, **18b**, **5a**, **5b**, **19a**, **19b**, **6a**, **6b**, the linear esterified tetramer **21**, and the diester of calix[4]arene **22**, together with 18-crown-6 and "silacrown" compounds. The study elucidated the dependence of extractability on the unit structure number, substituent, substituent at the p-position, ester number, cyclic structure, and so on.[26] The percent extraction of alkali metal picrates with 14 extractants, except two crown ether compounds, into CH$_2$Cl$_2$ at 20°C is listed in Table 3.3. They also concluded that the high extractability and selectivity are attributable to the size effect and the converging effect of the functional group compared with various types of derivatives. That is, they determined the X-ray diffraction (XRD) crystal structures of **4a** and **19a** in an effort to define cavity size and shape and total topology. Tetramer **4a** possesses a distorted cone or cup-like conformation in which the four ester groups are mutually *syn*, thus defining a central cavity with 3.10–3.28 Å and exhibiting high sodium extractability among five alkali metals. In contrast, the centrosymmetric hexamer **19a** has a different conformation. Three adjacent ester groups are *cis*, but the other three ester groups were inverted *anti*; consequently, it exhibited larger metal selectivity.

TABLE 3.3
Percent Extraction of Alkali Metal Picrates in CH_2Cl_2 at 20°C

Extractant	Li+	Na+	K+	Rb+	Cs+
4a	15.0	94.6	49.1	23.6	48.9
4b	6.7	85.7	22.3	9.8	25.5
18a	1.8	60.4	12.9	4.1	10.8
18b	1.1	34.2	4.8	1.9	4.6
5a	11.4	50.1	85.9	88.7	100.0
5b	1.7	10.3	29.1	41.2	54.8
19a	4.7	10.4	51.3	94.1	94.6
19b	2.6	6.7	25.2	77.7	94.6
6a	1.1	6.0	26.0	30.2	24.5
6b	0.9	8.3	25.5	29.8	20.1
20a	0.8	7.5	20.2	28.9	30.1
20b	0.4	4.1	12.1	17.5	27.0
21	2.6	3.9	5.4	6.8	7.1
22	0.0	0.0	0.0	0.0	0.0

Source: McKervey, M. A. et al., *J. Chem. Soc. Chem. Commun.*, 1985, 388–390.

It was reported that the more electron-dense amide[27-30] and ketone[31-33] derivatives of calix[4]arene showed higher extractability for alkali metal ions than ester derivatives. In particular, these derivatives showed high Na selectivity. Arnaud-Neu et al. reported on the extraction of alkali and alkaline earth metal ions with three amide derivatives of calix[4]arene having different alkyl branches on the nitrogen atom[29] and with 11 amide derivatives 23a–23k.[30] The percentage extraction of alkali and strontium picrates into CH_2Cl_2 at 20°C is listed in Table 3.4. All the tetraamide derivatives show a high extraction level for alkali cations, with sodium always being the best extracted. Amide derivatives are better extracting reagents than esters, but less selective due to high basicity of carbonyl oxygen atoms. The Na+/K+ selectivity, expressed by the separation factor *S*, is very low and slightly decreased with a decreasing number of carbon atoms in the substituents for alkyl derivatives, while the *S* values for the other amide derivatives with multiple bonds are high due to increasing the relative rigidity of the extraction reagents and electron-withdrawing character. It was found that the propargylic amide derivative with a triple bond showed high extractability and high Na selectivity. From this result, it was found that the effect of substituents on selectivity is also important. Thermodynamic parameters in a homogeneous methanol and acetonitrile phase for 23a, 23b, and 4a were estimated, and it was concluded that the lower Na/K selectivity of amides is mainly due to the high ΔH_c value for the K+ complex. The methyl ketone derivative of calix[4]arene, 24, showed a higher extraction strength than the methyl ester derivative 4b, while Na selectivity is less than for the ester derivative, as summarized in Table 3.5.

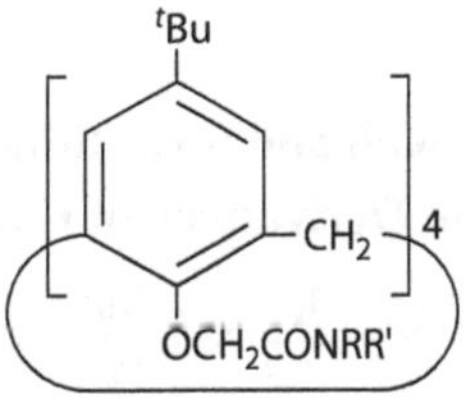

R = R' = Et   **23a**
R = R' = -[CH₂]₄-   **23b**
R = R' = Pr   **23c**
R = R' = Bu   **23d**
R = R' = Pentyl   **23e**
R = R' = Hexyl   **23f**
R = R' = Octyl   **23g**
R = R' = Allyl   **23h**
R = R' = Propagyl   **23i**
R = R' = CH₂Ph   **23j**
R = Et, R' = CH₂Ph   **23k**

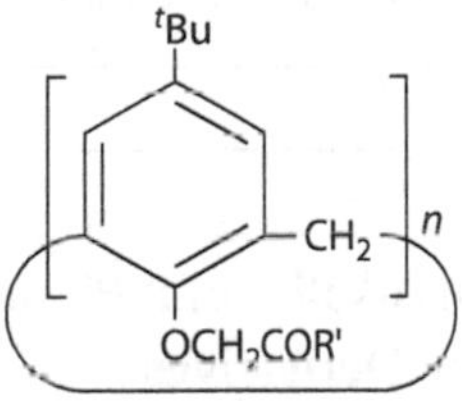

$n = 4$ R' = Me   **24**
$n = 6$ R' = Me   **25**
$n = 8$ R' = Me   **26**

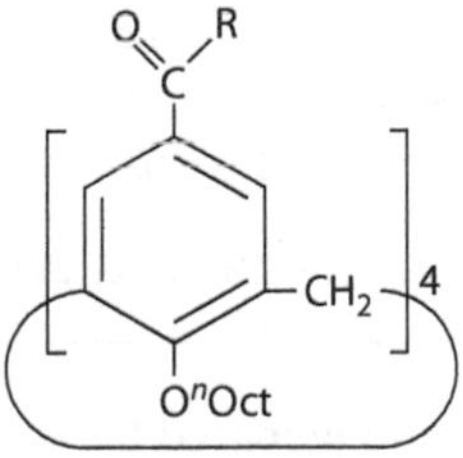

R = CH₃   **27a**
R = OCH₃   **27b**
R = NH₂   **27c**

Conner et al. carried out the extraction of alkali metal ions with cone-conformational calix[4]arene derivatives with ketone **27a**, ester **27b**, and amide **27c** groups at the upper rim. Extraction equilibrium constants for picrate salts are listed in Table 3.6. These showed lower extractability for Na⁺ than those at the lower rim, while showing comparable extractability for K⁺ and Cs⁺.[34] This is a typical rim-size effect on the extraction.

TABLE 3.4

Percent Extraction of Alkali and Strontium Picrates into CH_2Cl_2 at 20°C by Using Amide Derivatives of Calix[4]arene

Extractant	Li$^+$	Na$^+$	K$^+$	Rb$^+$	Cs$^+$	S(Na$^+$/K$^+$)	Sr^{2+}
23a	62.5	95.5	73.7	24.0	11.8	1.29	86.3
23b	47.8	91.1	57.5	16.1	11.2	1.58	72.1
23c	71.6	95.0	79.6	33.3	9.7	1.19	78.5
23d	67.7	92.0	78.4	37.8	14.8	1.17	74
23e	69.9	93.0	80.9	37.3	14.4	1.15	76.6
23f	71.2	87.6	77.2	36.9	26.2	1.13	68.4
23g	74.6	94.8	84.6	48.3	28.4	1.12	65.0
23h	40.1	91.5	63.7	16.0	6.4	1.44	53.4
23i	8.6	76.0	11.4	3.9	2.8	6.67	2.4
23j	36.5	79.8	56.1	14.4	11.5	1.42	27.9
23k	50.2	95.0	68.7	24.1	15.1	1.38	58.8

Source: Arnaud-Neu, F. et al., *J. Chem. Soc. Perkin Trans. 2*, 1995, 453–461.

TABLE 3.5

Percent Extraction of Alkali Metal Picrates into CH_2Cl_2 at 20°C by Using Ketone Derivatives

Extractant	Functional Group	n	Li$^+$	Na$^+$	K$^+$	Rb$^+$	Cs$^+$
4b	Ester	4	6.7	85.7	22.3	9.8	25.5
5b	Ester	6	1.7	10.3	29.1	41.2	54.8
6b	Ester	8	0.9	8.3	25.5	29.8	20.1
24	Ketone	4	31.4	99.2	84.1	53.7	83.8
25	Ketone	6	1.2	6.2	12.8	11.6	13.6
26	Ketone	8	0.7	9.9	25.1	20.8	15.3

Source: Arnaud-Neu, F. et al., *J. Am. Chem. Soc.*, 111, 1989, 8681–8691.

Since parent calix[4]arenes are prepared by using a sodium ion as a template, their derivatives on which functional groups are introduced generally exhibit sodium selectivity. On the contrary, some exceptional results were reported. Akabori et al. carried out the extraction and membrane transport of alkali metal ions with calix[4]arene derivatives with a phosphoric acid and three *O*-methyl groups, **28a**, **28b**, and **28c**.[35] They reported that **28a–c** exist as an equilibrium between the cone and partial cone conformations by inverting a methoxy benzene unit. Also, their mixed conformations rearrange only in the cone conformation during uptake of lithium and sodium ions, while they still exist as a mixture during potassium uptake. A more clear result for lithium selectivity was reported by Talanova et al. by using series of mono-ionizable calix[4]arene derivatives,

TABLE 3.6
Extraction Equilibrium Constants for Picrate Salts

Extractant	Functional Group	$10^{-5} K_e$			
		Li+	Na+	K+	Cs+
27a	Ketone at upper rim	0.122 ± 0.011	1.411 ± 0.088	0.686 ± 0.054	0.109 ± 0.008
27b	Ester at upper rim	0.152 ± 0.007	0.842 ± 0.078	1.295 ± 0.142	0.102 ± 0.014
27c	Amide at upper rim	0.062 ± 0.007	1.463 ± 0.029	1.084 ± 0.139	0.489 ± 0.066
4c	Ester at lower rim	0.056	11.3	0.1	0.11

Source: Conner, M. et al., *J. Org. Chem.*, 57, 1992, 3744–3746.
Note: $M^+ + Pic^- + L \rightleftharpoons MLPic: K_e$

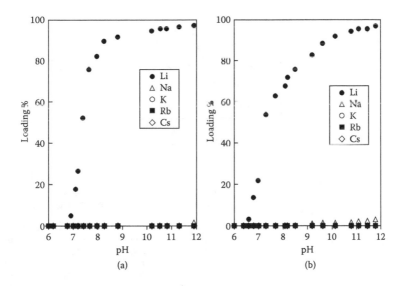

FIGURE 3.5 Effects of pH on loading percentages of alkali metal ions with (a) **29e** (mobile) and (b) **30e** (cone). (From Talanova, G. G. et al., *J. Chem. Soc. Perkin Trans. 2,* 2002, 1869–1874.)

conformationally mobile **29a–e** and cone-conformational **30a–e**.[36] The series of mono-ionizable compounds **29** selectively extracted Li+ from among five alkali metal ions. The Li extractabilities with **29a** and **29e**, and **30a–e** reached 100%, while Na and K extractabilities were less than 0.1%, and Rb and Cs were not detectable, as shown in Figure 3.5(a) and (b). Different from the results from Akabori et al.,[35] they proposed utilization of flexible ligands for extraction efficiency and selectivity. Since calix[4]arene is prepared by using sodium salt as the template ion, most of calix[4]arene compounds, especially in cone conformation, should exhibit sodium selectivity. However, both compounds exhibited extremely high Li selectivity. It is notable that coordinatively inert alkyl groups introduced affect selectivity change from sodium to smaller lithium different from the normal calix[4]arene selectivity.

In our current work, extraction of three alkali metal ions has been investigated with a dipropyl–diacetic acid crossed type of calix[4]anene in the cone conformation, **31**.[37] Lithium ion was simultaneously loaded onto a single molecule of **31** up to 200% in a single system, while sodium and potassium ions were loaded up to 150 and 100%, respectively. In a competitive system, sodium ion as the first ion and lithium ion as the second one were loaded up to 100%, respectively, whereas potassium ion was hardly extracted. This result also showed an obvious selectivity change based on a size effect by introducing alkyl groups. These results may suggest a creative concept for the design of extractant molecules with high separation efficiency by neither functional groups nor cyclic structure, but by the use of coordinatively inert alkyl branches.

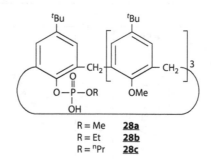

R = Me   **28a**
R = Et   **28b**
R = ^{n}Pr   **28c**

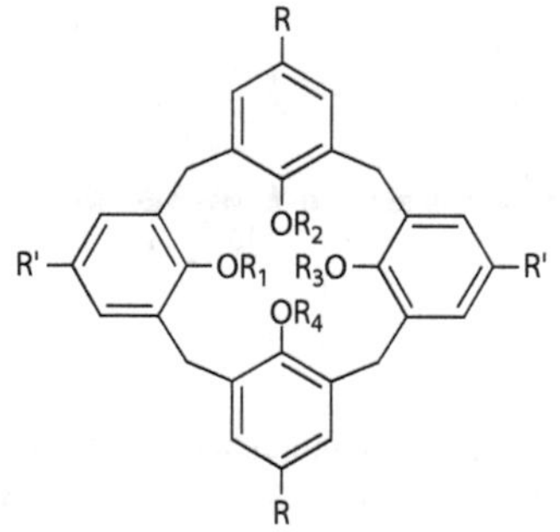

R = R' = ^{t}Bu		
$R_1 = R_2 = R_3$ = Me		
R_4 = OCH$_2$C(O)NHSO$_2$X		
X = CF$_3$		**29a**
Me		**29b**
Ph		**29c**
C$_6$H$_4$-4-NO$_2$		**29d**
R_4 = OCH$_2$COOH		**29e**

R = R' = ^{t}Bu		
$R_1 = R_2 = R_3 = ^t$Bu		
R_4 = OCH$_2$C(O)NHSO$_2$X		
X = CF$_3$		**30a**
Me		**30b**
Ph		**30c**
C$_6$H$_4$-4-NO$_2$		**30d**
R_4 = OCH$_2$COOH		**30e**

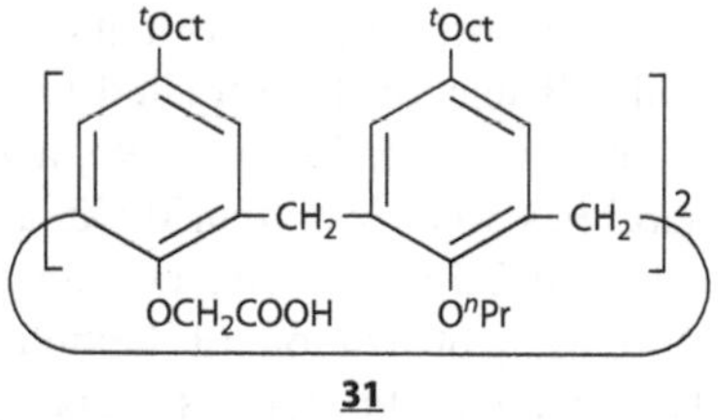

31

TABLE 3.7
Percent Extraction of Alkali Metal Picrates in CH_2Cl_2 at 25°C by Using Four Different Conformers of 4a and Two Different Conformers of 18a

Extractant	Conformation	Li+	Na+	K+	Cs+
4a	Cone	17.6	100.0	86.1	24.6
	Partial cone	5.2	62.1	94.3	49.9
	1,3-Alternate	0.0	22.1	70.0	54.0
	1,2-Alternate	1.5	88.8	100.0	98.9
18a	Cone	0.0	97.0	27.4	4.0
	1,3-Alternate	0.0	23.9	87.1	54.2

Source: Iwamoto, K., and Shinkai, S., J. Org. Chem., 57, 1992, 7066–7073.

Iwamoto et al. first reported the conformational effect.[38] They prepared four conformers of calix[4]arene tetraester 4a and 18a, by changing various preparation conditions, to investigate the extraction of alkali metal ions. The percent extraction of alkali metal picrates in CH_2Cl_2 at 25°C is listed in Table 3.7. The cone conformer showed Na selectivity, whereas the other conformers preferred K. The choice of the base for the template effect in the preparation of the calixarene is very important and is closely related to selectivity in the extraction of alkali metals. They reported on the alkyl-substituted calix[4]arene derivative, whose conformation was not inverted but fixed, as reported in their previous paper.[39] Subsequently, attention was paid to the conformation of the compounds used for extraction. Their work elucidating the relation between conformation and extractability made the detailed discussion possible afterward.

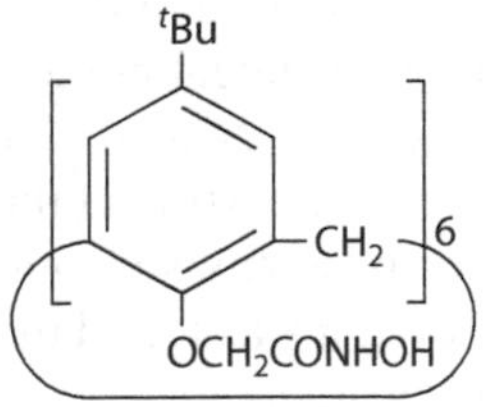

32

Uranyl extraction with super-uranophile calix[6]arene derivatives by Shinkai et al. is related to the molecular design for coordination geometry.[40–42] The authors focused on the uranyl ion, UO_2^{2+}, which prefers a pseudoplanar hexacoordinated binding structure, and used the unmodified and carboxyl derivatives of calix[6] arene, repectively, 2 and 14. The uranyl ion was effectively extracted with 14 at pH 8.0 ($\%E = 100$) and 10.0 ($\%E = 93.3$), while it was very poorly extracted with 2 at pH 8.0 ($\%E = 4.6$) and 10.0 ($\%E = 0.0$). In a further paper, Nagasaki and Shinkai

TABLE 3.8

Extraction of UO_2^{2+} in the Presence of Various Metal Cations at 30°C

Coexisting Metal Ions	($[M^{n+}]/[UO_2^{2+}]$) Ratio	E% of UO_2^{2+} 14	E% of UO_2^{2+} 30
None		100	100
Mg(II)	1000	100	100
Ni(II)	10	77	98
Zn(II)	12	51	96
Fe(III)	10		66
Fe(III)	1		100

Source: Nagasaki, T., and Shinkai, S., *J. Chem. Soc. Perkin Trans. 2*, 1991, 1063–1066.

reported on the selective extraction of the uranyl ion with the hexahydroxamic derivative of calix[6]arene **32**.[43] UO_2^{2+} was selectively extracted with **32** even in the presence of an excess amount of heavy metal ions apart from Fe^{3+}, while extraction with hexaacetic acid **14** was suppressed by the presence of other metal ions, as shown in Table 3.8.

There are many studies on extraction with calixarene derivatives with relatively soft functional groups. For example, Shinkai et al. reported on the extraction of alkali metal hydroxides with picric acid, silver and copper nitrates with picric acid, and uranyl acetate by using 2-pyridinocalix[6]arene **33** and calix[8]arene **34**, which showed good extractability for UO_2^{2+} consistent with the pseudoplanar hexacoordinate geometry, while showing poor extractability of the other ions, even for the soft metal ion Ag^+.[44] This is an example wherein the structural effect is more important than the functional group affinity.

Pappalardo et al. prepared four different conformers of *p-t*-butylcalix[4]arene tetra-2-pyridyl **35** and tetra-2-pyridine-*N*-oxide **36** derivatives to investigate the extraction of alkali metal ions.[45] The extractabilities were low due to the mismatch in HSAB theory between relatively soft functional groups and hard metal ions. From NMR spectra, however, it was found that Na^+ was surrounded by phenoxy oxygen and pyridyl nitrogen atoms.

p-t-Octylcalix[4]arene 2-, 3-, and 4-pyridyl derivatives **37–39** in the cone conformation and the corresponding monomers were prepared to investigate extraction of silver nitrate into chloroform in nitrate media by our group.[46] Comparison of calix[4]arene and the monomer derivatives (extractant concentrations were adjusted to be 5 and 20 mM in chloroform, respectively, so that the total concentrations of the functional group were the same) elucidated the cyclic structural effect of calix[4]arene, showing high silver extractability. Comparison of the compounds with nitrogen at different positions showed that **37** possesses the highest extractability due to its high electric charge density and matching of the size and geometry of the binding site for Ag^+. This was supported by the ^{1}H-NMR spectra of the 2-pyridyl complex with Ag^+, which showed that Ag^+ was surrounded not by phenoxy oxygen atoms but by pyridyl nitrogen atoms.

Yordanov et al. carried out the extraction of Pd^{2+}, Pt^{2+}, Cd^{2+}, Pb^{2+}, and Ni^{2+} with the calix[4]arene derivatives with sulfur functionality at lower and upper rims **41–43**.[47,48] They reported that a stable palladium chlorocomplex is readily and selectively extracted from hydrochloric acid media with p-t-butylcalix[4] arene tetrathiocarbamoyl derivative **40**. They also carried out the extraction of the chlorocomplexes of Sn^{4+}, Pd^{2+}, Pt^{2+}, and Au^{3+}, and for Hg^{2+} and Ag^+ with the thioether derivative **41**.[49] Since the functionalized compound at the lower rim of **40**, and the upper rim of **42** exhibited similar extraction ability to Pd^{2+}, the effect of high extraction ability is attributed to the sulfur functionality. They found that the extractabilities of Hg^{2+} and Ag^+ when t-butyl groups were present were higher than in their absence due to the difference in lipophilicity.

Silver nitrate extraction in nitric acid media into chloroform with hard oxygen-donating ketone[50,51] and amide[52] derivatives was reported by our group. High Ag^+ selectivity over Pd^{2+} and other precious metals was observed by using ketone derivatives of p-t-octylcalix[6]arene **44** and calix[4]arene **45**, while the corresponding monomer **46** had the opposite selectivity. Cone-conformational **45** showed the highest Ag extraction with 1:1 stoichiometry, as shown in Figure 3.6(a)–(c). This was attributed to effects of coordination size and converging functional group, together with the solvation mechanism, in which only a single anion was necessary for silver extraction. The Ag extractability with **44** and **45** from chloride media dropped to lower values than that from nitrate media. Similar behavior was observed for the tetraamide derivative of calix[4]arene **47**, while Ag extractability with the amide monomer **48** from chloride media was

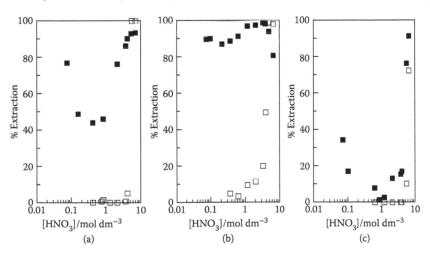

FIGURE 3.6 Effect of nitric acid concentration on metal extraction with ketone derivatives: (a) **42**, (b) **43**, and (c) **44** from nitrate media. Closed square: Ag(I); open square: Pd(II). (From Ohto, K. et al., *Anal. Chim. Acta*, 341, 1997, 275–283.)

higher than that from nitrate media. This is attributed to the fact that **47** extracts Ag as a cation from both media with 1:1 stoichiometry, while the tautomerized and anion-exchangeable **48** extracts silver ions as an anion, $AgCl_2^-$, from chloride media with 1:2 stoichiometry (Ag:**48**), and as a cation from nitrate media.

R : C(S)NMe$_2$ **40**
H **41**

R : C(S)NMe$_2$ **42**
Me **43**

$n = 6$ **44**
$n = 4$ **45**

47

46

48

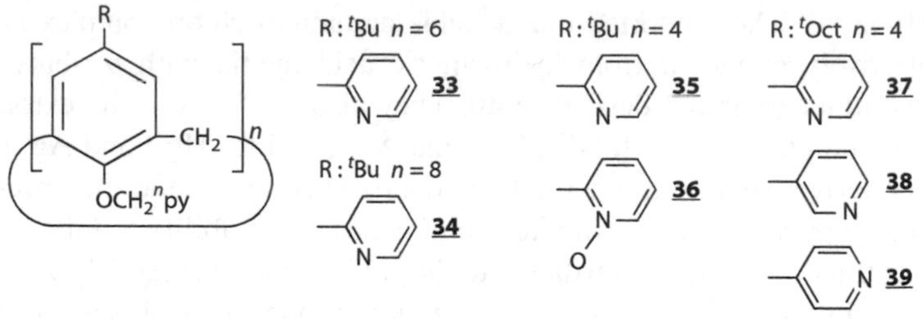

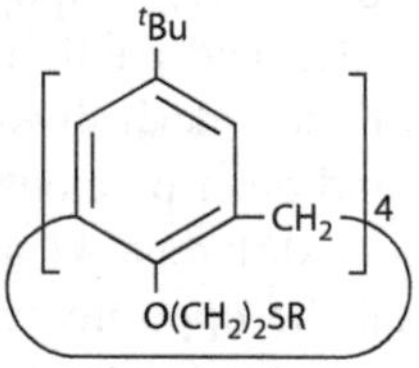

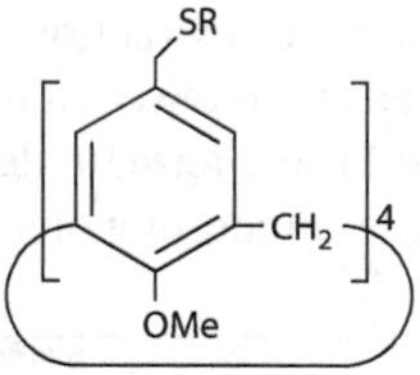

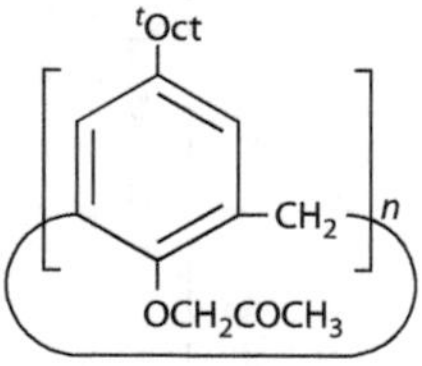

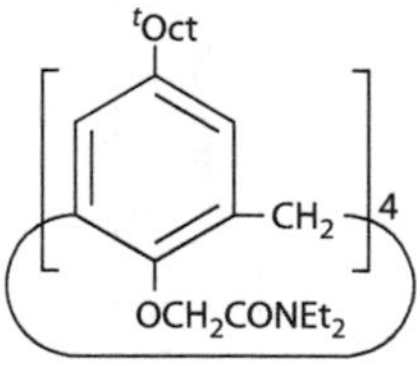

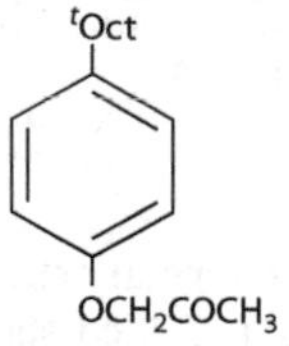

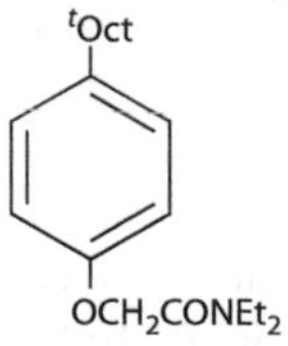

TABLE 3.9
Percent Extraction of Lanthanide and Actinides by Using Calix[4]arene Derivatives and the Corresponding Monomers with CMPO Functionality

Extractant	^{152}Eu(III)	^{237}Np[a]	^{239}Pu(IV)	^{241}Am(III)
49a	>99	67	99	>99
49b	>99	67	95	82
49c	>99	67	99	>99
49d		80	>99	>99
49e	>99	67	99	>99
50a	<0.1	33	25	<0.1
50b	54	47	96	64
50c	93	67	95	94
CMPO		46	96	54

Source: Arnaud-Neu, F. et al., *J. Chem. Soc. Perkin Trans.* 2, 1996, 1175–1182.

[a] Np ion is present in different oxidation states (IV–VI) like NpO^{2+}, NpO_2^+, and NpO_2^{2+} under this experimental condition.

Phosphine oxide and carbamoylmethylphosphine oxide are typical hard coordination ligands. Both compounds are commercially available as trioctylphosphine oxide (TOPO) and octylphenyl-*N*,*N*-diisobutylcarbamoylmethlyphosphine oxide (CMPO), and they also act as excellent ligands. In particular, the CMPO molecule has been proposed for use for actinide separation from nuclear waste solutions.[53–57] Arnaud-Neu et al. reported the extraction of Eu^{3+}, Np^{3+}, Th^{4+}, and Am^{3+} with a series of *O*-alkylcalix[4]arenes with diphenylcarbamoylmethylphosphine oxide at the upper rim.[58,59] They found that the calix[4]arene derivatives 49a–e showed higher extractability than the original CMPO or the linear derivatives 50a–c, as listed in Table 3.9.

Delmau et al. synthesized calix[4]arene with CMPO groups on the upper rim to investigate the extraction behavior of trivalent lanthanides (Ln) and Am^{3+} as an actinide (An).[60,61] The CMPO-like calixarene extractant 49a exhibited apparent light Ln selectivity, as shown in Figure 3.7, whereas it was found that the original CMPO showed rather poor mutual separation efficiency and middle Ln selectivity from the relationship between their ion radii and distribution ratio. The separation of Ln and An was, however, also poor, which is a normal result for hard-donor extractants. They examined the effect of solvents (octanol and chloroform) and the extraction media (HNO_3 and $NaNO_3$) on the distribution ratio. They found that two factors did not affect the separation efficiency, and the extraction system with octanol-$NaNO_3$ showed the highest distribution ratio among them. Furthermore, they investigated the effect of

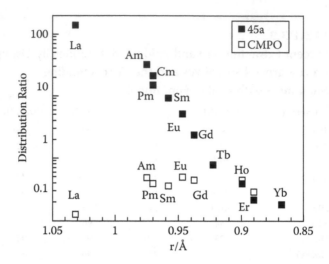

FIGURE 3.7 Relationship between ion radii and distribution ratio. Closed square: **49a**, open square: CMPO. (From Delmau, L. H. et al., *Chem. Commun.,* 1998, 1627–1628.)

alkyl chain length at the lower rim on distribution and found that the length did not affect separation efficiency and distribution ratio. Matthews et al. prepared some calix[4]arene derivatives with CMPO groups at the upper rim and alkoxyl groups at the lower rim to investigate the extraction of a series of Lns and Ans.[62] Among them, the derivative with a 1,2-dimethoxy-3,4-dipropoxy group showed the highest extraction ability. The separation of Am^{3+} and Cm^{3+} was proposed.

$$R = -C_5H_{11} \qquad \textbf{\underline{49a}}$$
$$-CH_2CH(C_2H_5)C_4H_9 \quad \textbf{\underline{49b}}$$
$$-C_{10}H_{21} \qquad \textbf{\underline{49c}}$$
$$-C_{12}H_{25} \qquad \textbf{\underline{49d}}$$
$$-C_{14}H_{29} \qquad \textbf{\underline{49e}}$$

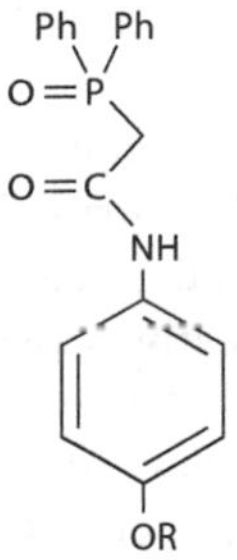

$$Ph \quad Ph$$
$$O = P$$
$$O = C$$
$$NH$$

$$OR$$

$$R = -C_5H_{11} \qquad \underline{\textbf{50a}}$$
$$-CH_2CH(C_2H_5)C_4H_9 \quad \underline{\textbf{50b}}$$
$$-C_{10}H_{21} \qquad \underline{\textbf{50c}}$$

Atamas et al. synthesized different calixarene derivatives of carbamoylmethylphosphine oxide **51** and diphosphine dioxide **52** to investigate the extraction of Ln^{3+} and Am^{3+}.[63] The extraction ability of both extractants is less than that of calix[4]arene with CMPO functions via amide bonds, but much higher than that of the original CMPO. Separation efficiencies of both extractants for Am/Eu separation and mutual separation of Lns were also poorer than that of the CMPO derivative through the amide. The result suggested that the CMPO through the amide is better for extraction ability and separation efficiency. Schmidt et al. prepared the *N*-methylated CMPO-like *p-t*-butylcalix[4]arene **53** to investigate Lns and Am extraction.[64] Although the extraction and the separation efficiencies were drastically depressed by *N*-methylation due to the stoichiometrical change from 2:1 (secondary amide) to 1:1 (*N*-methylated amide), mutual separation of Lns was improved. Distribution ratios for cation extraction at 3 mol dm^{-3} HNO_3 are summarized in Table 3.10.

There have been reports on work wherein carbamoylmethylphosphine oxide or diphosphine oxide groups were introduced on the lower rim of calix[4]arene. Arnaud-Neu et al. carried out the extraction of alkali metals, alkaline earth metals, Th^{4+}, and Eu^{3+} with *p-t*-butylcalix[4]arene having phosphine oxides with different spacer lengths **54a** ($n = 1$) and **54b** ($n = 2$) on the lower rim.[65] A big effect of the shortening of the carbon chain bearing the phosphine oxide groups was observed. It would be attributed to not only the chelate effect with phenoxy oxygen atoms but also the size effect of the coordination site. Compared with

TABLE 3.10
Distribution Ratios for Cation Extraction at 3 mol dm^{-3} HNO_3

Extractant	La(III)	Ce(III)	Nd(III)	Sm(III)	Eu(III)	Am(III)
53	2.1	1.8	0.72	0.21	0.14	0.87
49a	145		135	60	45	156

Source: Schmidt, C. et al., *Org. Biomol. Chem.*, 1, 2003, 4089–4096.

commercially available TOPO and CMPO, both derivatives showed extremely high extraction ability.

Barboso et al. introduced CMPO groups with three different methylene spacers on *p-t*-butylcalix[4]arene at the lower rim **55a–c** for the extraction of thorium and the three Ln classes (light, middle, heavy).[66] The CMPO derivative with a shorter spacer length showed Th selectivity over the Lns, while the longer derivative showed poor Th selectivity, but higher extractability of the Lns. The derivative with CMPO at a narrow lower rim showed light to middle Ln selectivity, while the derivative with a wide upper rim obviously showed large light Ln selectivity.

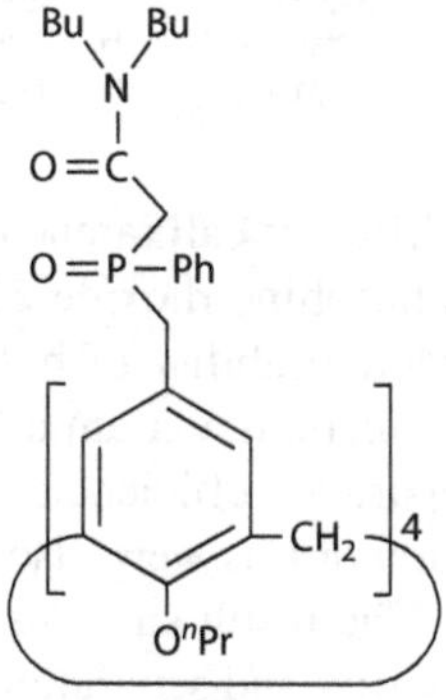

51

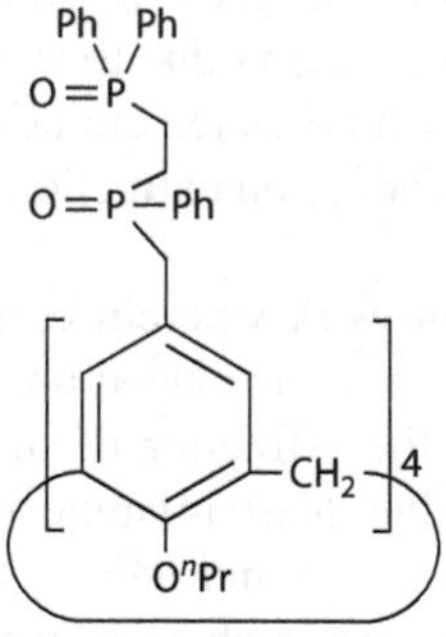

52

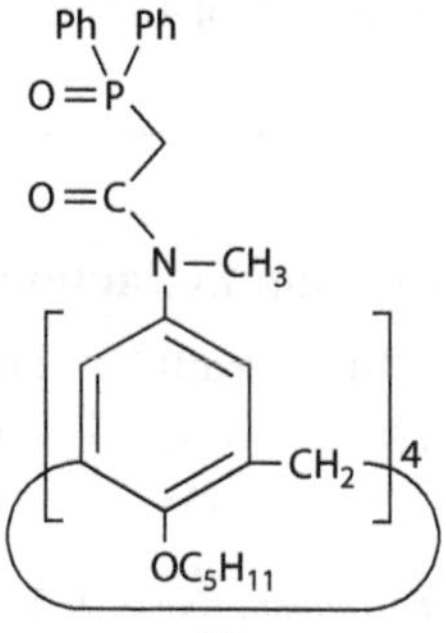

53

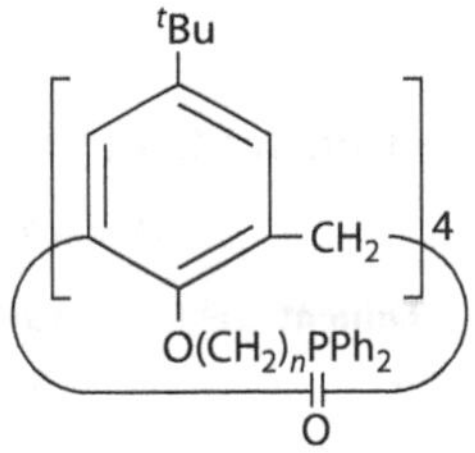

$n = 1$ **54a**
2 **54b**

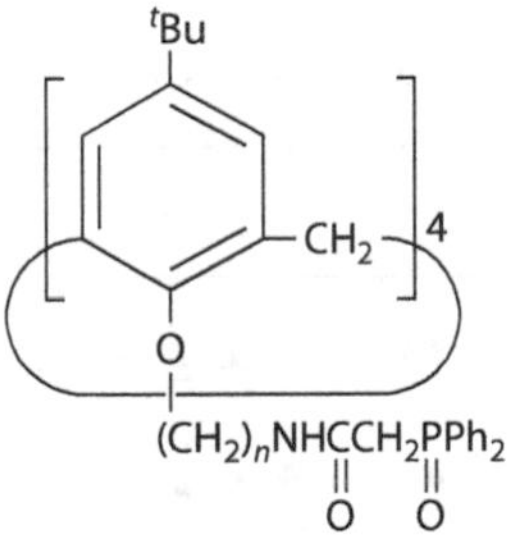

$n = 2$ **55a**
3 **55b**
4 **55c**

There are other reports on rare earth (RE) extraction with calixarene compounds. Ludwig et al. prepared *p-t*-butylcalix[4]arene tetraacetic acid **13**, *p-t*-butylcalix[6]arene hexaacetic acid **14**, and propylene-methylene alternately bridged calix[4]arene tetraacetic acid with a larger cavity **56** to investigate the extraction of a series of trivalent Lns.[67,68] They determined the extraction reaction by slope analysis for the dependencies on pH and extractant concentration. The cyclic hexamer showed high extractability and is comparable to the tetramer with a wider rim. They concluded that the extraction ability is related to cavity size, number of carboxylic acid groups, and flexibility of conformation. Furthermore, they first found that coexistence of Na$^+$ in the Ln extraction system caused a change in the extractability and selectivity, as described later.

The more lipophilic *p-t*-octylcalix[4]arene tetraacetic acid **57** in the cone conformation, *p-t*-octylcalix[6]arene hexaacetic acid **58**, the corresponding monomer **59**, and the corresponding linear trimer **58** were employed to investigate the extraction of a series of trivalent RE metals by our group.[69,70] The aggregation constant and aggregation number of the cyclic compounds **57** and **58** in toluene and chloroform were estimated. The dimerization constants of **57–59** in toluene and chloroform are listed in Table 3.11. Hexamer **58** is less aggregative than the tetramer **57**. It may be related to the conformational rigidity. The extraction reaction was also determined by slope analysis similar to the procedure by Ludwig et al. The extraction ability and the separation efficiency of calixarene extractants for trivalent REs were discussed using the relationship between the reciprocal

TABLE 3.11

Dimerization Constants of the Extractants

Extractant	Diluent	Dimerization Constant K_d/dm³ mol⁻¹
58	Toluene	4.54×10^{-3}
58	Chloroform	1.21×10^{-3}
57	Toluene	7.00×10^{-2}
57	Chloroform	4.78×10^{-3}
59	Chloroform	1.03×10^{-4}

Source: Ohto, K. et al., *Anal. Sci.*, 11, 1995, 893–902.

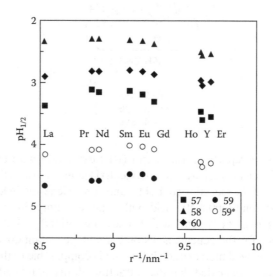

FIGURE 3.8 Relationship between $pH_{1/2}$ and sequence of trivalent rare earth ions in HNO_3–glycine media. (From Arnaud-Neu, F. et al., *Chem. Eur. J.*, 5, 1999, 175–186; HCl-HEPES media, unpublished data.)

number of their ion radii and the pH values for 50% extraction ($pH_{1/2}$) shown in Figure 3.8. The enhanced extractability and separation efficiency by the multiple effects of converging functional group and size fitting of calix[4]arene in cone conformation were explained.

Ludwig et al. also carried out the extraction of a series of Ln, transition, and heavy metal ions in perchlorate media with a carboxylic acid derivative of calix[4] arene with *n*-octadecyl groups, **61**,[71] into chloroform containing 5% 2-octanol as a modifier. Due to the higher hydrophobicity compared with **13** with *t*-butyl groups at the *p*-position, **61** showed a 20 times higher distribution ratio for Lns than **13**. Furthermore, they reported the favorable separation by the extraction of Lns and

Am with an amide–acetic acid crossed derivative of *p-t*-octylcalix[6]arene **62**.[72] The compound **62** showed a higher separation factor for Am/Eu (118) compared with **14** as a reference compound, which had a separation factor of 66.

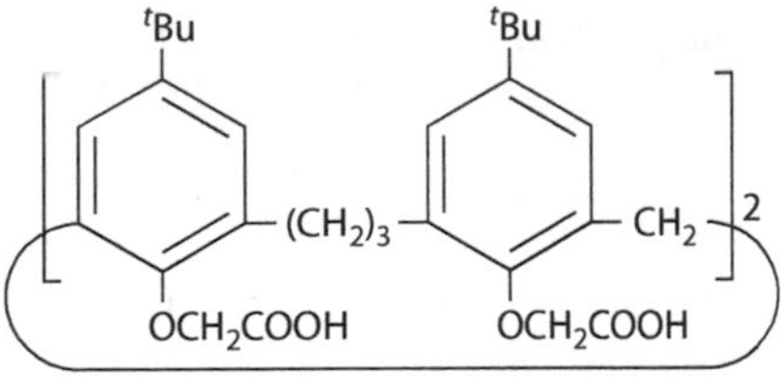

^{t}Bu ^{t}Bu

$-(CH_2)_3-$ $-CH_2-\Big]_2$

OCH$_2$COOH   OCH$_2$COOH

56

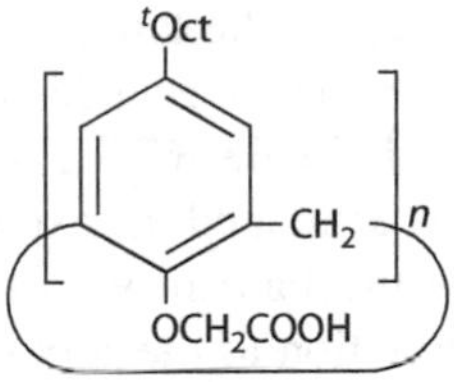

^{t}Oct ^{t}Oct

$-CH_2-\Big]_n$

OCH$_2$COOH   OCH$_2$COOH

$n=4$ **57**   **59**
$n=6$ **58**

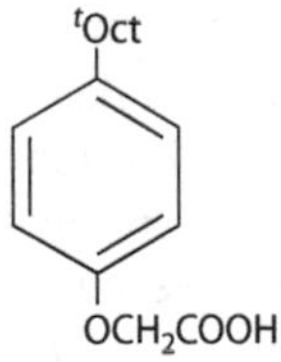

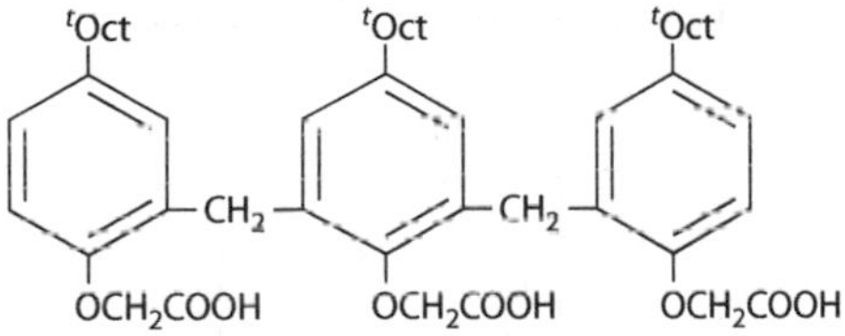

^{t}Oct ^{t}Oct ^{t}Oct

$-CH_2-$ $-CH_2-$

OCH$_2$COOH OCH$_2$COOH OCH$_2$COOH

60

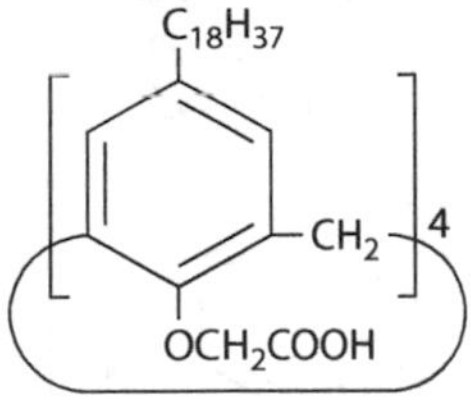

$C_{18}H_{37}$

$-CH_2-\Big]_4$

OCH$_2$COOH

61

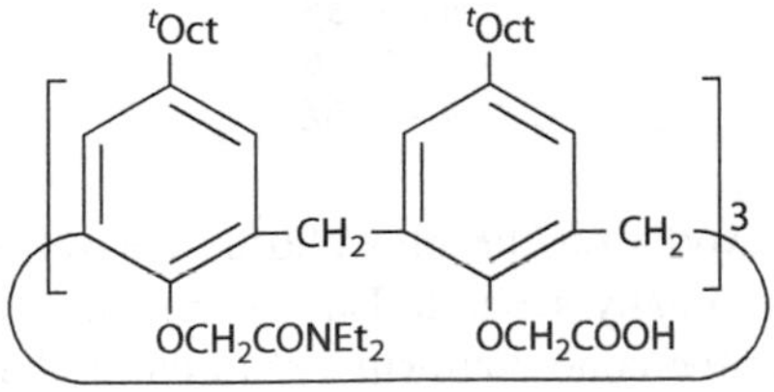

^{t}Oct ^{t}Oct

$-CH_2-$ $-CH_2-\Big]_3$

OCH$_2$CONEt$_2$ OCH$_2$COOH

62

TABLE 3.12

Comparison of log K_{ex1} Values of Am^{3+} Extraction into CHCl$_3$

Extractant	64	58	67	65	66	14	63
Substituent at p-position	^{t}Bu	^{t}Oct	H	H	H	^{t}Bu	^{t}Bu
n, number of aryl moieties	7	6	7	5	6	6	5
log K_{ex1}	−3.59	−4.14 >	−4.25 >	−4.33 >	−4.59 >	−4.90 >	−6.0 >

Source: Ludwig, R. et al., *Radiochim. Acta*, 88, 2000, 335–343.

Note: $M^{3+} + H_nL_{org} \rightleftharpoons MH_{n-3}L_{org} + 3H^+$: K_{ex1}

Ludwig et al. also systematically investigated the lanthanide extraction with p-t-butylcalixarene and debutylated calixarene acetic acid derivatives with different unit numbers from 5 to 7, using **14, 63, 64, 58, 65–67**.[73] They also determined the extraction reaction by slope analysis and estimated the extraction equilibrium constants. It is just a simple ion-exchange mechanism with three protons with a 1:1 stoichiometry. They concluded debutylation at the p-position affected less extractability due to flexible cavity size, less preorganization, and less hydrophobicity. A comparison of the logarithm of the extraction equilibrium constants of calixarene compounds is summarized in Table 3.12.

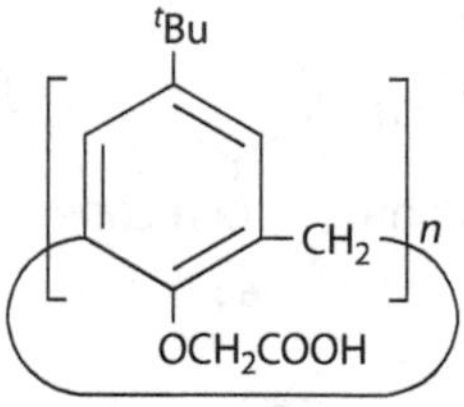

n = 5 **63**
n = 7 **64**

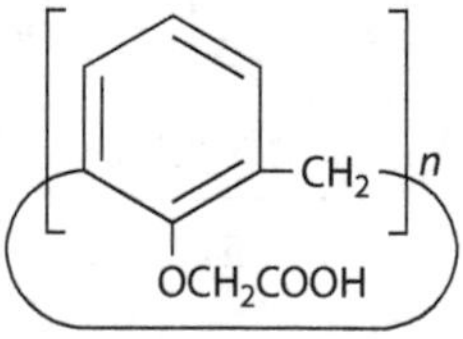

n = 5 **65**
n = 6 **66**
n = 7 **67**

There are several works on the phosphorus derivatives of the extractants, which also aimed to mutually separate Lns described earlier. The extraction of trivalent RE ions with the cone-conformational calix[4]arene derivative having a phosphonic acid group at the upper rim, **68**, was carried out by our group.[74]

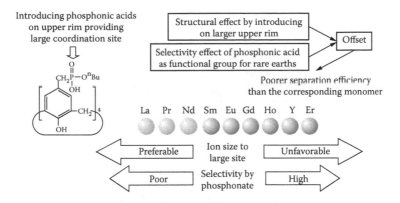

FIGURE 3.9 Typical offset of two effects base on the cyclic structure and the functionality of the calix[4]arene type extractants. (From Ohto, K. et al, in *Proceedings of ISEC'93, Solvent Extraction in the Process Industries*, Vol. 1, SCI, London, 1993, 364–369).

Since the binding site at the upper rim is much larger than the ion diameter of RE^{3+}, the separation efficiency of **68** for RE^{3+} is poorer than that of the corresponding monomeric analogue **69**, although the extraction ability of **68** is comparable to that of **69**. This is a typical offset shown in Figure 3.9, which the effect of the cyclic structure adversely affects to offset the effect based on the functional group affinity. This negative result led to the strategy to design the extractant molecule shown in Figure 3.4.

Oshima et al. carried out the extraction of a series of Lns with *p-t*-octylcalix[6] arene with octylphosphoric acids at the lower rim of **70** together with the corresponding monomer **71**.[75] The compound **70** possesses extremely high extraction ability compared with **69** and the hexaacetic acid derivative **14**. The $pH_{1/2}$ value of Ho^{3+} for **70** is −0.8, while for **14** it is 2.6.

Jurecka et al. carried out the extraction of Ln^{3+} and Th^{4+} by using *p-t*-butylcalix[4]arene having phosphonic acid groups of different spacer lengths on the lower rim, **72–74**.[76] All phosphonic acid extractants showed, as expected, high heavy RE selectivity. The extraction efficiencies decrease with increasing length of the methylene spacer, although the dependence was not monotonic, and an unusually high selectivity was observed for Yb^{3+}. Rohovec et al. also prepared a calix[4] arene derivative with four dibasic phosphonic acids and a propylene spacer length in a cone conformation, **75**.[77] They found that **75** exhibited lighter RE selectivity.

68

69

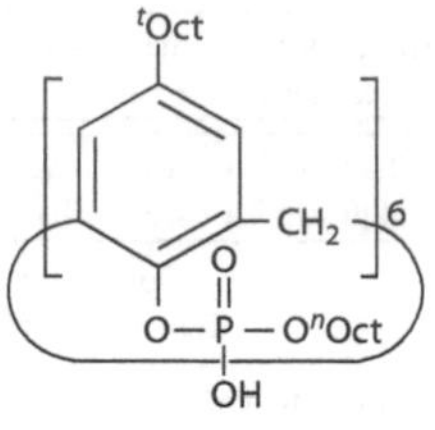

70 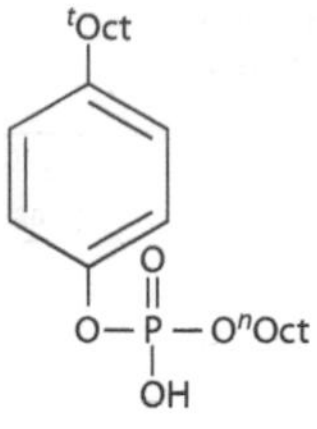   71

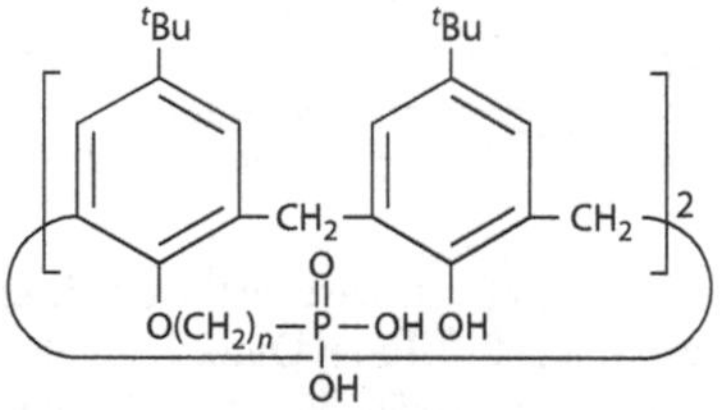

n = 1 **72**
n = 2 **73**
n = 3 **74**

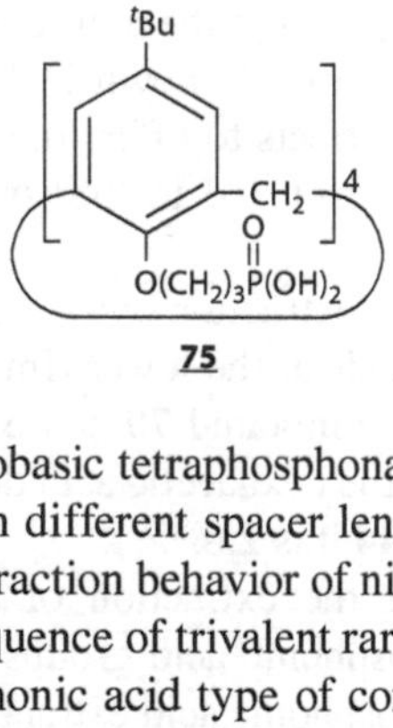

75

In our current works, monobasic tetraphosphonate derivatives of calix[4]arene in the cone conformation with different spacer lengths, **76**[78] and **77**[79], have been prepared to investigate the extraction behavior of nine trivalent REs. The selectivity as given by $pH_{1/2}$ over a sequence of trivalent rare earth ions for **57**, **76–78**, and a representative organophosphonic acid type of commercial extractant, PC-88A, is shown in Figure 3.10. On comparison of the extraction ability of **76** to **57**, it may be remarked that the nature of the functional group is a truly significant factor in molecular design. Both extractants possess high extraction efficiency; particularly, **76** exhibited extremely high extraction ability. Similar to the results reported by Jurecka et al.,[76] the extraction ability and the separation efficiency were strongly related to the length of the methylene spacer. The difference is related to the absence or presence of the participation of the phenoxy oxygen atoms in the extraction.[78] That is, chelation of the phosphonic acid and the phenoxy oxygen atoms enhances the extraction strength, but depresses the selectivity. Extractant **76** exhibited higher extraction strength than **77**, whereas it exhibited poorer separation efficiency among the various REs. There are some conceivable reasons, and one of them is related to chelation as follows: (1) mismatch in the distance between the chelating atoms and the ionic radii of the rare earth metals, (2) unsuitableness of phenoxy oxygen atom to heavier REs for multiplier effect with phosphonic acid,

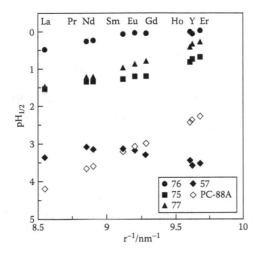

FIGURE 3.10 Relationship between $pH_{1/2}$ and sequence of trivalent rare earth ions for **57, 75–77** and PC-88A. (From Ohto, K. et al., *J. Inclusion Phenom. Macrocycl. Chem.*, DOI 10.1007/s10847-012-0255-0.)

and (3) significant enhancement of extraction ability. In addition, 1:2 metal:ligand stoichiometry, which is similar to that of the acyclic ligand complex, depressing separation efficiency and selection of unbranched alkyl groups as side chains should also be carefully considered. As shown in Figure 3.10, the prepared extractants **77** and **78** still exhibited poorer separation efficiency than the best commercial extractant in RE separation, PC-88A. Based on the concepts mentioned above, better extractants in separation efficiency among REs are still needed.

$n = 1$ **76**
$n = 3$ **77**

78

Followed by our strategy to design the extractants illustrated in Figure 3.4, the focus next shifted to light RE selectivity of quaternary ammonium compounds under special conditions with 10 M nitrate salt in an apolar diluent.[80] Such conditions led to not only an increased nitrate concentration as counteranion, but also a salting-out effect. Calix[4]arenes with quaternary ammonium groups at the upper rim, **79** and **80**, together with the corresponding monomer, **81**, were prepared to investigate extraction behavior of trivalent rare earth metals in nitrate media.[81] Although the prepared extractants exhibited light RE selectivity as expected, the detailed evaluation and comparison were not carried out, because **79** and **80** exhibited poor extractability under the limited experimental conditions due to poor solubility in toluene.

Bartsch et al. synthesized a series of di-ionizable p-t-butylcalix[4]arenes with N-(X)sulfonylcarboxamide groups at the lower rim (e.g., **82–83**) to investigate the extraction of Pb^{2+},[82] Hg^{2+},[83,84] Li^+ and Ba^{2+},[85] and Pb^{2+}, Hg^{2+}, and Ag^+.[86] They explained the effect of X-substituents and conformation related to the preorganization of the calixarene on the extractability. For Pb selectivity, **82a** suffered no suppression of Pb extraction in the presence of the same amount of K, Cs, Cu, Co, Ni, Zn, and Cd, although it was affected by the presence of Na, Sr, and Hg.

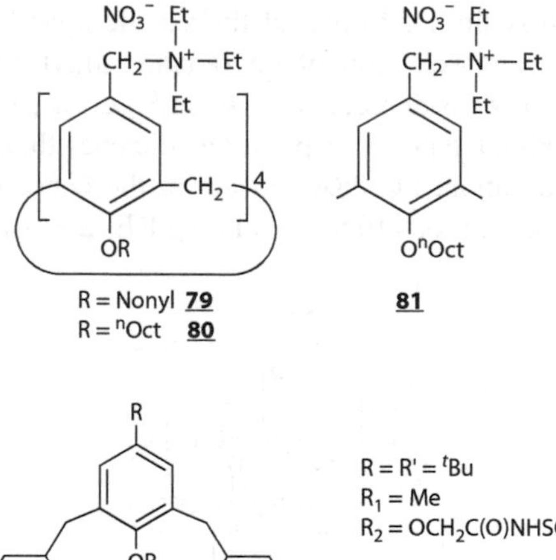

R = Nonyl **79**
R = "Oct **80**

81

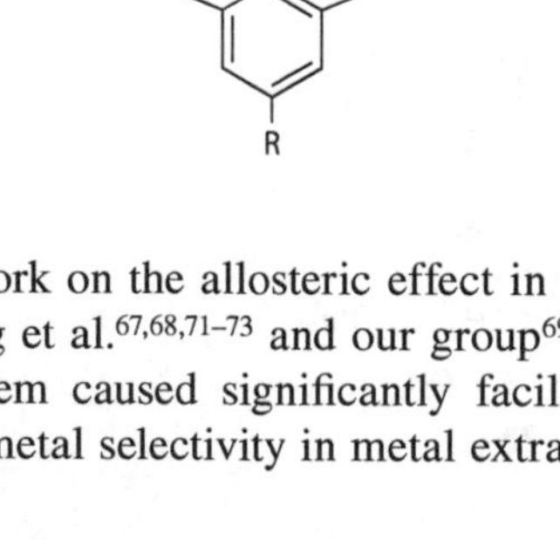

R = R' = 'Bu
R_1 = Me
R_2 = OCH$_2$C(O)NHSO$_2$X

X = CF$_3$ **82a**
    Me   **82b**
    Ph   **82c**
    C$_6$H$_4$-4-NO$_2$ **82d**

R = R' = H   **83**
R_1 = Me
R_2 = OCH$_2$C(O)NHSO$_2$CF$_3$

Next, the work on the allosteric effect in the extraction of metal ions is discussed. Ludwig et al.[67,68,71–73] and our group[69,87–89] found that Na addition to the extraction system caused significantly facilitated extractability and a drastic change of the metal selectivity in metal extraction with calix[4]arene tetraacetic

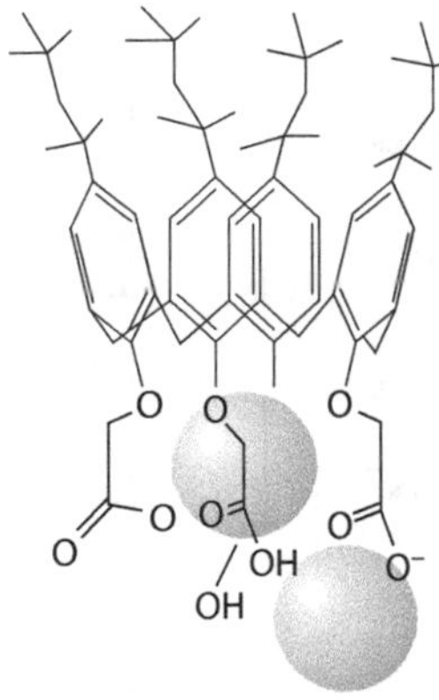

FIGURE 3.11   Self-coextraction of two sodium ions with **57**. (From Ohto, K. et al., *Chem. Lett.*, 27(7), 1998, 631–632; Ohto, K. et al., *J. Inclusion Phenom. Macrocycl. Chem.*, 65, 2009, 111–120.)

acid derivatives. The "self-co-extraction mechanism" of Na$^+$ with p-t-octylcalix[4] arene tetraacetic acid **57**,[90] in which two Na$^+$ ions are simultaneously (vs. sequentially) extracted by a single molecule of **57**, was proposed by our group as shown in Figure 3.11. The first sodium ion is surrounded by four phenoxy oxygen atoms, three of four carbonyl oxygen atoms, and one of four carboxyl groups. After uptake of the first Na$^+$ ion, the extraction of the second Na$^+$ ion is facilitated due to the electron outflow from the rest of the carbonyl oxygen atoms. Consequently, it was concluded that the first and second Na$^+$ ions were simultaneously extracted with **57**.[91] The second Na$^+$ extraction, which was facilitated, was just a simple ion-exchange reaction, different from the specific first Na$^+$ uptake. Because the extraction of coexisting metal ions was also enhanced by Na$^+$ addition, as shown in Figure 3.12, the co-extraction mechanism is reasonable. In our current work, p-t-octylcalix[4]arene-bearing diacetic acid and dicarboxylic acid with different methylene spacers at distal positions, **84–86**, were prepared to investigate alkali metal extraction and estimate extraction equilibrium constants for a numerical evaluation of the allosteric effect.[92] Stepwise extraction equilibrium constants of alkali metals with the extractants, **57**, **84–86**, are listed in Table 3.13.

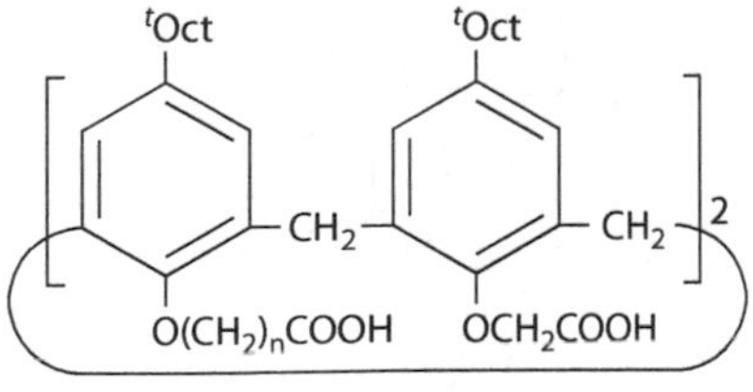

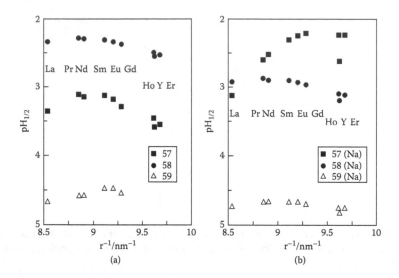

FIGURE 3.12   Relationship between $pH_{1/2}$ with **57–59**, and RE sequence in the absence (a) and presence (b) of Na^+. (From Ohto, K. et al., *Anal. Sci.*, 11, 1995, 893–902; Ohto, K. et al., *Polyhedron*, 16, 1997, 1655–1661.)

Pb^{2+} extraction with **57** was also carried out by our group.[93] Two Pb^{2+} ions were also co-extracted with a single molecule of **57**. The Pb^{2+} extraction mechanism, however, was different from that for Na^+, as it was not simultaneous but stepwise. Each mechanism was independently elucidated by measurements of the percent shift of the aryl peak with ^{1}H-NMR spectroscopy and of the percent metal loading with atomic absorption spectrophotometry (AAS) before and after extraction, as shown in Figure 3.13(a) and (b). In the mechanism, the extraction of the second Pb^{2+} ion was also facilitated by the first Pb^{2+} extraction, which is also an allosteric effect.

Another typical example of the allosteric effect in extraction is the research by Yamato et al.,[94] who prepared 1,3-alternate conformational thiacalix[4]arene with two 2-pyridyl groups and a crown-4-bridge, **87**. They determined by NMR spectra that the extractant discharged Li^+, which was extracted with the hard crown moiety by size fitting after Ag^+ uptake by the soft 2-pyridyl moieties. They concluded that the cyclic structure of the crown ring at the opposite site became wider after Ag^+ extraction, and consequently the size discrimination in favor of Li^+ was reduced.

87

TABLE 3.13
Stepwise Extraction Equilibrium Constants of Alkali Metals with the Extractants 57, 84–86

Extractant		Li$^+$	Na$^+$	K$^+$	$\beta_{Na/Li}$	$\beta_{K/Na}$	$\beta_{K/Li}$
			(a) Individual System				
57	K_{ex1}	2.85×10^{-4}	0.316	1.05×10^{-3}	1110	0.00332 (301)	3.68
	K_{ex2}	5.44×10^{-4}	0.791	1.35×10^{-3}	1450	0.00171 (586)	2.48
84	K_{ex1}	1.42×10^{-4}	0.068	4.53×10^{-5}	479	0.000666 (1500)	0.319
							(3.14)
	K_{ex2}	2.52×10^{-4}	1.12×10^{-3}	1.82×10^{-5}	4.44	0.0163 (61.5)	0.0722
							(13.8)
85	K_{ex1}	1.34×10^{-4}	0.019	2.90×10^{-5}	142	0.00153 (655)	0.216
							(4.62)
	K_{ex2}	2.46×10^{-4}	7.50×10^{-4}	5.20×10^{-7}	3.05	0.000693 (1440)	0.00211
							(473)
86	K_{ex1}	9.10×10^{-5}	0.011	4.52×10^{-6}	121	0.000411 (2430)	0.0500
							(20.1)
	K_{ex2}	1.21×10^{5}	3.40×10^{-4}	2.14×10^{-7}	28.1	0.000629 (1590)	0.0177
							(56.5)
			(b) Competitive System				
57	K_{ex1}		0.316				
	K_{ex2}	9.97×10^{-3}	0.451	0.188	4520	0.417	1890
84	K_{ex1}		0.068				
	K_{ex2}	2.36×10^{-5}	2.08×10^{-4}	6.98×10^{-3}	8.81	33.6	296
85	K_{ex1}		0.019				
	K_{ex2}	2.80×10^{-5}	2.03×10^{-5}	6.10×10^{-4}	0.725	30	21.8
86	K_{ex1}		0.011				
	K_{ex2}	3.48×10^{-4}	1.82×10^{-4}	2.52×10^{-4}	0.523	1.38	0.72

Source: Yoneyama, T. et al., *Talanta*, 88, 2012, 121–128.
Note: The values in parentheses represent the inverse.

Regardless of the cyclic structure of calixarenes, they are generally called calixpodands (acyclic ligand), which are modified at both rims. Calixcrowns are unique compounds combining the advantages of calixarene and crown ethers. There are many reports on extraction using calixcrowns. The first compound **17** was reported by Ungaro et al.[25] Then, Arduini et al. synthesized the cone-conformational 1,2-proximal-*p-t*-butylcalix[4]biscrown-5 derivative **88** to investigate alkali metal extraction.[95] It was found that the extractant exhibited Rb$^+$ and K$^+$ selectivity over Na$^+$ and Cs$^+$. Asfari et al. prepared a double-calixcrown compound possessing two calixcrown-5 moieties **89** to investigate alkali metal ions.[96] Among the alkali ions, it was found that K$^+$ and Rb$^+$ were selectively extracted. Casnati et al. prepared a series of cone and 1,3-alternate conformational calix[4]

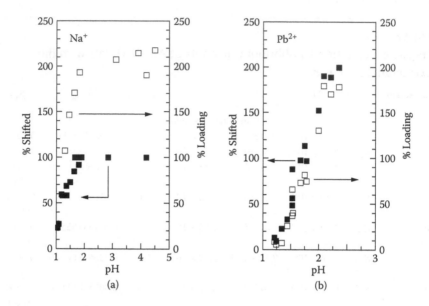

FIGURE 3.13 Effect of pH on %Shifted and %Loading with **57**. (a) Na+, (b) Pb2+. (From Ohto, K. et al., *J. Inclusion Phenom. Macrocycl. Chem.*, 65, 2009, 111–120; Ohto, K. et al., *Anal. Chim. Acta*, 387, 1999, 61–69.)

arene crown-5[97] and -6[98–100] ethers to carry out the extraction of alkali metals. They found that they were very good ionophores compared with valinomycin, an antibiotic substance. The compounds **90a** and **90b** exhibited high Cs+/Na+ selectivity in the alkali picrate extraction into dichloromethane. (The degrees of extraction of Na+ and Cs+ for **4a** were 2.4 and 65.4%, and those for **4b** were 2.6 and 63.5%, respectively.) When they were incorporated in nitrophenyl octyl ether (NPOE) membranes, they exhibited extremely high Cs+/Na+ selectivities through the membrane transport of 11,750 and 13,650, respectively, even under the 1:1000 concentration ratio of Cs+/Na+.[99] Although the early targets for calix-crown compounds were a series of alkali and alkaline earth metal ions, since some of them showed Cs+ selectivity, the target has been shifted to Cs+, because Cs removal from nuclear waste solutions is an urgent task. Cs removal has been one of the hottest research topics for extraction with calixarene compounds. Various 1,3-alternate derivatives of calix[4]biscrown-6 with excellent Cs+/Na+ selectivity were also prepared and were used to investigate the extraction from acid solution for reprocessing application by Dozol's and Vicens' groups.[101–106] Dozol et al. investigated Cs extractive removal in 1 M nitric acid into *o*-NPOE as diluent and found extremely high Cs+/Na+ selectivity of >28,500, 10,500, and >33,000 for **90a**, **90b**, and **90c**, respectively.[102] They also applied such excellent extractants to a supported liquid–membrane system.[101–103,107] Moyer et al. introduced use of calixcrowns for alkaline nuclear waste cleanup, particularly in Cs removal.[108] Among calixcrowns, calix[4]arene-bis-(*t*-octylbenzocrown-6), called BobCalixC6, **91**, exhibited high Cs selectivity over Na due to its preorganized structure. Extensive tests for optimization of Cs removal and thermal and radiation stability were carried out. Further works on degradation of the extractants and solvent modifier **92** by radiation and addition of the modifier for solubility improvement and demulsification were performed.[109]

88

R : ^iPr **90a**
^nPr **90b**
^nOct **90c**

89

91

92

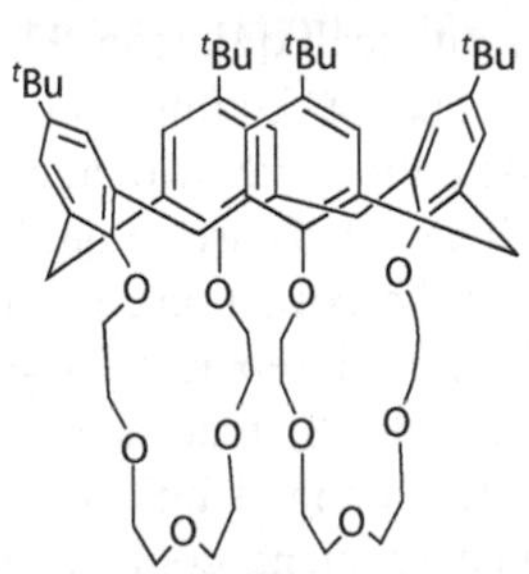

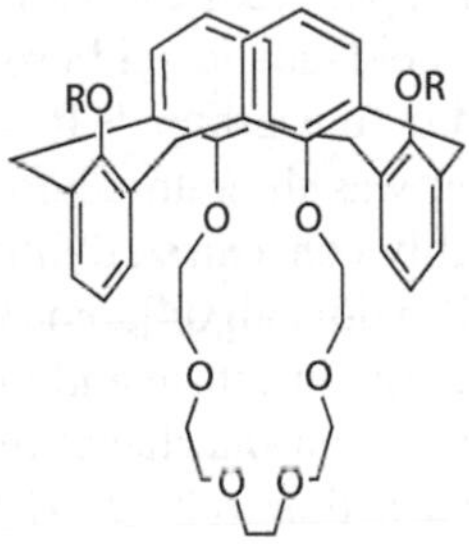

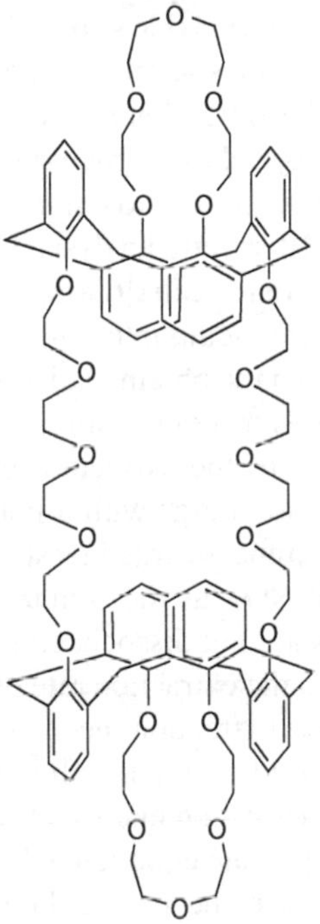

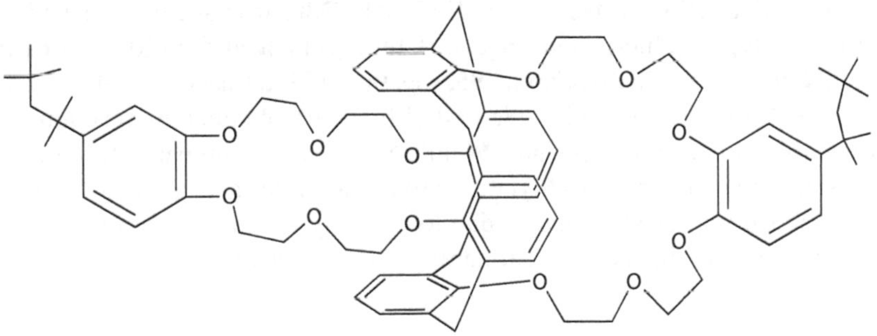

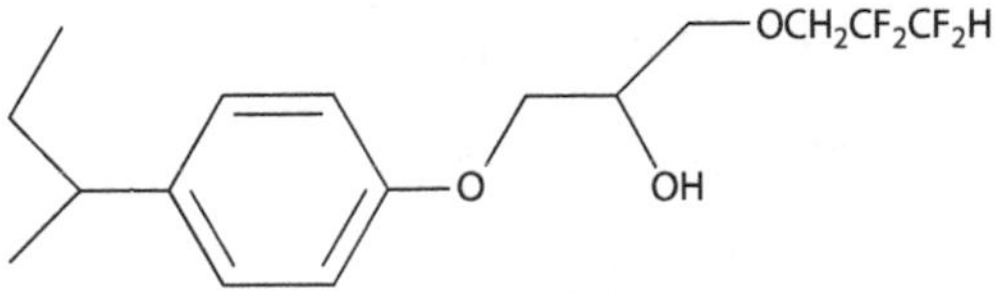

Whereas calixarene compounds exhibit excellent properties, they can also have several disadvantages, such as poor solubility, extraction rate, and so on, depending on the nature of the particular system examined. Several works on the extraction rate were reported, including Izatt et al.[22] and Akabori et al.,[35] described earlier. Yaftian et al. reported the extraction of alkali ions[110] and Ag+[111] with amide and phosphine oxide crossed p-t-butylcalix[4]arene 93. They estimated transport rate and mass-transport coefficients at feeding and receiving phases, and found that the extraction rate with the calixarene extractant was much faster than that with dicyclohexyl-18-crown-6. They also carried out batchwise experiments to determine the 1:1 stoichiometry of the extraction complex by using slope analysis and NMR spectra analysis. Ludwig et al. reported UO_2^{2+} extraction with p-t-butylcalix[4]arene, calix[6]arene, and calix[8]arene acetic acid derivatives 13–15 from nitrate media into toluene-octanol and estimated the diffusion coefficients.[71] Although the obtained diffusion coefficients could not be compared due to the different solvent composition for three extractants, the work may contribute to the elucidation of the extraction mechanism from the viewpoint of kinetics. The extraction rate of Ag+ with a butyric acid–acetic acid crossed derivative of p-t-octylcalix[4]arene 84 was investigated by our group.[112] It was elucidated that the extraction of NO_3^- as the counteranion was the rate-determining step, since carboxylic acids are not dissociated in highly concentrated nitric acid media. He et al. investigated the extraction rate of Nd^{3+} with calix[4]arene tetraacetic acid 13 and elucidated the extraction mechanism at the interface and the rate-controlling step.[113] The deduced rate-controlling step was an interfacial reaction. They also determined the activation energy of the extraction to be 21.41 kJ mol^{-1} and proposed an extraction rate equation to be $-d[Nd^{3+}]/dt = k_f [Nd^{3+}][H_4A]^{0.727} [H^+]^{-0.978}$.

Work to improve the poor solubility of the extractants has been reported. p-Debutylated, p-t-butyl, p-t-amyl, and p-t-octylcalix[4]arene tetraacetic acid derivatives were prepared by our group to investigate the solubility in various diluents, such as aliphatic hexane and EXXSOL D80, aromatic toluene, and chloroform as a typical chlorinated solvent.[114] Tetraacetic acid derivatives of calix[4]arene 94 and p-t-amylcalix[4]arene 95, together with tetraacetic acid derivatives of p-t-alkylcalix[4]arene with mixed alkyl branches obtained by reaction of the various mixed phenol compounds, 96 and 97, were also prepared to investigate their solubility. The mixed alkylated calix[4]arene tetraacetic acid derivatives showed much improved solubility due to not only low purity as a mixture, but also the symmetric breaking of calixarene, as listed in Table 3.14.

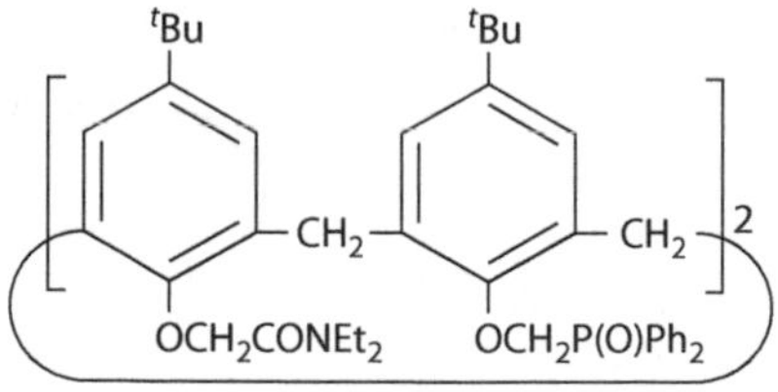

93

TABLE 3.14
Maximum Solubility of Tetraacetic Acid Derivatives of Calix[4]arene with Different Alkyl Chains in Various Diluents

Extractant	Substituent at p-position	Solubility in Organic Diluents, mol m³			
		CHCl$_3$	Toluene	Hexane	EXXSOL D80
94	Debutylated	<0.56	<0.54	<0.074	<0.093
13	^{t}Bu	50.1	9.3	0.08	<0.12
95	^{t}Am	57.9	15.3	0.08	<0.13
57	^{t}Oct	63.0	18.4	0.78	<0.054
96	^{t}Bu and ^{t}Am	62.7	13.5	<0.19	<0.10
97	^{t}Am and ^{t}Oct	265.0	61.1	<0.19	<0.080

Source: Ohto, K. et al., *Solvent Extr. Ion Exch.*, 17, 1999, 1309–1325.

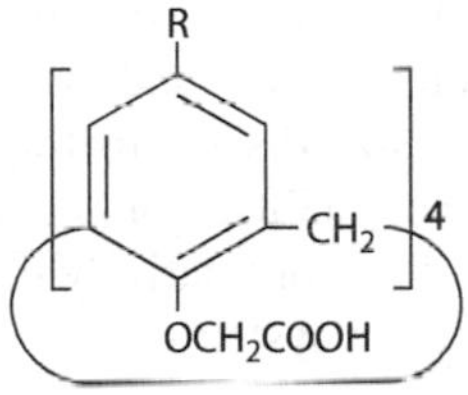

R = H   **94**
R = ^{t}Am   **95**

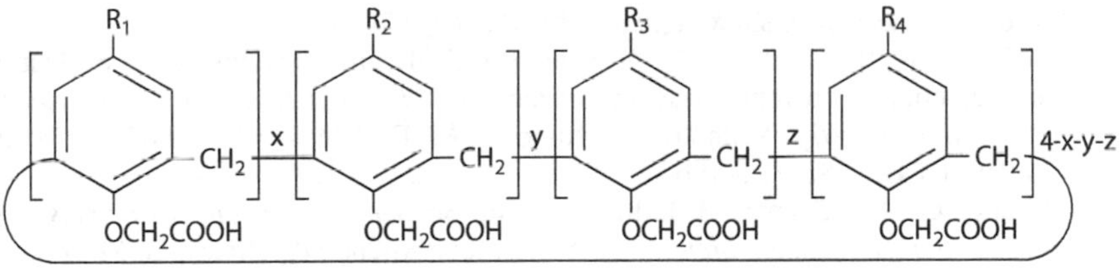

R$_1$, R$_2$, R$_3$, R$_4$ = ^{t}Bu or ^{t}Am   **96**
R$_1$, R$_2$, R$_3$, R$_4$ = ^{t}Am or ^{t}Oct   **97**

Works using a reversed-micelle system[115] and an ionic liquid[116–120] have been also reported to improve the poor solubility in organic solvents. The reversed micelle prepared by cationic cetyltrimethyl ammonium bromide (CTAB) or anionic bis-2-ethylhexyl sulfosuccinate (AOT) facilitates the solubilization of **13** in isooctane. The solubility of **13** proportionally increased with the increased concentrations of both surfactants. The relationship between the solubility of **13** and the AOT concentration was expressed by the equation (**13**) in reversed micelle = 0.108 [AOT]. Shimojo et al. described that the employment of 1-octyl-3-methylimidazolium hexafluorophosphate as an ionic liquid resulted in the enhanced solubility of less lipophilic calixarene extractants and in enhanced silver extraction by 230-fold compared with the conventional chloroform system.[117] Luo et al. prepared a series of new hydrophobic and protic alkylammonium ionic liquids with bis(trifluoromethylsulfonyl)imide or bis(perfluoroethylsulfonyl)imide as conjugated anions to apply the Sr and Cs extraction with BOBCalixC6, **91**, or dicyclohexano-18-crown-6.[120] They concluded that the observed enhancement in the extraction efficiency can be attributed to the greater hydrophilicity of the cations of the protic ionic liquids.

3.5 CONCLUSIONS

The introduction of calixarenes as extractants has opened a new realm of separation chemistry with many exciting possibilities for new selectivity and control of behavior not possible with other classes of extractants. An extensive foundation of understanding has been laid, with some notable successes in real application, but in fact, we have only scratched the surface. Calixarene compounds have specific discriminating properties for guest metal ions due to the structural effect of the metal binding site of a certain size. Further potential to design the improved extractants for metal separation will be found by optimizing the separation properties of calixarenes, such as size-discriminating effect, conformation effect, converging effect of multifunctionality, complementarity effect together with allosteric effect, and the effect by coordinatively inert alkyl branching, which are hardly observed in other extractant systems.

REFERENCES

1. Inoue, K., Nakashio, F. Industrial chelating extractants. Their developments and recent advances. *Kagaku Kougaku* 1982, 46, 164–171. (In Japanese).
2. Sudderth, R. B., Kordosky, G. A. Some practical considerations in the evaluation and selection of solvent extraction reagents. In *Chemical Reagents in the Mineral Processing Industry*, Malhotra, D., Aime, S. M. E., Riggs, W. F., Eds., S. M. E., Littleton, CO, 1986, 181–196.
3. Ritcey, G. M. Chapter 3 Extractants. In *Solvent Extraction—Principles and Applications to Process Metallurgy*, Vol. 1, revised 2nd ed., Ritcey, G. M., Ed., G. M. Ritcey & Associates, Ottawa, 2006, 69–157.
4. Pedersen, C. J. Cyclic polyethers and their complexes with metal salts. *J. Am. Chem. Soc.* 1967, 89, 2495–2496.
5. Pedersen, C. J. Cyclic polyethers and their complexes with metal salts. *J. Am. Chem. Soc.* 1967, 89, 7107–7136.
6. Cram, D. J., Cram, J. M. Host–guest chemistry: complexes between organic compounds simulate the substrate selectivity of enzymes. *Science* 1974, 183, 803–809.
7. Kyba, E. P., Helgeson, R. C., Madan, K., Gokel, G. W., Tarnowski, T. L., Moore, S. S., Cram, D. J. Host–guest complexation. 1. Concept and illustration. *J. Am. Chem. Soc.* 1977, 99, 2564–2571.
8. Lehn, J.-M. Cryptates: Inclusion complexes of macropolycyclic receptor molecules. *Pure Appl. Chem.* 1978, 50, 871–892.
9. Cram, D. J. Preorganization—From solvents to spherands. *Angew. Chem. Int. Ed. Engl.* 1986, 25, 1039–1057.

10. Pearson, R. G. Hard and soft acids and bases. *J. Am. Chem. Soc.* 1963, 85, 3533–3539.
11. Oda, R., Shono, T., Tabushi, I. Application to metal separation. In *Kagaku Zoukan 74, Crown Ether No Kagaku*, Kagaku Doujin, Tokyo, 1978, 110–119. (In Japanese).
12. Cram, D. J., Kaneda, T., Helgeson, R. C., Brown, S. B., Knobler, C. B., Maverick, E. K., Trueblood, N. Host–guest complexation. 35. Spherands, the first completely preorganized ligand systems. *J. Am. Chem. Soc.* 1985, 107, 3645–3668.
13. Izatt, R. M., Pawlak, K., Bradshaw, J. S. Thermodynamic and kinetic data for macrocycle interactions with cations and anions. *Chem. Rev.* 1991, 91, 1721–2085.
14. Takeshita, M., Shinkai, S. Recent topics on functionalization and recognition ability of calixarenes: the "third host molecule." *Bull. Chem. Soc. Jpn.* 1995, 68, 1088–1097.
15. Zinke, A. Ziegler, E. The hardening process of phenol-formaldehyde resins. X. *Chem. Ber.* 1944, 77B, 264–272.
16. Kaemmerer, H., Happel, G., Caesar, F. Spectroscopic analysis of a cyclic tetrameric compound from p-cresol and formaldehyde. *Makromol. Chem.* 1972, 162, 179–197.
17. Gutsche, C. D., Muthukrishnan, R. Calixarenes. 1. Analysis of the product mixtures produced by the base-catalyzed condensation of formaldehyde with para-substituted phenols. *J. Org. Chem.* 1978, 43, 4905–4906.
18. Gutsche, C. D., Dhawan, B., No, K. H., Muthukrishnan, R. Calixarenes. 4. The synthesis, characterization, and properties of the calixarenes from p-tert-butylphenol. *J. Am. Chem. Soc.* 1981, 103, 3782–3792.
19. Gutsche, C. D. Calixarenes. *Aldrichim. Acta.,* 1995, 28, 3–9.
20. Ohto, K. Review of the extraction behavior of metal cations with calixarene derivatives. *Solvent Extr. Res. Dev. Jpn.* 2010, 17, 1–18.
21. Eigen, M. Fast elementary steps in chemical reaction mechanisms. *Pure Appl. Chem.* 1963, 6, 97–115.
22. Izatt, R. M., Lamb, J. D., Hawkins, R. T., Brown, P. R., Steven, R., Christensen, J. J. Selective M+-H+ coupled transport of cations through a liquid membrane by macrocyclic calixarene ligands. *J. Am. Chem. Soc.* 1983, 105, 1782–1785.
23. Chang, S.-K., Cho, I. New metal cation-selective ionophores derived from calixarenes. *Chem. Lett.* 1984, 13, 477–478.
24. Chang, S.-K., Cho, I. New metal cation-selective ionophores derived from calixarenes: Their syntheses and ion-binding properties. *J. Chem. Soc. Perkin Trans. 1* 1986, 211–214.
25. Ungaro, R., Pochini, A. New ionizable ligands from p-tert-butylcalix[4]arene. *J. Inclusion Phenom.* 1984, 2, 199–206.
26. McKervey, M. A., Seward, E. M., Ferguson, G., Ruhl, B., Harris, S. J. Synthesis, X-ray crystal structures, and cation transfer properties of alkyl calixaryl acetates, a new series of molecular receptors. *J. Chem. Soc. Chem. Commun.* 1985, 388–390.
27. Calestani, G., Ugozzoli, F., Arduini, A., Ghidini, E., Ungaro, R. Molecular inclusion in functionalized macrocycles. Part 14. Encapsulated potassium cation in a new calix[4]arene neutral ligand. Synthesis and X-ray crystal structure. *J. Chem. Soc. Chem. Commun.* 1987, 344–346.
28. Arduini, A., Ghidini, E., Pochini, A., Ungaro, R., Andreetti, G. D., Calestani, G., Ugozzoli, F. Molecular inclusion in functionalized macrocycles. Part 15. p-tert-Butylcalix[4]arene tetraacetamide: A new strong receptor for alkali cations. *J. Inclusion Phenom.* 1988, 6, 119–134.
29. Arnaud-Neu, F., Schwing-Weill, M.-J., Ziat, K., Sremin, S., Harris, S. J., McKervey, M. A. Selective alkali and alkaline earth cation complexation by calixarene amides. *New. J. Chem.* 1991, 15, 33–37.

30. Arnaud-Neu, F., Barrett, G., Fanni, S., Marrs, D., McGregor, W., McKervey, M. A., Schwing-Weill, M. J., Vetrogen, V., Wechsler, S. Extraction and solution thermodynamics of complexation of alkali and alkaline-earth cations by calix[4]arene amides. *J. Chem. Soc. Perkin Trans. 2* 1995, 453–461.

31. Ferguson, G., Kaitner, B., McKervey, M. A., Seward, E. M. Synthesis, X-ray crystal structure, and cation transfer properties of a calix[4]arene tetraketone. A new versatile molecular receptor. *J. Chem. Soc. Chem. Commun.* 1987, 584–585.

32. Schwing, M.-J., Arnaud-Neu, F., Marques, E. Cation binding properties of alkyl calixaryl derivatives. A new family of molecular receptors. *Pure Appl. Chem.* 1989, 61, 1597–1603.

33. Arnaud-Neu, F., Collins, E. M., Deasy, M., Ferguson, G., Harris, S. J., Kaitner, B., Lough, A. J., McKervey, M. A., Marques, E., Ruhl, B. L., Schwing-Weill, M. J., Seward, E. M. Synthesis, X-ray crystal structures, and cation-binding properties of alkyl calixaryl esters and ketones, a new family of macrocyclic molecular receptors. *J. Am. Chem. Soc.* 1989, 111, 8681–8691.

34. Conner, M., Janout, V., Regen, S. L. Synthesis and alkali metal binding properties of "upper rim" functionalized calix[4]arenes. *J. Org. Chem.* 1992, 57, 3744–3746.

35. Akabori, S., Itabashi, H., Shimura, H., Inoue, M. Syntheses of phosphate diesters having p-tert-butylcalix[4]arene and alkyl group as ester moieties, and their selective lithium cation transport abilities through liquid membranes. *Chem. Commun.* 1997, 2137–2138.

36. Talanova, G. G., Talanov, V. S., Hwang, H.-S., Eliasi, B. A., Bartsch, R. A. New mono-ionizable, Li^+-selective calix[4]arenes. *J. Chem. Soc. Perkin Trans. 2* 2002, 1869–1874.

37. Sadamatsu, H., Yoneyama, T., Kawakita, H., Morisada, S., Ohto, K. Unpublished data.

38. Iwamoto, K., Shinkai, S. Synthesis and ion selectivity of all conformational isomers of tetrakis[(ethoxycarbonyl)methoxy]calix[4]arene. *J. Org. Chem.* 1992, 57, 7066–7073.

39. Iwamoto, K., Araki, K., Shinkai, S. Conformations and structures of tetra-O-alkyl-p-tert-butylcalix[4]arenes. How is the conformation of calix[4]arenes immobilized? *J. Org. Chem.* 1991, 56, 4955–4962.

40. Shinkai, S., Koreishi, H., Ueda, K., Manabe, O. A new hexacarboxylate uranophile derived from calix[6]arene. *J. Chem. Soc. Chem. Commun.* 1986, 233–234.

41. Shinkai, S., Shirahama, Y., Satoh, H., Manabe, O. Selective extraction and transport of uranyl ion with calixarene-based uranophiles. *J. Chem. Soc. Perkin Trans. 2*, 1989, 1167–1171.

42. Araki, K., Hashimoto, N., Otsuka, H., Nagasaki, T., Shinkai, S. Molecular design of a calix[6]arene-based super-uranophile with C_3 symmetry. High UO_2^{2+} selectivity in solvent extraction. *Chem. Lett.* 1993, 829–832.

43. Nagasaki, T., Shinkai, S. Synthesis and solvent extraction studies of novel calixarene-based uranophiles bearing hydroxamic groups. *J. Chem. Soc. Perkin Trans. 2* 1991, 1063–1066.

44. Shinkai, S., Otsuka, T., Araki, K., Matsuda, T. (2-Pyridylmethoxy)calixarenes: New versatile ionophores for metal extraction. *Bull. Chem. Soc. Jpn.* 1989, 62, 4055–4057.

45. Pappalardo, S., Ferguson, G., Neri, P., Rocco, C. Synthesis and complexation studies of regioisomers and conformational isomers of p-tert-butylcalix[4]arene bearing pyridine or pyridine N-oxide pendant groups at the lower rim. *J. Org. Chem.* 1995 60, 4576–4584.

46. Ohto, K., Higuchi, H., Inoue, K. Solvent extraction of silver with pyridinocalix[4]arenes. *Solvent Extr. Res. Dev. Jpn.* 2001, 8, 37–46.

47. Yordanov, A. T., Mague, J. T., Roundhill, D. M. Synthesis of heavy metal ion selective calix[4]arenes having sulfur containing lower-rim functionalities. *Inorg. Chem.* 1995, 34, 5084–5087.

48. Yordanov, A. T., Falana, O. M., Koch, H. F., Roundhill, D. M. (Methylthio)methyl and (N,N-dimethylcarbamoyl)methyl upper-rim-substituted calix[4]arenes as potential extractants for Ag(I), Hg(II), Ni(II), Pd(II), Pt(II), and Au(III). *Inorg. Chem.* 1997, 36, 6468–6471.

49. Yordanov, A. T., Whittlesey, R. B., Roundhill, D. M. Lower rim 2-methylthioethoxy substituted calix[4]arenes as shape selective complexants for mercury and silver. *Supramol. Chem.* 1998, 9, 13–15.

50. Ohto, K., Murakami, E., Shiratsuchi, K., Inoue, K., Iwasaki, M. Solvent extraction of silver(I) and palladium(II) ions with ketone derivative of calix[4]arene from highly acidic nitrate media. *Chem. Lett.* 1996, 25, 173–174.

51. Ohto, K., Murakami, E., Shinohara, T., Shiratsuchi, K., Inoue, K., Iwasaki, M. Selective extraction of silver(I) over palladium(II) with ketonic derivatives of calixarenes from highly concentrated nitric acid. *Anal. Chim. Acta* 1997, 341, 275–283.

52. Ohto, K., Yamaga, H., Murakami, E., Inoue, K. Specific extraction behavior of amide derivative of calix[4]arene for silver(I) and gold(III) ions from highly acidic chloride media. *Talanta* 1997, 44, 1123–1130.

53. Horwitz, E. P., Kalina, D. G., Diamond, H., Kaplan, L., Vandegrift, G. F., Leonard, R. A., Steindler, M. J., Schulz, W. W. TRU decontamination of high-level Purex waste by solvent extraction using a mixed octyl(phenyl)-N,N-diisobutylcarbamoylmethylphosphite oxide/TBP/NPH (TRUEX) solvent. In *Proceedings of the International Symposium on Actinide/Lanthanide*, September 1984, Choppin, G. R., Navratil, J. D., Schulz, W. W., Eds. World Scientific, Singapore, 1984, 43–69.

54. Horwitz, E. P., Schulz, W. W. The TRUEX process: A vital tool for disposal of U.S. defense nuclear waste. In *New Separation Chemistry Techniques for Radioactive Waste and Other Specific Applications*, Cecille, L., Casarci, M., Pietrelli, L., Eds. Elsevier, New York, 1991, 21–29.

55. Musikas, C., Schulz, W. W., Liljenzin, J.-O. Solvent extraction in nuclear science and technology. In *Solvent Extraction Principles and Practice*, 2nd ed., revised and expanded, Rydberg, J., Cox, M., Musikas, C., Choppin, G. R., Eds. Marcel Dekker, New York, 2004, 507–557.

56. Boehmer, V. CMPO-substituted calixarenes. In ACS Symposium Series 757, *Calixarenes for Separations*, Lumetta, G. L., Rogers, R. D., Gophalan, A. S., Eds. ACS, Washington, DC, 2000, 135–149.

57. Arnaud-Neu, F., Barboso, S., Byrne, D., Charbonniere, L. J., Schwing-Weill, M. J., Ulrich, G. Binding of lanthanides(III) and Thorium(IV) by phosphorylated calixarenes. In ACS Symposium Series 757, *Calixarenes for Separations*, Lumetta, G. L., Rogers, R. D., Gophalan, A. S., Eds. ACS, Washington, DC, 2000, 150–164.

58. Malone, J. F., Marrs, D. J., McKervey, M. A., O'Hagen, P., Thompson, N., Walker, A., Arnaud-Neu, F., Mauprivez, O., Schwing-Weill, M.-J. Calix[n]arene phosphine oxides. A new series of cation receptors for extraction of europium, thorium, plutonium, and americium in nuclear waste treatment. *J. Chem. Soc. Chem. Commun.* 1995, 2151–2153.

59. Arnaud-Neu, F., Boehmer, V., Dozol, J.-F., Gruetter, C., Jakobi, R. A., Kraft, D., Mauprivez, O., Rouquette, H., Schwing-Weill, M.-J., Simon, N., Vogt, W. Calixarenes with diphenylphosphorylacetamide functions at the upper rim. A new class of highly efficient extractants for lanthanides and actinides. *J. Chem. Soc. Perkin Trans. 2* 1996, 1175–1182.

60. Delmau, L. H., Simon, N., Schwing-Weill, M.-J., Arnaud-Neu, F., Dozol, J.-F., Eymard, S., Tournois, B., Boehmer, V. "CMPO-substituted" calix[4]arenes, extractants with selectivity among trivalent lanthanides and between trivalent actinides and lanthanides. *Chem. Commun.* 1998, 1627–1628.

61. Delmau, L. H., Simon, N., Schwing-Weill, M.-J., Arnaud-Neu, F., Dozol, J.-F., Eymard, S., Tournois, B., Gruetter, C., Musigmann, C., Tunayar, A., Boehmer, V. Extraction of trivalent lanthanides and actinides by "CMPO-like" calixarenes. *Sep. Sci. Technol.* 1999, 34, 863–876.

62. Matthews, S. E., Saadioui, M., Boehmer, V., Barboso, S., Arnaud-Neu, F., Schwing-Weill, M.-J., Carrera, A. G., Dozol, J.-F. Conformationally mobile wide rim carbamoylmethylphosphine oxide (CMPO)-calixarenes. *J. Prakt. Chem.* 1999, 341, 264–273.

63. Atamas, L., Klimchuk, O., Rudzevich, V., Pirozhenko, V., Kalchenko, V., Smirnov, I., Babain, V., Efremova, T., Varnek, A., Wipff, G., Arnaud-Neu, F., Roch, M., Saadioui, M., Boehmer, V. New organophosphorus calix[4]arene ionophores for trivalent lanthanide and actinide cations. *J. Supramol. Chem.* 2003, 2, 421–427.

64. Schmidt, C., Saadioui, M., Boehmer, V., Host, V., Spirlet, M.-R., Desreux, J. F., Brisach, F., Arnaud-Neu, F., Dozol, J.-F. Modification of calix[4]arenes with CMPO-functions at the wide rim. Synthesis, solution behavior, and separation of actinides from lanthanides. *Org. Biomol. Chem.* 2003, 1, 4089–4096.

65. Arnaud-Neu, F., Browne, J. K., Byrne, D., Marrs, D. J., McKervey, M. A., O'Hagen, P., Schwing-Weill, M.-J., Walker, A. Extraction and complexation of alkali, alkaline earth, and f-Element cations by calixaryl phosphine oxides. *Chem. Eur. J.* 1999, 5, 175–186.

66. Barboso, S., Carrera, A. G., Matthews, S. E., Arnaud-Neu, F., Boehmer, V., Dozol, J.-F., Rouquette, H., Schwing-Weill, M.-J. Calix[4]arenes with CMPO functions at the narrow rim. Synthesis and extraction properties. *J. Chem. Soc. Perkin Trans. 2* 1999, 719–724.

67. Ludwig, R., Inoue, K., Yamato, T. Solvent extraction behaviour of calixarene-type cyclophanes towards trivalent La, Nd, Eu, Er, and Yb. *Solvent Extr. Ion Exch.* 1993, 11, 311–330.

68. Ludwig, R., Inoue, K., Shinkai, S., Gloe, K. Solvent extraction behaviour of *p*-tert-butylcalix[n]arene carboxylic acid derivatives towards trivalent lanthanides and sodium. In *Proceedings of ISEC'93, Solvent Extraction in the Process Industries*, Vol. 1, Logsdail, D. H., Slater, M. J., Eds. SCI, London, 1993, 273–278.

69. Ohto, K., Yano, M., Inoue, K., Yamamoto, T., Goto, M., Nakashio, F., Nagasaki, T., Shinkai, S. Extraction of rare earths with new extractants of calixarene derivatives. In *Proceedings of ISEC'93, Solvent Extraction in the Process Industries*, Vol. 1, Logsdail, D. H., Slater, M. J., Eds. SCI, London, 1993, 364–369.

70. Ohto, K., Yano, M., Inoue, K., Yamamoto, T., Goto, M., Nakashio, F., Shinkai, S., Nagasaki, T. Solvent extraction of trivalent rare earth metal ions with carboxylate derivatives of calixarenes. *Anal. Sci.* 1995, 11, 893–902.

71. Ludwig, R., Gauglitz, R. Calixarene type extractants for metal ions with improved properties. In *Proceedings of ISEC'96, Value Adding through Solvent Extraction*, Vol. 1, Shallcross, D. C., Paimin, R., Prvcic, L. M., Eds. University of Melbourne, Melbourne, 1996, 365–369.

72. Ludwig, R., Kunogi, K., Dung, N., Tachimori, S. A calixarene-based extractant with selectivity for Am[III] over Ln[III]. *Chem. Commun.* 1997, 1985–1986.

73. Ludwig, R., Lentz, D., Nguyen, T. K. D. Trivalent lanthanide and actinide extraction by calixarenes with different ring sizes and different molecular flexibility. *Radiochim. Acta* 2000, 88, 335–343.

74. Ohto, K., Ota, H., Inoue, K. Solvent extraction of rare earths with a calix[4]arene compound containing phosphonate groups introduced onto upper rim. *Solvent Extr. Res. Dev. Jpn.* 1997, 4, 167–182.

75. Oshima, T., Yamamoto, T., Ohto, K., Goto, M., Nakashio, F., Furusaki, S. A calix-arene-based phosphoric acid extractant for rare earth separation. *Solvent Extr. Res. Dev. Jpn.* 2001, 8, 194–204.

76. Jurecka, P., Vojtisek, P., Novotny, K., Rohovec, J., Lukes, I. Synthesis, characterisation, and extraction behaviour of calix[4]arene-based phosphonic acids. *J. Chem. Soc. Perkin Trans. 2* 2002, 1370–1377.

77. Matulkova, I., Rohovec, J. Synthesis, characterization, and extraction behaviour of calix[4]arene with four propylene phosphonic acid groups on the lower rim. *Polyhedron* 2005, 24, 311–317.

78. Ohto, K., Takedomi, A., Chetry, A. B., Morisada, S., Kawakita, H., Oshima, T. The effect of phenoxy oxygen atoms on extremely high extraction ability and less separation efficiency of trivalent rare earth elements with tetraphosphonic acid derivative of calix[4]arene. *J. Inclusion Phenom. Macrocycl. Chem.*, DOI 10.1007/s10847-012-0255-0.

79. Ohto, K., Matsufuji, T., Yoneyama, T., Tanaka, M., Kawakita, H., Oshima, T. Preorganized, cone-conformational calix[4]arene possessing four propylenephosphonic acids with high extraction ability and separation efficiency for trivalent rare earth elements. *J. Inclusion Phenom. Macrocycl. Chem.* 2011, 71, 489–497.

80. Komasawa, I., Hisada, K., Kiyamura, M. Extraction and separation of rare-earth elements by tri-n-octylmethylammonium nitrate. *J. Chem. Eng. Jpn.* 1990, 23, 308–315.

81. Ohto, K., Ishii, H., Kawakita, H., Harada, H., Inoue, K. Solvent extraction of rare earth and precious metals with quaternary ammonium type of calix[4]arene. *J. Ion Exch.* 2007, 18, 240–245.

82. Talanova, G. G., Hwang, H.-S., Talanov, V. S., Bartsch, R. A. Calix[4]arenes with a novel proton-ionizable group: synthesis and metal ion separations. *Chem. Commun.* 1998, 419–420.

83. Talanova, G. G., Hwang, H.-S., Talanov, V. S., Bartsch, R. A. Calix[4]arenes with hard donor groups as efficient soft cation extractants. Remarkable extraction selectivity of calix[4]arene *N*-(*X*)sulfonylcarboxamides for Hg^{II}. *Chem. Commun.* 1998, 1329–1330.

84. Talanova, G. G., Elkarim, N. S. A., Talanov, V. S., Bartsch, R. A. A calixarene-based fluorogenic reagent for selective mercury(II) recognition. *Anal. Chem.* 1999, 71, 3106–3109.

85. Talanova, G. G., Talanov, V. S., Gorbunova, M. G., Hwang, H.-S., Bartsch, R. A. Effect of upper rim para-alkyl substituents on extraction of alkali and alkaline earth metal cations by di-ionizable calix[4]arenes. *J. Chem. Soc. Perkin Trans. 2* (2002) (12), 2072–2077.

86. Talanova, G. G., Talanov, V. S., Hwang, H.-S., Park, C., Surowiec, K., Bartsch, R. A. Rigid versus flexible: How important is ligand "preorganization" for metal ion recognition by lower rim-functionalized calix[4]arenes? *Org. Biomol. Chem.* 2004, 2, 2585–2592.

87. Ohto, K., Shiratsuchi, K., Inoue, K., Goto, M., Nakashio, F., Shinkai, S., Nagasaki, T. Extraction behavior of copper(II) ion by calixarene carboxylates derivatives pre-organized by sodium ion. *Solvent Extr. Ion Exch.* 1996, 14, 459–478.

88. Ohto, K., Yano, M., Inoue, K., Nagasaki, T., Goto, M., Nakashio, F., Shinkai, S. Effect of coexisting alkaline metal ions on the extraction selectivity of lanthanide ions with calixarene carboxylate derivatives. *Polyhedron* 1997, 16, 1655–1661.

89. Ohto, K., Shioya, Higuchi, A., H., Oshima, T., Inoue, K. Effect of coexisting sodium ion on extractive separation of metal ions with calix[4]arene tetracarboxylic acid. *ARS Sep. Acta* 2002, 1, 61–70.

90. Ohto, K., Ishibashi, H., Inoue, K. Self-coextraction of sodium ions with calix[4] arene tetracarboxylate. *Chem. Lett.* 1998, 27(7), 631–632.

91. Ohto, K., Ishibashi, H., Kawakita, H., Inoue, K., Oshima, T. Allosteric coextraction of sodium and metal ions with calix[4]arene derivatives 1. Role of the first-extracted sodium ion as an allosteric trigger for self-coextraction of sodium ions with calix[4] arene tetracarboxylic acid. *J. Inclusion Phenom. Macrocycl. Chem.* 2009, 65, 111–120.

92. Yoneyama, T., Sadamatsu, H., Kuwata, S., Kawakita, H., Ohto, K. Allosteric coextraction of sodium and metal ions with calix[4]arene derivatives 2: First numerical evaluation for the allosteric effect on alkali metal extraction with crossed carboxylic acid type calix[4]arenes. *Talanta* 2012, 88, 121–128.

93. Ohto, K., Fujimoto, Y., Inoue, K. Stepwise extraction of two lead ions with a single molecule of calix[4]arene tetracarboxylic acid. *Anal. Chim. Acta* 1999, 387, 61–69.

94. Peres-Casas, C., Rahman, S., Begum, N., Xi, Z., Yamato, T. Allosteric bindings of thiacalix[4]arene-based receptors with 1,3-alternate conformation having two different side arms. *J. Inclusion Phenom. Mol. Rec. Chem.* 2008, 60, 173–185.

95. Arduini, A., Casnati, A., Dodi, L., Pochini, A., Ungaro, R. Selective 1,2-functionalization of calix[4]arenes at the lower rim. Synthesis of a new type of bis-calixcrown ether. *J. Chem. Soc. Chem. Commun.* 1990, 1597–1598.

96. Asfari, Z., Abidi, R., Arnaud-Neu, F., Vicens, J. Synthesis and complexing properties of a double-calix[4]arene crown ether. *J. Inclusion Phenom. Mol. Rec. Chem.* 1992, 13, 163–169.

97. Casnati, A., Pochini, A., Ungaro, R., Bocchi, C., Ugozzoli, F., Egberink, R. J. M., Struijk, H., Lugtenberg, R., Jong, F. D., Reinhoudt, D. N. 1,3-Alternate calix[4] arenecrown-5 conformers: New synthetic ionophores with better K^+/Na^+ selectivity than valinomycin. *Chem. Eur. J.* 1996, 2, 436–445.

98. Ungaro, R., Casnati, A., Ugozzoli, F., Pochini, A., Dozol, J.-F., Hill, C., Rouquette, H. 1,3-Dialkoxycalix[4]arene crown-6 in 1,3-alternate conformation: Cesium-selective ligands using the cation-arene effect. *Angew. Chem. Int. Ed. Engl.* 1994, 33, 1506–1509.

99. Casnati, A., Pochini, A., Ungaro, R., Ugozzoli, F., Arnaud-Neu, F., Fanni, S., Schwing-Weill, M. J., Egberink, R. J. M., Jong, F. D., Reinhoudt, D. N. Synthesis, complexation, and membrane transport studies of 1,3-alternate calix[4]arene-crown-6 conformers: A new class of cesium selective ionophores. *J. Am. Chem. Soc.* 1995, 117, 2767–2777.

100. Bocchi, C., Careri, M., Casnati, A., Mori, G. Selectivity of calix[4]arene-crown-6 for cesium ion in ISE: Effect of the conformation. *Anal. Chem.* 1995, 67, 4234–4238.

101. Asfari, Z., Wenger, S., Vicens, J. Calixcrowns and related molecules. *J. Inclusion Phenom. Mol. Rec. Chem.* 1994, 19, 137–148.

102. Hill, C., Dozol, J.-F., Lamare, V., Rouquette, H., Eymard, S., Tournois, B., Vicens, J., Asfari, Z., Bressot, C., Ungaro, R., Casnati, A. Nuclear waste treatment by means of supported liquid membranes containing calixcrown compounds. *J. Inclusion Phenom. Mol. Rec. Chem.* 1994, 19, 399–408.

103. Asfari, Z., Bressot, C., J. Vicens, Hill, C., Dozol, J.-F., Rouquette, H., Eymard, S., Lamare, V., Tournois, B. Doubly crowned calix[4]arenes in the 1,3-alternate conformation as cesium-selective carriers in supported liquid membranes. *Anal. Chem.* 1995, 67, 3133–3139.

104. Haverlock, T. J., Bonnesen, P. V., Sachleben, R. A., Moyer, B. A. Applicability of a calixarene-crown compound for the removal of cesium from alkaline tank waste. *Radiochim. Acta* 1997, 76, 103–108.

105. Kim, J. S., Suh, I. H. K., Jong, K., Cho, M. H. Selective sensing of cesium ions by novel calix[4]arene bis(dibenzocrown) ethers in an aqueous environment. *J. Chem. Soc. Perkin Trans. 1* 1998, 2307–2312.

106. Lamare, V., Dozol, J.-F., Fuangswasdi, S., Arnaud-Neu, F., Thuery, P., Nierlich, M., Asfari, Z., Vicens, J. A new calix[4]arene-bis(crown ether) derivative displaying an improved cesium over sodium selectivity: Molecular dynamics and experimental investigation of alkali-metal ion complexation. *J. Chem. Soc. Perkin Trans. 2* 1999, 271–284.

107. Hill, C., Dozol, J.-F., Rouquette, H., Eymard, S., Tournois, B. Study of the stability of some supported liquid membranes. *J. Membrane Sci.* 1996, 114, 73–80.

108. Moyer, B. A., Birdwell, J. F., Bonnesen Jr., P. V., Delmau, L. H. Use of macrocycles in nuclear-waste cleanup: A real world application of a calixcrown in cesium separation technology. In *Macrocyclic Chemistry: Current Trends and Future Perspectives*, Gloe, K. Ed. Springer, Dordrecht, 2005, 383–405.

109. Mincher, B. J., Modolo, G., Mezyk, S. P. Review article: The effects of radiation chemistry on solvent extraction: 2. a review of fission-product extraction. *Solvent Extr. Ion Exch.* 2009, 27, 331–353.

110. Yaftian, M. R., Burgard, M., Matt, D., Wieser, C., Dieleman, C. B. Multifunctional calix[4]arenes containing pendant amide and phosphoryl groups: Their use as extracting agents and carriers for alkali cations. *J. Inclusion Phenom.* 1997, 27, 127–140.

111. Yaftian, M. R., Burgard, M., Bachiri, A. E., Matt, D., Wieser, C., Dieleman, C. B. Calix[4]arenes with pendant amide and phosphine oxide functionalities: Use as extractive agents and carriers for silver (I) ions. *J. Inclusion Phenom. Mol. Recognit. Chem.* 1997, 29, 137–151.

112. Ohto, K., Terada, M., Shiratsuchi, K., Cierpiszewski, R., Inoue, K. A solvent extraction kinetics study of silver ion with a crossed type carboxylic acid derivative of calix[4]arene. *Solvent Extr. Res. Dev. Jpn.* 2007, 14, 145–150.

113. He, W., Wuping, L., Wang, W., Li, D., Niu, C. Mass transfer kinetics of neodymium(III) extraction by calix[4]arene carboxylic acid using a constant interfacial area cell with laminar flow. *J. Chem. Technol. Biotechnol.* 2008, 83, 1314–1320.

114. Ohto, K., Tanaka, H., Ishibashi, H., Inoue, K. Solubility in organic diluents and extraction behavior of calix[4]arene carboxylates with different alkyl chains. *Solvent Extr. Ion Exch.* 1999, 17, 1309–1325.

115. Goto, M., Shinohara, K., Shimojo, K., Oshima, T., Kubota, F. Solubilization of calixarenes in an aliphatic organic solvent by reverse micelles. *J. Chem. Eng. Jpn.* 2002, 35, 1012–1016.

116. Visser, A. E., Swatloski, R. P., Hartman, D. H., Huddleston, J. G., Rogers, R. D. Calixarenes as ligands in environmentally benign liquid–liquid extraction media, aqueous biphasic systems and room temperature ionic liquid. In ACS Symposium Series 757, *Calixarenes for Separations*, Lumetta, G. L., Rogers, R. D., Gophalan, A. S., Eds. ACS, Washington, DC, 2000, 223–236.

117. Shimojo K., Goto, M. First application of calixarenes as extractants in room-temperature ionic liquids. *Chem. Lett.* 2004, 33, 320–321.

118. Luo, H., Dai, S., Bonnesen, P. V., Buchanan, A. C., Holbrey, J. D., Bridges, N. J., Rogers, R. D. Extraction of cesium ions from aqueous solutions using calix[4]arene-bis(tert-octylbenzo-crown-6) in ionic liquids. *Anal. Chem.* 2004, 76, 3078–3083.

119. Shimojo K., Goto, M. Solvent extraction and stripping of silver ions in room-temperature ionic liquids containing calixarenes. *Anal. Chem.* 2004, 76, 5039–5044.

120. Luo, H., Yu, M., Dai, S. Solvent extraction of Sr^{2+} and Cs^+ based on hydrophobic protic ionic liquids. *Z. Naturforsch. A Phys. Sci.* 2007, 62, 281–291.

4 Protein Extraction by the Recognition of Lysine Residues Using Macrocyclic Molecules

Tatsuya Oshima
Department of Applied Chemistry, Faculty of
Engineering, University of Miyazaki, Miyazaki, Japan

CONTENTS

4.1 INTRODUCTION

Liquid–liquid extraction has been attracting attention as a useful separation method for biological products in the past decades, as it is well suited to large-scale operation as well as continuous operation mode, and can be designed to be selective through the control of conditions, systems, and extractants.[1–4] Additionally, specific extractants that show affinity to targeted biomolecules would conclusively enhance the selectivity to acquire a prominent separation system.

The extraction of biomacromolecules such as proteins into a hydrophobic phase has been challenging for a number of reasons. Proteins are hydrophilic macromolecules, and their surface is generally surrounded by hydrophilic or ionic residues to be dissolved in aqueous media. If we want to extract proteins into organic media, whole protein molecules should become hydrophobic enough to be transferred into the hydrophobic media via the complexation with some extractants or surfactants. Among the various approaches for effecting extraction of proteins, reversed-micelle systems have proven successful. A protein is transferred from an aqueous phase to the water pools of reversed micelles in an organic phase via interactions between the surfactant and the protein.[5,6] Various proteins can be transferred into a reversed-micelle solution, with the major driving force for protein extraction being the electrostatic interactions between the charged polar heads in the surfactant and the oppositely charged groups in the protein. Hydrophobic interactions are also a dominant factor. However, the separation factor for proteins in a typical reversed-micelle system is generally not so high because simple electrostatic interactions between the protein and the surfactants govern the protein transfer.

The selective extraction of proteins in reversed-micelle systems by incorporating an affinity ligand that discriminates the target protein from contaminating proteins has provided a major advance over the simple surfactant systems. Various biological interactions, including lectin-sugar, enzyme-inhibitor, antigen-antibody, protein-protein, and DNA-protein interactions, are available for affinity extraction. For example, concanavalin A has been selectively extracted over other proteins using an affinity reversed-micelle extraction system incorporating an alkylglucoside as a ligand via lectin–sugar interactions.[7–10] The extraction of lysozyme (Lyso) and bovine serum albumin (BSA) into reversed micelles was effected by the addition of the triazine dye, Cibacron Blue.[11,12] Trypsin can be separated from a mixture of proteins by reversed micelles containing an alkylated trypsin inhibitor through their protein–protein affinity interactions.[13,14] Thus, the specificity of the ligand for a target protein dominates the selectivity in protein extraction.

Recently, the macrocyclic classes of compounds calixarenes and crown ethers (Figure 4.1) were found to be effective extractants for proteins. The extraction systems were based on the formation of a hydrophobic supramolecular complex between multiple macrocyclic molecules and a protein. Calixarenes and crown

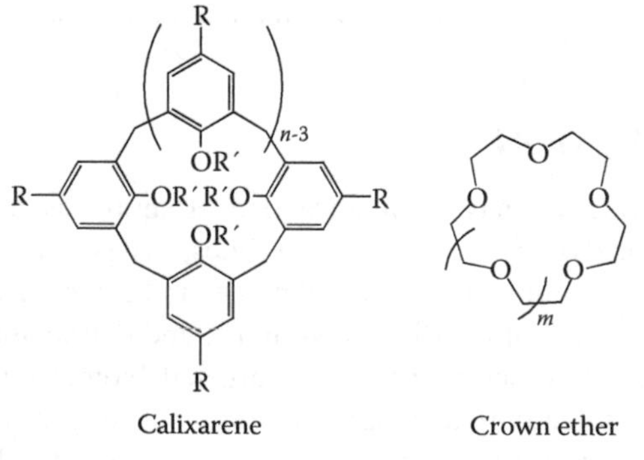

Calixarene   Crown ether

FIGURE 4.1   Molecular structures of calixarene ($n = 4$, 6, and 8 typically) and crown ether ($m = 0$–3 typically).

ethers are known to be host molecules for various ammonium compounds, which can be included into their cavities via hydrogen bonding and electrostatic interactions. In a similar manner, the macrocyclic compounds can complex with the protonated amino group of lysine residues and the terminal amino group in a protein. The resulting supramolecular complex becomes hydrophobic enough to be transferred into the hydrophobic phase in a liquid–liquid extraction system.

The liquid–liquid extraction of protein using calixarene derivatives was first reported in 2002.[15] Other reports of protein extraction by calixarenes for various uses followed.[16] Exciting possibilities for controlling selectivity owing to the structural versatility and synthetic accessibility of calixarenes have become apparent. In this chapter, protein extraction by the recognition of lysine residues using the macrocyclic molecules calixarenes and crown ethers as extractants is summarized from the recent literature. A novel selective extraction system for proteins can be developed according to the differences in the distribution of amino acid residues on the surface of proteins. An extraction system is also useful for studying protein refolding. Additionally, novel extraction systems for proteins using crown ethers have been studied in ionic liquid systems and aqueous two-phase systems. Moreover, complexation with macrocyclic molecules is a novel modification technique to enhance the biological activity of a protein of interest. Various enzymatic oxidations have been developed using supramolecular complexes between macrocyclic molecules and proteins.

4.2 PROTEIN EXTRACTION USING CALIXARENES

4.2.1 RECOGNITION OF BIOMOLECULES USING CALIXARENES

A calixarene is a cyclic oligomer that is formed by phenolic units linked with methylene bridges.[17–21] The cyclic calixarene oligomers are easily synthesized via a single-step condensation of phenolic units with formaldehyde. The cavity-shaped architecture of calixarenes is available for the inclusion of a targeted species. Additionally, various functional groups for a particular interaction with a guest species can be introduced to the upper or lower rim of the calixarene molecule regioselectively (Figure 4.2). Furthermore, the regioisomers and conformational isomers of the calixarene derivatives display different affinities for the target. Calixarene derivatives bearing various functional groups have been developed for complexation with metal ions, alkali ions, alkali earth ions, transition metal ions, and f-elements.[22–24]

Calixarenes are also known to be attractive platforms for the recognition of biomolecules. Various calixarene derivatives have been developed for complexation with biomolecules such as amines, amino acids, peptides, nucleosides, and saccharides, for applications with chromogenic and fluorogenic receptors, ion-selective electrodes, and liquid–liquid extraction.[25,26] Liquid–liquid extraction of biomolecules using calixarene derivatives has been studied since the 1990s. The binding selectivity achieved from the liquid–liquid extraction of butyl-ammonium picrates using calix[6]arene hexaethyl ester is greater than that of

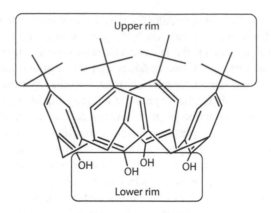

FIGURE 4.2   Upper rim and lower rim of calixarene.

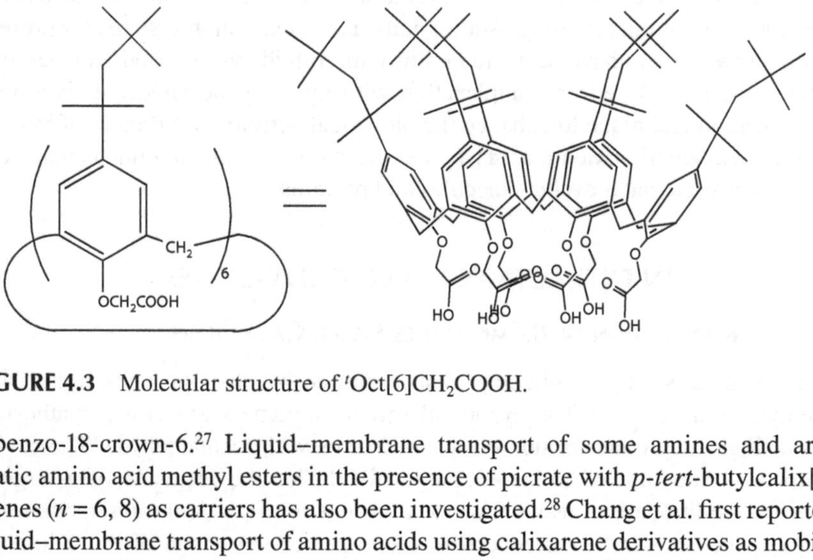

FIGURE 4.3   Molecular structure of ^{t}Oct[6]CH$_2$COOH.

dibenzo-18-crown-6.[27] Liquid–membrane transport of some amines and aromatic amino acid methyl esters in the presence of picrate with *p-tert*-butylcalix[*n*] arenes (*n* = 6, 8) as carriers has also been investigated.[28] Chang et al. first reported liquid–membrane transport of amino acids using calixarene derivatives as mobile carriers.[29,30] The complexation between calix[6]arene ester derivative and amino acid ester is considered to be based on the tripodal hydrogen bondings resulting in the formation of an inclusion complex. Chiral separation of amino acids was also attempted using a calix[4]arene analog having chiral pendant groups in a liquid– membrane transport experiment.[31] The chiral calix[4]arene derivatives show the chiral selectively *L*-amino acid esters in bulk liquid–membrane transport.

Calix[6]arene derivatives are also known to be effective host molecules for amino compounds. The *p-tert*-octylcalix[6]arene hexacarboxylic acid derivative (abbreviated as ^{t}Oct[6]CH$_2$COOH) (shown in Figure 4.3) exhibits one of the highest extractabilities for amino acid esters and biologically important amino compounds.[32–34] The calix[6]arene derivative has six carboxylic acid groups that interact with the protonated amino group via electrostatic interaction through a proton-exchange reaction. The pseudocavity size of ^{t}Oct[6]CH$_2$COOH, which is formed by carbonyl groups, is optimal for the inclusion of the protonated amino group. Furthermore, the C_6 symmetry of ^{t}Oct[6]CH$_2$COOH is stereochemically favorable for the interaction with the amino group.

The extractability of a series of calix[n]arene carboxylic acid derivatives (^{t}Oct[n]CH$_2$COOH (n = 4, 6, 8)) for tryptophan methyl ester (Trp-OMe), as well as a monomer analog (^{t}Oct[1]CH$_2$COOH) in chloroform, was compared under the same operational conditions (Figures 4.4 and 4.5). The order of the extractabilities was as follows:[32]

^{t}Oct[6]CH$_2$COOH > ^{t}Oct[8]CH$_2$COOH >> ^{t}Oct[4]CH$_2$COOH, ^{t}Oct[1]CH$_2$COOH

The calix[6]arene exhibited the highest extractability in the series because its size provided the best fit to the target. As the cavity of calix[8]arene is a little too large for the formation of optimal interactions with the amino acid, the extractability of the calix[8]arene ^{t}Oct[8]CH$_2$COOH is less. The extractability of the calix[4]

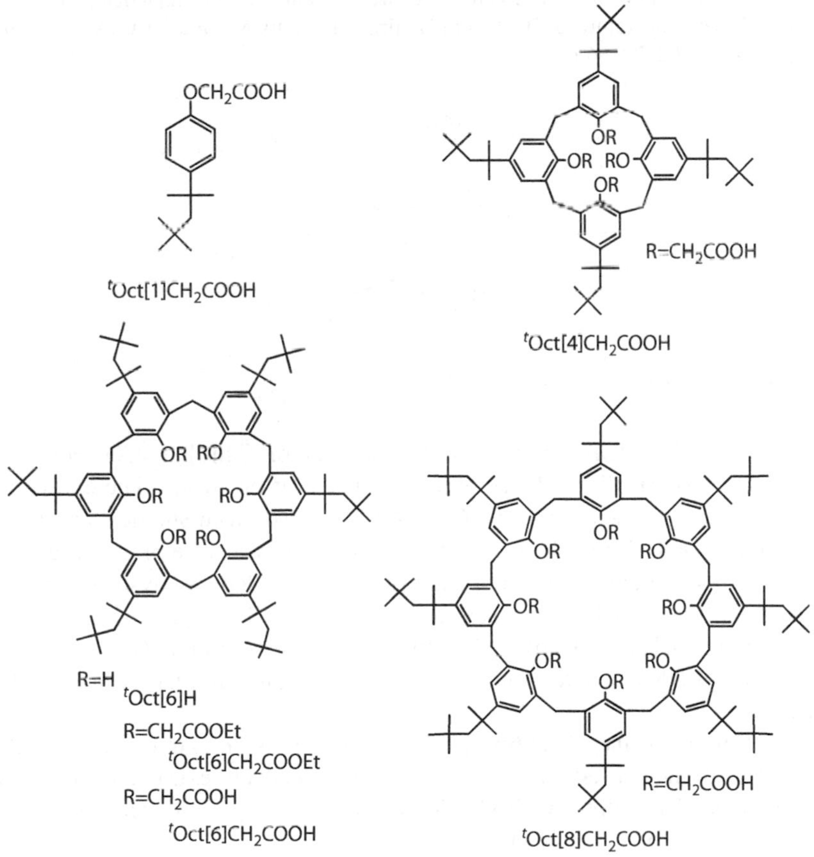

FIGURE 4.4 Molecular structures of ^{t}Oct[n]CH$_2$COOH (n = 1, 4, 6, and 8).

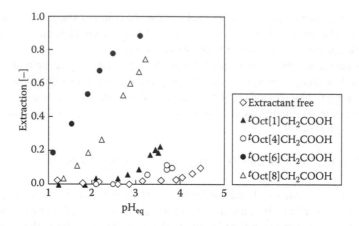

FIGURE 4.5 Extraction profiles of Trp-OMe using ^{t}Oct[n]CH$_2$COOH ($n = 1, 4, 6, 8$) as a function of pH. [Trp-OMe]$_{ini}$ = 1 mM, [^{t}Oct[n]CH$_2$COOH]$_{ini}$ = 0 (Extractant free), 5 mM ($n = 4, 6, 8$) or 30 mM ($n = 1$); diluent, chloroform; volume, 5 mL aqueous phase, 5 mL organic phase; temperature, 30°C. (From Oshima, T., Goto, M., and Furusaki, S., *J. Incl. Phenom.*, 43, 2002, 77–86.)

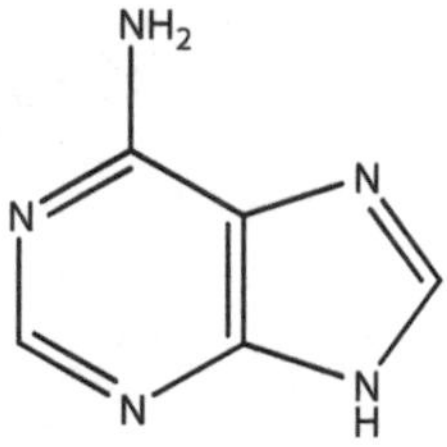

FIGURE 4.6 Molecular structure of adenine.

arene ^{t}Oct[4]CH$_2$COOH is much smaller than that of ^{t}Oct[6]CH$_2$COOH because the cavity is too small for the complete inclusion of the target. The extractability of the monomer analog ^{t}Oct[1]CH$_2$COOH is also much smaller than that of ^{t}Oct[6]CH$_2$COOH. Thus, the size of the macrocyclic structure of calixarene will determine its extractability. From the results of the slope analysis and the Job method analysis, calix[6]arene was found to form a 1:1 complex with Trp-OMe. Furthermore, inclusion of Trp-OMe into the cavity of ^{t}Oct[6]CH$_2$COOH was confirmed by ^{1}H-NMR (nuclear magnetic resonance) and circular dichroism (CD) spectra.

In a similar manner, ^{t}Oct[6]CH$_2$COOH can act as an extractant for nucleobases.[35] In particular, adenine (Figure 4.6) is selectively extracted through the formation of a 1:1 complex with ^{t}Oct[6]CH$_2$COOH via a proton-exchange reaction as shown in the following equation:

$$HA^+ + H_6R \ (org) \rightleftharpoons (HA)(H_5R) \ (org) + H^+$$

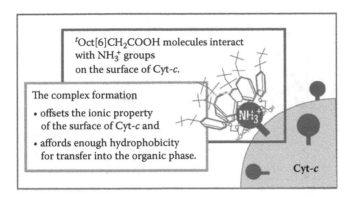

'Oct[6]CH$_2$COOH molecules interact
with NH$_3^+$ groups
on the surface of Cyt-c.

The complex formation

• offsets the ionic property
  of the surface of Cyt-c and

• affords enough hydrophobicity
  for transfer into the organic phase.

NH$_3^+$

Cyt-c

FIGURE 4.7 Schematic illustration of the complexation between cytochrome c and calix[6]arene carboxylic acid derivative. (From Oshima, T., Goto, M., and Furusaki, S., *Biomacromolecules*, 3, 2002, 438–444.)

where H$_6$R denotes 'Oct[6]CH$_2$COOH, and HA$^+$ denotes a protonated adenine.

Furthermore, 'Oct[6]CH$_2$COOH can also function as an extractant for separating dopamine over other catecholamines, adrenaline, and noradrenaline.[36] The extraction reaction was confirmed to be a proton-exchange reaction to form the 1:1 complex.

Thus, the calix[6]arene carboxylic acid 'Oct[6]CH$_2$COOH forms a stable complex with an amino group in various biomolecules, suggesting a general approach for extraction. On the basis of the same strategy, the ε-amino group of a lysine residue in various proteins would therefore be a probable target for 'Oct[6]CH$_2$COOH. Figure 4.7 shows a schematic illustration of the complexation between a cationic protein, cytochrome c (Cyt-c), and 'Oct[6]CH$_2$COOH.[15,16] As shown in the figure, the calix[6]arene molecules include the ε-amino groups of the lysine residues through a proton-exchange reaction. The complexation between the calix[6] arene molecules and the lysine residues via electrostatic interactions alters the protein molecule into a more hydrophobic state. Cyt-c has 19 lysine residues and simultaneously binds with multiple calix[6]arene molecules. As a result of plural proton-exchange reactions between the lysine residues and 'Oct[6]CH$_2$COOH, the resulting hydrophobic complex is transferred into the organic phase.

4.2.2 PROTEIN EXTRACTION USING CALIXARENES

The extraction behavior of Cyt-c using 'Oct[6]CH$_2$COOH as an extractant has been studied.[15,37] Figure 4.8 shows the extraction of Cyt-c using calix[n]arene carboxylic acid derivatives ('Oct[n]CH$_2$COOH (n = 4, 6, 8)) and the monomer analog ('Oct[1]CH$_2$COOH) in chloroform.[38] The data are plotted with the absorbance of the Soret band of extracted Cyt-c in the organic phase as the ordinate and the molar ratio between the extractant ('Oct[n]CH$_2$COOH) and Cyt-c as the abscissa. The calix[6]arene 'Oct[6]CH$_2$COOH exhibited the highest extractability for protein from all the extractants examined, and Cyt-c could be quantitatively

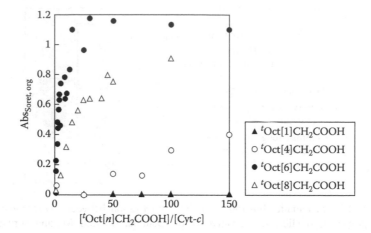

FIGURE 4.8 Dependency of the extraction of Cyt-c on the ratio [^{t}Oct[n]CH$_2$COOH]/[Cyt-c]. [Cyt-c]$_{ini}$ = 10 μM; pH$_{ini}$ = 6.2; chloroform diluent; 5 mL aqueous phase; 5 mL organic phase; temperature 30°C. (From Oshima, T. et al., *J. Ion Exch.*, 14, 2003, 373–376.)

extracted in the presence of 20-fold molar equivalent of the extractant. The extraction strength of the calix[8]arene ^{t}Oct[8]CH$_2$COOH was second to ^{t}Oct[6] CH$_2$COOH, while that of the calix[4]arene ^{t}Oct[4]CH$_2$COOH was much lower. Furthermore, the monomer analog (^{t}Oct[1]CH$_2$COOH) did not extract Cyt-c at all. Namely, the order of extractability for Cyt-c is as follows, which agrees with that obtained for amino acids:[31]

$$^t\text{Oct[6]CH}_2\text{COOH} > {}^t\text{Oct[8]CH}_2\text{COOH} > {}^t\text{Oct[4]CH}_2\text{COOH} > {}^t\text{Oct[1]CH}_2\text{COOH}$$

^{t}Oct[6]CH$_2$COOH shows the strongest affinity for protonated amino groups because of its ideal cavity size, the C$_3$ symmetry, and the preorganized carboxylic acid groups for inclusion. Cyt-c is extracted using the calix[6]arene molecules through the complexations with the –NH$_3^+$ groups of the lysine residues in Cyt-c. ^{t}Oct[8]CH$_2$COOH is also able to include the protonated amino group; however, the recognition is not strict because the cavity size is too large. By contrast, it is difficult for the –NH$_3^+$ group to be included in ^{t}Oct[4]CH$_2$COOH because of its small cavity, resulting in the lower extractability for Cyt-c. The hydrophobicity of the complex between the protein and the extractant is also a key factor for the extraction. As the complexation of Cyt-c with the monomer analog ^{t}Oct[1] CH$_2$COOH does not afford sufficient hydrophobicity for extraction, a precipitate is formed at the oil–water interface. As the extractability of *p-tert*-butylcalix[6] arene hexacarboxylic acid derivative (^{t}Bu[6]CH$_2$COOH) was similar to that of ^{t}Oct[6]CH$_2$COOH, the difference in the alkyl chain length at the upper rim does not affect the extractability of Cyt-c.[15]

Cyt-c is not extracted into the organic solution with the nonsubstituted calix[6] arene (^{t}Oct[6]H) or the ethyl ester derivative (^{t}Oct[6]CH$_2$COOEt) under the same conditions, likely because the proton-exchange reaction drives the complexation between the protein and the organic phase calixarene. Since ^{t}Oct[6]CH$_2$COOH does not form a molecular assembly like reversed micelles under the experimental conditions, the water content in the organic phase does not change appreciably before or after the extraction. As the complexation proceeds through the proton-exchange reaction, the extraction depends on the pH, with more protein being extracted under the higher pH condition. The complexation reaction can be expressed by a proton-exchange reaction as follows:

$$n\text{H}_6\text{R (org)} + (\text{NH}_3^+)_n\text{-protein} \rightleftharpoons (\text{H}_5\text{R} \cdot \text{NH}_3^+)_n\text{-protein (org)} + n\text{H}^+$$

where H$_6$R and (NH$_3^+$)$_n$-protein denote ^{t}Oct[6]CH$_2$COOH and Cyt-c, respectively. The number of the calixarene molecules binding to the protein molecule was not confirmed; however, the binding of seven calixarene molecules was previously detected from matrix-assisted laser desorption ionization time-of-flight mass spectrometry (MALDI-TOF MS) spectra (unpublished data).

Structural changes of Cyt-c extracted with ^{t}Oct[6]CH$_2$COOH were investigated by UV-vis and circular dichroism (CD) spectroscopy. For the extracted Cyt-c in the organic solution, the Soret band is shifted from the native one, and the ligand to metal charge transfer (LMCT) band from the sulfur atom of Met 80 (the axial ligand) to heme iron(III) at 695 nm has disappeared. The disappearance of the LMCT band indicates the cleavage of the coordination bond of Met 80 to heme iron.[39,40] The Q band changes to a broader pattern. The CD spectrum of Cyt-c extracted with the calixarene is also different from the native one in water. These spectral changes at the Soret and Q bands indicate that the heme environment of the extracted Cyt-c is different from that of the native Cyt-c.

The complexation between calixarene and Cyt-c has been studied by other researchers for various uses. Shuang and coworkers reported that carboxyphenyl-modified calix[4]arenes (Figure 4.9) can complex with Cyt-c.[41] These calixarenes feature substitution at the upper rim, the lower rim being terminated by n-propoxy groups. The cavity size of calix[4]arene is smaller than that of calix[6]arene; however, the upper rim of the calix[4]arene provides a wider cavity, which is possible to incorporate the –NH$_3^+$ groups of the lysine residues in Cyt-c. Binding affinity to Cyt-c in N, N-dimethyl formamide (DMF) using tetrakis-carboxyphenylcalix[4]

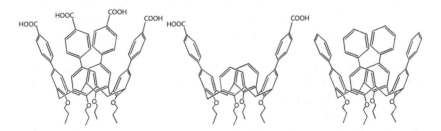

FIGURE 4.9 Molecular structures of upper-rim carboxyphenyl-modified tetrapropoxy calix[4]arene and its analogs. (From Oshima, T. et al., *J. Ion Exch.*, 14, 2003, 373–376.)

arene is higher than that obtained using either the bis-carboxylphenylcalix[4]arene or tetrakis-phenylcalix[4]arene as structural analogs, depending on the number of carboxyl groups. The fluorescence of the calix[4]arene derivative is quenched due to the Förster resonance energy transfer (FRET) upon the binding with Cyt-c.

A biosensor based on supported lipid films incorporating the calix[6]arene for detection of Cyt-c using electrochemical impedance spectroscopy (EIS) was studied by Hianik and coworkers.[42] 'Oct[6]CH₂COOH was incorporated into soybean phosphatidylcholine (SBPC) liposomes at different molar ratios (SBPC: 'Oct[6]CH₂COOH = 10:1, 30:1, and 100:1). The lipid layer containing the calix-arenes was formed on the surface of a gold electrode covered with an octadecane-thiol monolayer. The enhanced content of calixarenes in the liposomes resulted in an increase of sensor response. The developed biosensor allowed good discrimi-nation between Cyt-c and lysine with a limit of detection for Cyt-c of close to 10 nM. An excess amount of lysine was shown to cause interference in the deter-mination of Cyt-c. However, a comparison of the sensor response to Cyt-c and lysine demonstrated a large difference in sensitivity between these compounds. While Cyt-c can be detected at the nM concentration level, the effect of lysine can be measured only at much higher concentrations (mM range). The dissociation constant for Cyt-c was approximately 4.2×10^4 times lower than that of lysine, which provides clear evidence of the much higher affinity of Cyt-c for the sensor surface.

4.2.3 SEPARATION OF PROTEINS USING CALIXARENE

As the extraction of Cyt-c using 'Oct[6]CH₂COOH is based on complexation between 'Oct[6]CH₂COOH and a protonated amino group of a lysine residue, calix[6]arene shows extraction selectivity for lysine-rich proteins. Therefore, a separation system for lysine-rich proteins can be developed by liquid–liquid extraction using 'Oct[6]CH₂COOH.

A schematic illustration of a separation system for a lysine-rich protein using 'Oct[6]CH₂COOH is shown in Figure 4.10.[43] The separation process consisted of (1) selective extraction of a Lys-rich protein using the calix[6]arene and (2) recovery of the extracted lysine-rich protein. Cyt-c and lysozyme (abbreviated as Lyso) have similar molecular weights and isoelectric points (pIs), but differ in their number of Lys residues, since Cyt-c has 19 lysine residues, while Lyso only has 6. 'Oct[6]CH₂COOH quantitatively extracted the Lys-rich Cyt-c based on recognition of the ε-amino groups in the Lys residues. By contrast, Lyso was difficult to extract and tended to aggregate with the extractant, resulting in the formation of a precipitate at the oil–water interface. Thus, the calix[6]arene can selectively extract a lysine-rich protein based on the difference in the number of lysine residues on the surface of the protein.

Following extraction, the next step in the separation and purification of pro-teins is to strip the protein into a fresh aqueous solution. As the extraction pro-ceeds by the proton-exchange reaction, the extracted protein should be stripped using an aqueous acidic solution. However, when Cyt-c was extracted with 'Oct[6]

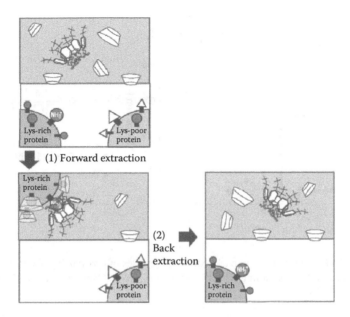

FIGURE 4.10   Conceptual illustration of the separation system for the lysine-rich protein using ^{t}Oct[6]CH$_2$COOH.

CH$_2$COOH, the protein was not recovered from the chloroform solution by contact with an aqueous acidic solution alone. It therefore appears that transfer of Cyt-c into the organic phase is an irreversible process. Addition of an alcohol can help the stripping of proteins. Figure 4.11 shows the extraction profile of Cyt-c using ^{t}Oct[6]CH$_2$COOH as well as the back-extraction profile by contacting a fresh acidic solution in an isooctane system containing 10 vol% of 1-octanol as a co-solvent.[38] The stripping experiment was examined with or without the addition of 15 vol% of ethanol in a stripping solution. As shown in Figure 4.11, the extracted Cyt-c could be quantitatively recovered from the isooctane solution using an aqueous solution containing ethanol at a pH of less than 2. Thus, the organic solvent as well as the co-solvent should be selected carefully for the back-extraction. For instance, the extracted protein can be effectively stripped from isooctane containing 10 vol% 1-octanol as a co-solvent, but not from polar organic solvents such as chloroform.[43,44] The addition of co-solvents such as alcohol that distribute into both the aqueous and organic phases is effective for controlling the gap of hydrophilicity/lipophilicity of the aqueous and organic phases.

The separation of Cyt-c and Lyso by solvent extraction using ^{t}Oct[6]CH$_2$COOH was examined under the optimal conditions, namely, selective extraction of Cyt-c at pH 6.2 in the presence of 50 mM sodium ions, followed by back-extraction to an aqueous solution at pH 1.5 containing 15 vol% of ethanol. Figure 4.12 shows the liquid chromatograms for a separation test of Cyt-c and Lyso by solvent extraction. In the forward extraction stage, an aqueous solution containing Cyt-c and Lyso (Figure 4.12(a)) was placed in contact with an organic solution

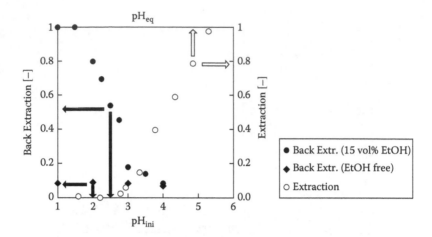

FIGURE 4.11 Extraction and back-extraction profiles of Cyt-*c* using *Oct[6]CH₂COOH in isooctane media (containing 10 vol% 1-octanol). (From Oshima, T. et al., *J. Ion Exch.*, 14, 2003, 373–376.)

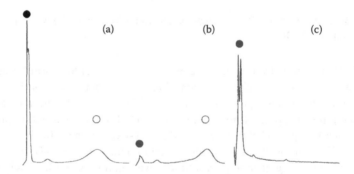

FIGURE 4.12 Liquid chromatograms of a separation between Cyt-*c* (filled circles) and Lyso (open circles) with *Oct[6]CH₂COOH. (a) Before the extraction (aqueous phase); (b) after the extraction (aqueous phase); (c) after back-extraction (stripping aqueous phase).

containing the calix[6]arene. After phase separation, 93% of the Cyt-*c* in the aqueous solution was extracted by *Oct[6]CH₂COOH, while Lyso remained in the solution (Figure 4.12(b)). In the subsequent stripping stage, 97% of the extracted Cyt-*c* was recovered in the aqueous solution, whereas Lyso was not detected (Figure 4.12(c)). Namely, the calix[6]arene can separate cationic proteins that are similar in molecular weight and pI by discriminating the numbers of lysine residues on the surfaces of the protein molecules.

The tertiary structure of Cyt-*c* after the back-extraction is denatured due to the acidity and the presence of the alcohol in the aqueous stripping solution, but it recovers after neutralization. The activity of Cyt-*c* as an electron-transfer protein was partly retained (67%) through the forward extraction and back-extraction after neutralization.

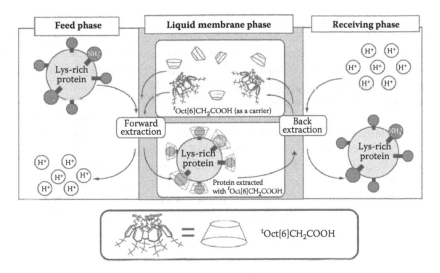

FIGURE 4.13 Schematic illustration of liquid–membrane transport of a protein using ^{t}Oct[6]CH$_2$COOH as a carrier.

Moreover, the liquid–membrane transport of Cyt-c, using ^{t}Oct[6]CH$_2$COOH as a carrier, has been demonstrated, with a view to continuous separation of this specific protein.[45] Figure 4.13 shows a schematic illustration of the liquid–membrane transport of a protein using ^{t}Oct[6]CH$_2$COOH as a carrier. The transport of the protein occurred as follows: (1) extraction of Cyt-c using ^{t}Oct[6]CH$_2$COOH from the feed phase into the liquid–membrane phase through a proton-exchange reaction, followed by (2) diffusion of the complex toward the opposite interface through the bulk liquid–membrane phase and (3) stripping of the extracted Cyt-c into the receiving phase with hydrogen ions.

The rate of Cyt-c transfer into the membrane phase increased with increasing concentrations of ^{t}Oct[6]CH$_2$COOH. The back-extraction of the extracted Cyt-c into the receiving phase, however, was relatively slow. In the presence of 2.0×10^{-3} M of ^{t}Oct[6]CH$_2$COOH, 97% of Cyt-c was extracted from the feed phase and 77% of Cyt-c was recovered into the receiving phase after 60 hours. Transport of Cyt-c was examined using a series of calix[n]arene carboxylic acid derivatives (n = 4, 6, 8). The rate of forward extraction of Cyt-c using ^{t}Oct[4]CH$_2$COOH was much smaller and was not recoverable. The forward extraction rates with ^{t}Oct[6]CH$_2$COOH and ^{t}Oct[8]CH$_2$COOH were similar; however, when Cyt-c was extracted with ^{t}Oct[6]CH$_2$COOH, it was not recovered into the receiving phase, and a precipitate formed at the interface between the liquid–membrane phase and the receiving phase. By contrast, Cyt-c successfully permeated membranes in the presence of ^{t}Oct[6]CH$_2$COOH.

As the extraction is based on a proton-exchange mechanism, the transport rate is governed by carrier concentration and the pH gradient between the feed and the receiving phases. The forward extraction rate from the feed phase became saturated when the pH in the feed phase was greater than 4.0. Moreover, the extracted

Cyt-c was difficult to strip from the receiving phase when the pH value of this phase was higher than 2.5. Adding an alkali salt to the aqueous solution is necessary to suppress precipitation; however, alkali ions are competitively extracted by ^{t}Oct[6]CH$_2$COOH. The transport rate of Cyt-c was not affected by adding 10 mM NaCl, but the rate decreased when more than 50 mM NaCl was added, as excess sodium ions inhibited transport.

Liquid–membrane separation of Cyt-c from a mixture of cationic proteins was demonstrated under optimal conditions. Figure 4.14 shows the time course of Cyt-c and Lyso transport through the liquid–membrane by ^{t}Oct[6]CH$_2$COOH. As expected, the Lys-rich protein Cyt-c was selectively extracted by ^{t}Oct[6] CH$_2$COOH. After 37 hours, 99% of Cyt-c was extracted from the feed phase, while 73% of Lyso remained in the feed phase. Moreover, 79% of Cyt-c was recovered into the receiving phase, whereas only 7% of Lyso appeared in the receiving

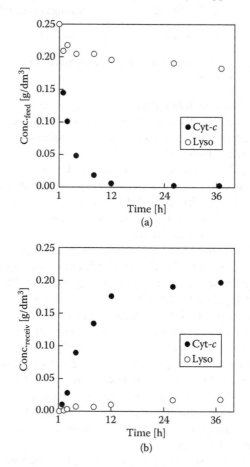

FIGURE 4.14 Transport profiles for the transport of Cyt-c and Lyso through liquid–membranes using ^{t}Oct[6]CH$_2$COOH in (a) the feed phase and (b) the receiving phase. (From Oshima, T. et al., *J. Membr. Sci.*, 307, 2008, 284–291.)

phase, indicating that these cationic proteins were successfully and specifically separated through the liquid–membrane system using ^{t}Oct[6]CH$_2$COOH.

^{t}Oct[6]CH$_2$COOH is also available for the extraction of the major serum antibody immunoglobulin G (IgG). Martinez-Aragon et al. reported the extraction of the very large protein (MW = 135,000 Da) IgG into isooctane containing 10 vol% 1-octanol as a co-solvent.[46] Increasing the shaking or stirring speed led to a higher amount of precipitation of IgG; however, there was no precipitation at a shaking speed of 100 rpm in the presence of 0.05 mM ^{t}Oct[6]CH$_2$COOH. The extraction behavior at high pH was not expected due to the negative net charge of the IgG. However, the highest extraction was obtained at pH 10.95, which is above the pI of IgG (45% at 100 rpm). Therefore, it is likely that the electrostatic interactions between the lysine group of IgG and the calix[6]arene are not the only driving force for this host-guest interaction. Salt concentration is also important for controlling the IgG extraction as well as to prevent precipitation. The minimum amount of NaCl to solubilize IgG and prevent denaturation is 150 mM. On the other hand, IgG extraction into an organic phase is higher at lower salt concentrations. Therefore, the extraction is highest in the presence of 150 mM NaCl. From the results of the kinetics study, the limiting step in the extraction process is likely the interaction of the calix[6]arene molecules with the lysine residues of IgG. It is also possible to back-extract IgG from an organic phase to an aqueous phase without denaturation using a pH shift.

4.2.4   Protein Refolding via Liquid–Liquid Extraction Using Calixarene

The expression of recombinant proteins in bacteria on the basis of recombinant DNA technologies often results in the formation of insoluble aggregates called inclusion bodies. For the regeneration of protein into the active form, the insoluble aggregates containing the misfolded proteins should be solubilized using denaturants. The solubilized protein is then refolded by decreasing the concentration of the denaturant. The proteins must be isolated from the solution containing the denaturant and unfolded protein in order to prevent the reaggregation of the folded intermediate. Shimojo et al. studied protein refolding by means of a liquid–liquid extraction process using the calix[6]arene carboxylic acid derivative ^{t}Oct[6]CH$_2$COOH.[47] Misfolded protein in inclusion bodies was dissolved and denatured in an aqueous solution containing urea as a denaturant. The denatured protein was extracted using ^{t}Oct[6]CH$_2$COOH from the aqueous solution into an organic solution. Formation of a supramolecular complex between the protein and multiple ^{t}Oct[6]CH$_2$COOH molecules denatured the protein molecules from each other, which suppressed the generation of aggregates due to protein–protein interactions. Later, the extracted protein was recovered into a denaturant-free aqueous solution and spontaneously refolded into a biologically active conformation.

As a case study, Cyt-*c* was extracted from an aqueous solution containing 8 M urea at pH 4.5–6.0. As no aggregation of Cyt-*c* was observed, the Cyt-*c* molecules seem to have been successfully isolated using ^{t}Oct[6]CH$_2$COOH. The extracted

Cyt-c was recovered into an acidic solution with a pH less than 2.5 containing 30 vol% of 1-butanol as a co-solvent without formation of aggregates. Analysis using UV-vis, CD, and fluorescence spectroscopy confirmed that the structure of the refolded protein was identical to that of the native one. In particular, the far-UV CD spectrum of refolded Cyt-c was very similar to that of native Cyt-c, suggesting that the refolded Cyt-c possessed a native-like secondary structure. From the result of the reduction kinetics of refolded Cyt-c using ascorbic acid, refolded Cyt-c was found to regain around 72% of its native activity as an electron-transfer protein. These results suggest that the protein extraction system using calixarene provided a chaperone-like function for a denatured protein. Furthermore, the system incorporating ^{t}Oct[6]CH$_2$COOH is expected be useful for simultaneously handling protein refolding and the separation of proteins.

4.2.5 Enzymatic Reaction of Extracted Proteins in Organic Media

Enzymatic reactions in organic media provide several advantages, including that synthesis is favored over hydrolysis, enzyme stability may be enhanced, enzyme selectivity is more tunable, and the conditions favor water-insoluble substrates.[48,49] In particular, a homogeneous enzymatic reaction in a nonaqueous medium with an enzyme dissolved by means of various methods is more attractive than the heterogeneous reaction using an insoluble enzyme because of the possibility for higher catalytic activity. Cyt-c complexed with ^{t}Oct[6]CH$_2$COOH exhibits peroxidase activity in organic media.[36] When the heme environment at the active site (namely, the sixth coordination position of the heme) is altered by complexation with modifiers like calixarenes, the substrate for peroxidation can more easily access the active site of Cyt-c compared to its native state.

The oxidation of 2,6-dimethoxyphenol (2,6-DMP) as a reaction model (Scheme 4.1) proceeds in the presence of 500 nM Cyt-c (2.5×10^{-5}-fold equivalent with respect to the substrate) in chloroform. Similarly, the oxidation of o-phenylenediamine (o-PDA) is also facilitated in the presence of Cyt-c complexed with the calixarene. The enhancement of the N-demethylase activity of Cyt-c by the complexation with the calixarene is also confirmed by N-demethylation of N,N-dimethylaniline (Scheme 4.2).[50] The enzymatic activity in organic media is influenced by various factors. Since enzymes in organic media "memorize" the ionogenic state corresponding to the pH in the last aqueous solution they were exposed to, the pH dependence of the catalytic activity generally agrees with that in water. The peroxidase activity of Cyt-c increases with an increase in the

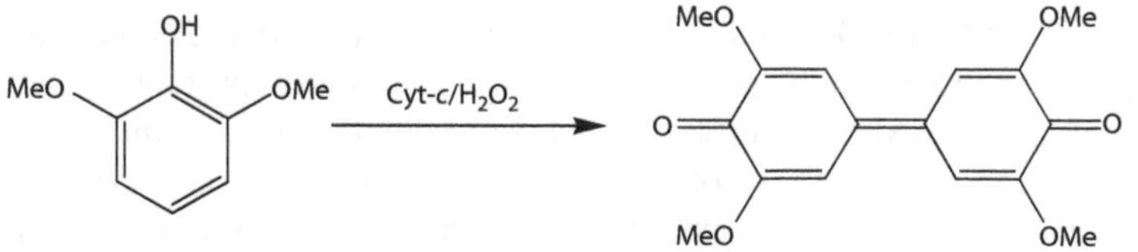

SCHEME 4.1 Enzymatic oxidation of 2,6-DMP using the calixarene · Cyt-c complex.

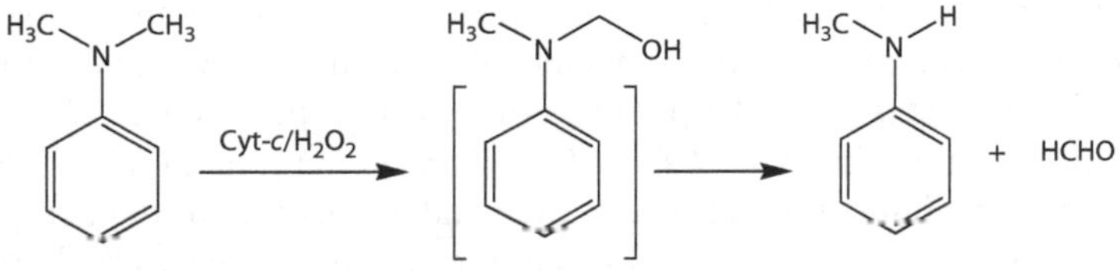

SCHEME 4.2 Enzymatic N-demethylation of N,N-dimethylaniline using the calixarene · Cyt-c complex.

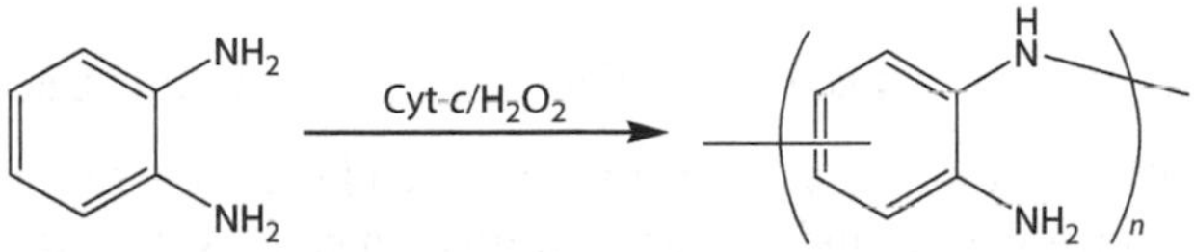

SCHEME 4.3 Enzymatic polymerization of o-PDA using the calixarene · Cyt-c complex.

extraction pH at the extraction operation because it tends to denature under the lower pH values. The oxidation activity also depends on the kind of organic solvent. More hydrophobic diluents such as chloroform, toluene, and p-xylene are favorable for the reaction. In hydrophilic organic diluents such as acetonitrile and tetrahydro-furan (THF), enzymes generally adopt an inactivate state due to the loss of essential water. The oxidation rate of the aromatic compounds increases with increasing concentrations of the Cyt-c as the catalyst and the hydrogen peroxide substrate. However, Cyt-c is inactivated in the presence of an excess amount of hydrogen peroxide.

Enzymatic polymerization using an isolated enzyme via a nonbiosynthetic pathway provides a synthetic strategy for useful polymers that are difficult to produce with the aid of conventional chemical catalysts.[51] The enzymatic polymerization of o-PDA (Scheme 4.3) by adding Cyt-c and hydrogen peroxide was demonstrated under optimized conditions.[52] During the experiments, both Cyt-c and hydrogen peroxide were periodically injected to regenerate the oxidation reaction. After evaporation of the mixed solution, the oxidative product was obtained as a black powder that was insoluble in toluene and ethanol, while it was soluble in THF and DMF. The polymeric material of o-PDA, with a molecular weight of more than 5000 Da, is obtained in chloroform or hexane. The average molecular weight of the polymer was around 7000 Da. The spectroscopic analysis of the polymer product by Fourier transform infrared (FT-IR) and ^{1}H-NMR showed that it was made up by various bonds formed between the monomer units.

Solvent extraction of the major hemeprotein hemoglobin using p-$tert$-butylcalix[n]arene carboxylic acid derivatives (n = 4, 6, 8) was also investigated by Karmali and coworkers.[53] The order of hemoglobin extractabilities at pH 5.1 was calix[6]arene > calix[4]arene > calix[8]arene. Namely, all the tested calixarene derivatives functioned as extractants for the large protein hemoglobin (MW = 64,500 Da). Hemoglobin extraction with calixarenes takes place at initial pH values higher than the pI, which suggests that protein–calixarene interactions are not limited to ionic interactions. The complex of hemoglobin and calix[6]arene exhibits pseudoperoxidase activity in chloroform, sufficient to catalyze the oxidation of syringaldazine in the presence of hydrogen peroxidase. The complex exhibited a specific activity of 9.92×10^{-2} U mg protein^{-1} at an initial pH of 7.5 in an organic medium. The half-life of the catalytic activity of the complex was between 1.96 and 2.64 days. Hemoglobin in the organic phase was back-extracted into aqueous solutions at alkaline pH.

4.3 PROTEIN EXTRACTION USING CROWN ETHERS

4.3.1 DISSOLUTION OF PROTEINS IN ORGANIC MEDIA USING CROWN ETHERS

Since the 1970s, various studies have been conducted on the complexation between 18-crown-6 ether and primary ammonium cations.[54–58] As the structure of 18-crown-6 is reasonable for the inclusion of a protonated amino group due to the cavity size and preorganized oxygen available for cation–dipole interactions, it has been used as a complexing reagent for biologically important amino compounds. Similarly, 18-crown-6 and its derivatives can also act as complexing reagents for the recognition of protein lysine residues resulting in formation of a supramolecular complex. The resulting n:1 (crown ether:protein) types of supramolecular complex are more hydrophobic compared with the native protein and can be dissolved in organic solution.

Odell et al. reported that proteins can be made soluble in organic solvents by complexation with crown ethers and cryptands.[59,60] 18-Crown-6 and cryptand[2.2.2] are more effective for dissolving proteins than smaller or larger macrocycles. Reinhoudt et al. studied the effect of crown ethers on an enzyme-catalyzed reaction in apolar organic solvents.[61,62] 18-Crown-6 and dicyclohexano-18-crown-6 (abbreviated DCH18C6) (Figure 4.15) enhanced the rate of the α-chymotrypsin-catalyzed transesterification of N-acetyl-L-phenylalanine ethyl ester with propan-1-ol in n-octane.

Tsukube et al. also studied the chemical activation of Cyt-c in methanol via complexation with crown ethers.[63–65] A lariat ether, an alcohol-armed 18-crown-6 derivative, exhibited the highest solubilization efficiency for Cyt-c. The supramolecular complex between Cyt-c and the lariat ether worked as a catalyst in the oxidation of pinacyanol chloride with hydrogen peroxide. The supramolecular complex between Cyt-c and 18-crown-6 promoted asymmetric oxidation of organic sulfoxide in methanol at lower temperatures. From the results of

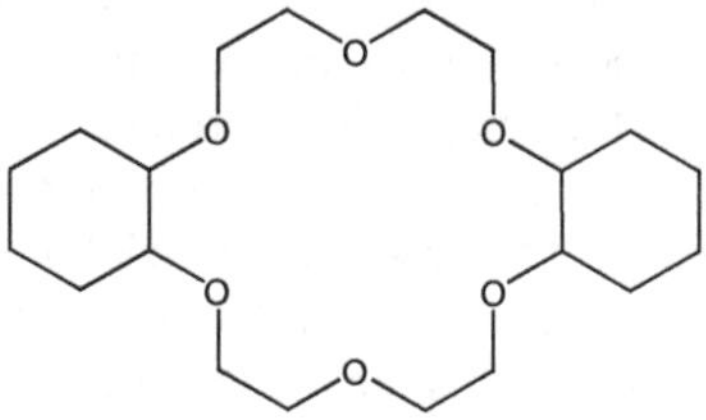

FIGURE 4.15 Molecular structure of DCH18C6.

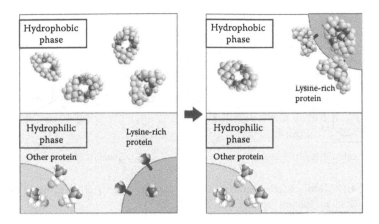

FIGURE 4.16 Schematic illustration of protein extraction using a crown ether in an ionic liquid extraction system and aqueous two-phase system.

electrospray ionization mass spectrometry (ESI-MS), UV, CD, and Raman spectroscopic characterizations, binding of four or five 18-crown-6 molecules to a Cyt-*c* was confirmed. The complexation between lysine residues and 18-crown-6 molecules has also been confirmed by ESI-MS.[66–68] The most intense peak of a mixture of 18-crown-6 and tetralysine is a quadruply charged peptide with four crown ethers attached. Namely, there was a one-to-one correlation between the number of 18-crown-6 molecules and lysine residues.

Crown ether functions like calixarene as an extractant for proteins in an ionic liquid extraction system as well as an aqueous two-phase system (ATPS). A schematic illustration of the selective extraction of a lysine-rich protein using the crown ether DCH18C6 is shown in Figure 4.16. Through multiple complexation interactions between the crown ether molecules and the amino groups, the lysine-rich protein surface became more hydrophobic, and in the presence of other proteins, the crown ether-Cyt-*c* complex was selectively extracted into the more hydrophobic phase.

4.3.2 EXTRACTION OF PROTEINS IN AN IONIC LIQUID SYSTEM

Liquid–liquid extraction systems have many attractive properties, such as negligible volatility, nonflammability, high thermal stability, and controllable hydrophobicity.[69,70] Crown ethers can function as extractants in an ionic liquid system. Armstrong and coworkers reported that the positive form of some amino acids can be complexed by dibenzo-18-crown-6 and extracted into an ionic liquid phase.[71] Subsequently, Pletnev and coworkers reported that hydrophilic amino acids were quantitatively extracted into ionic liquid containing DCH18C6.[72]

Shimojo and Goto reported the extraction of Cyt-*c* from an aqueous phase into ionic liquids.[73,74] Cyt-*c* was extracted into a hydroxyl group-containing ionic liquid ([C$_2$OHmim][Tf$_2$N]) (Figure 4.17) using DCH18C6. The Cyt-*c* extraction increased with increasing concentrations of DCH18C6, and quantitative extraction was achieved when the DCH18C6 concentration was about 1000-fold larger

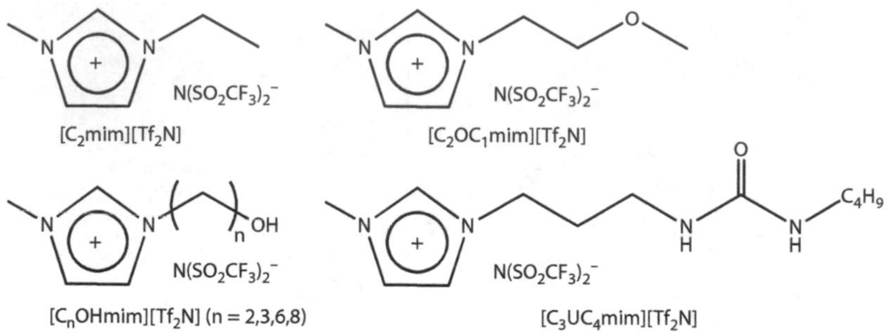

FIGURE 4.17 Molecular structures of ionic liquids for extraction of proteins. (From Shimojo, K. et al., *Biomacromolecules*, 7, 2006, 2–5.)

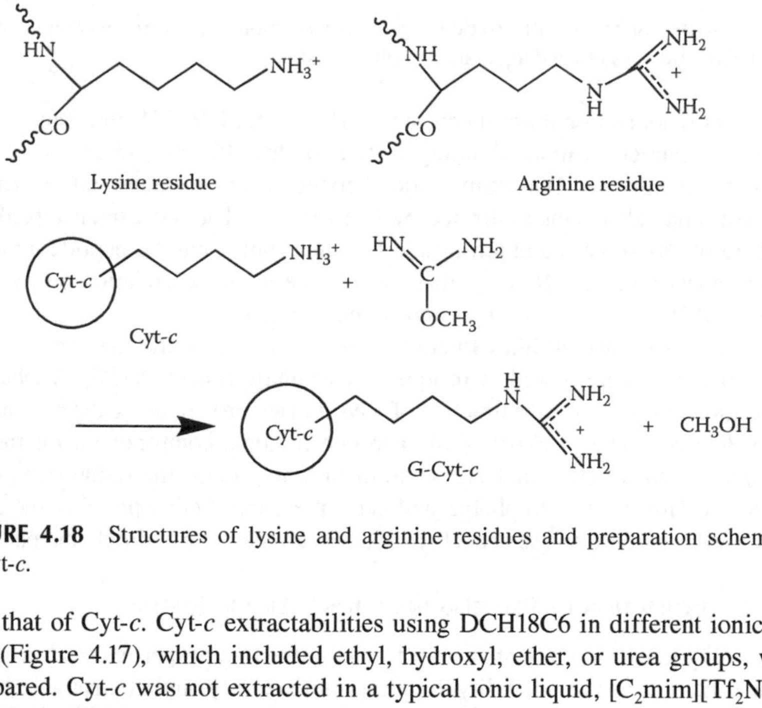

FIGURE 4.18 Structures of lysine and arginine residues and preparation scheme of G-Cyt-*c*.

than that of Cyt-*c*. Cyt-*c* extractabilities using DCH18C6 in different ionic liquids (Figure 4.17), which included ethyl, hydroxyl, ether, or urea groups, were compared. Cyt-*c* was not extracted in a typical ionic liquid, [C$_2$mim][Tf$_2$N]. In [C$_n$OHmim][Tf$_2$N] (n = 2, 4, 6, and 8), the degree of Cyt-*c* extraction decreased considerably as the alkyl spacer between the hydroxyl group and imidazolium was elongated. These results indicate that increasing the hydrophobicity of ionic liquids reduced the degree of Cyt-*c* extraction. Extraction of four heme proteins (Cyt-*c*, guanidylated Cyt-*c* (G-Cyt-*c*), myoglobin (Mb), and horseradish peroxidase (HRP)) into an ionic liquid using DCH18C6 was conducted to elucidate the participation of the protein amino group on the extraction. Lys-rich proteins Cyt-*c* and Mb have 19 lysine residues as well as 2 arginine residues (Figure 4.18). By contrast, the Arg-rich protein HRP has 6 lysine residues as well as 21 arginine residues. G-Cyt-*c* is a chemically modified protein in which 9.7 lysine residues of Cyt-c are transformed into homoarginine moieties (Figure 4.18). Cyt-c was quantitatively extracted into [C_2OHmim][Tf_2N] using DCH18C6, and Mb was moderately extracted (extractability = 64.8%). On the other hand, the degrees of extraction for G-Cyt-c (extractability = 4.3%) and HRP (extractability = 15.0%) were relatively low. The results suggested that DCH18C6 molecules could bind lysine residues, resulting in the extraction of the Lys-rich proteins.

The structural changes in Cyt-c dissolved in ionic liquids by complexation with DCH18C6 were probed using UV-vis, CD, and resonance Raman spectroscopy. The complete disappearance of the LMCT band for Cyt-c complexed with DCH18C6 in [C_2OHmim][Tf_2N] indicated the cleavage of the coordination bond of Met 80 to heme iron via the complexation with DCH18C6. The CD spectrum for the Cyt-c-DCH18C6 complex in [C_2OHmim][Tf_2N] was characterized by the disappearance of the negative Cotton effect for native Cyt-c and a concomitant increase in the intensity of the positive Cotton effect at 404 nm, suggesting a change in heme–polypeptide interactions in the vicinity of the heme active site and an increase in the planarity of the ferric heme moiety. From the results, it is likely that the sixth ligand Met 80 in the heme group of the complex between Cyt-c and DCH18C6 in the ionic liquid is replaced by other amino acid residues, and that a nonnatural, six-coordinate, low-spin ferric heme structure is induced. The supramolecular complex in [C_2OHmim][Tf_2N] has peroxidase activity and can accelerate the initial reaction rate relative to that achieved by the native Cyt-c. The k_{cat} (catalytic constant) and K_m (Michaelis constant) for the Cyt-c-DCH18C6 complex in [C_2OHmim][Tf_2N] is around 3- and 15-fold lower than the value for native Cyt-c, respectively, which results in a 5.2-fold enhancement of the k_{cat}/K_m value.

4.3.3 EXTRACTION OF PROTEINS IN AN AQUEOUS TWO-PHASE SYSTEM

An aqueous two-phase system (ATPS), also commonly known as an aqueous biphasic system, is formed by combining certain inorganic salts and aqueous solutions of water-soluble polymers in specific concentrations.[1–4] A mixture of certain water-soluble polymers or a polymer and a salt in an aqueous solution forms two immiscible phases, and the resulting biphasic medium can be used for separation of various materials. Extraction of biomolecules in ATPS generally depends on physicochemical properties such as isoelectric point, surface hydrophobicity, and molecular weight. Affinity extraction based on the interactions between a targeted molecule and a specific affinity ligand is effective for enhancing the selectivity in ATPS.[75,76] Recently, DCH18C6 was confirmed to be effective for controlling the distribution of Cyt-c in Li_2SO_4- polyethylene glycol (PEG) ATPS.[77]

In preliminary experiments, the extraction tests for Cyt-c using crown ethers were carried out in the commonly used dextran-PEG ATPS. However, the distribution of Cyt-c was not influenced by crown ethers in dextran-PEG. As the result of a number of trials, an extraction system for protein separation was realized using DCH18C6 as the ligand in Li_2SO_4-PEG ATPS. Lithium sulfate (Li_2SO_4) was selected to form the salt-rich phase because Li^+, as the smallest alkali metal ion, has much lower affinity for DCH18C6 than the other alkali ions.

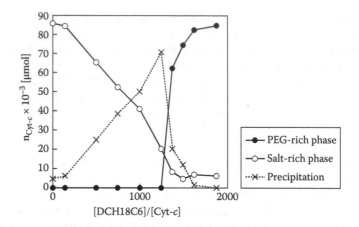

FIGURE 4.19   The effect of DCH18C6:Cyt-*c* molar ratio on the extraction of Cyt-*c* using DCH18C6 in Li$_2$SO$_4$/PEG ATPS. [Cyt-*c*] = 10 µM, [LiClO$_4$] = 15 mM, [urea] = 7.5 mM, pH$_{ini}$ = 5.0. (From Oshima, T., Suetsugu, A., and Baba, Y., *Anal. Chim. Acta*, 674, 2010, 211–219.)

Figure 4.19 shows the dependence of the amount of Cyt-*c* extracted on the DCH18C6:Cyt-*c* molar ratio in Li$_2$SO$_4$-PEG ATPS at pH 5.0. The amount of Cyt-*c* in the salt-rich phase decreased with increasing concentrations of DCH18C6 in the system. However, Cyt-*c* did not appear in the PEG-rich phase, and Cyt-*c* precipitated at the interface until the DCH18C6:Cyt-*c* molar ratio reached 1375:1, whereupon the amount of precipitated Cyt-*c* started to decrease. Cyt-*c* was almost quantitatively extracted into the PEG-rich phase in the presence of DCH18C6 at a concentration 1500 times that of Cyt-*c*. The lysine-rich protein Cyt-*c* was distributed in the hydrophobic PEG-rich phase by the multiple interactions between the DCH18C6 molecules and the lysine residues. The binding constant for Cyt-*c* and DCH18C6 in the aqueous medium, based on cation–dipole interactions, would be relatively small compared with that for Cyt-*c* and the calix[6]arene carboxylic acid ^{t}Oct[6]CH$_2$COOH in the organic medium based on electrostatic interactions. Hence, a larger relative concentration of DCH18C6 is required for the quantitative extraction of Cyt-*c* in ATPS.

Extractabilities of different crown ethers and their structural analogs (Figure 4.20) for Cyt-*c* in Li$_2$SO$_4$-PEG ATPS were compared to study the dominant structural factors necessary for extraction. Of the various crown ethers and their analogs that were investigated, only DCH18C6 was able to extract Cyt-*c* as shown in Figure 4.21. The ligands differed in cavity size and the functional groups in the side chain. The podand PGM ether is an acyclic molecule that has the same number of oxygen atoms as DCH18C6. The smaller crown ethers HM12C4, HM15C5, B15C5, 12C4, and 15C5 had no influence on the extraction of Cyt-*c*. The amount of Cyt-*c* in the salt-rich phase decreased somewhat by adding B18C6, DCH24C8, and 18C6, but Cyt-*c* was not observed in the PEG-rich phase and precipitated. From these results, it is likely that the extraction of Cyt-*c* is based

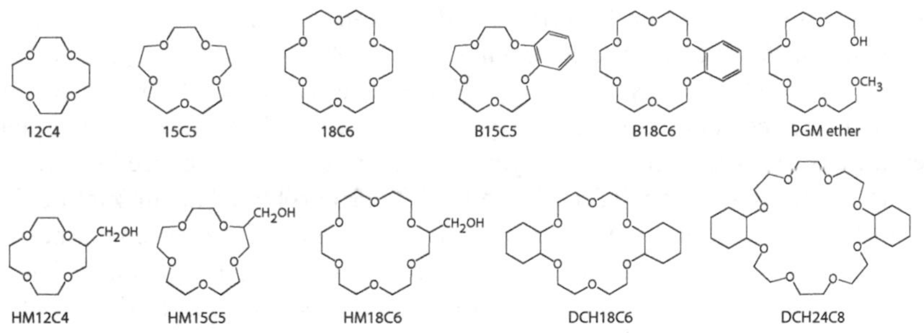

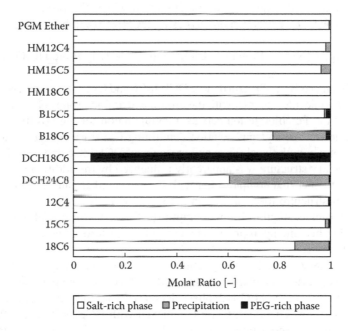

FIGURE 4.20 Molecular structures of crown ether derivatives and their analogs for Cyt-*c* extraction in ATPS.

FIGURE 4.21 Extraction of Cyt-*c* in the presence of crown ethers and their analogs in Li_2SO_4 = PEG ATPS. [Cyt-*c*] = 10 μM, [crown ether] = 18.75 mM, [$LiClO_4$] = 15 mM, [urea] = 15 mM, pH_{ini} = 5.0. (From Oshima, T., Suetsugu, A., and Baba, Y., *Anal. Chim. Acta*, 674, 2010, 211–219.)

on the complexation between DCH18C6 molecules and the protonated amino groups of Cyt-*c*. To study the effect of the complexation between DCH18C6 and a protonated amino group, extraction of various amino compounds in Li_2SO_4-PEG ATPS has also been examined.[78] The number of primary amino groups of the guest molecules as well as their hydrophobicity is a dominant factor in enhancing the extraction based on the complexation with DCH18C6. From the results of the slope analysis, the number of DCH18C6 molecules complexing with an amino compound agreed with the number of primary amino groups. Thus, the complexation between DCH18C6 and a primary amino group promotes the extraction of the guest amino compound into the PEG-rich phase in the ATPS.

In the present ATPS system using DCH18C6 as a ligand, addition of perchlorate anion to the solution is necessary to extract Cyt-c into the PEG-rich phase. Perchlorate ion is recognized as a hydrophobic anion and is likewise favorable for extraction of Cyt-c using DCH18C6 in Li_2SO_4-PEG ATPS. The amount of Cyt-c (10 µM) extracted into the PEG-rich phase increased with increasing concentrations of $LiClO_4$. In the presence of more than 15 mM $LiClO_4$, precipitation was not observed, and Cyt-c was quantitatively extracted into the PEG-rich phase. Namely, Cyt-c is transferred into the PEG-rich phase by forming a complex with DCH18C6 and perchlorate ion. As the complexation is based on cation-dipole interactions between the protein and the crown ether, the extraction proceeds under weakly acidic and neutral conditions in which the protein is positively charged. The Cyt-c extraction process using DCH18C6 can be represented by

$$n\text{CE (org)} + (NH_3^+)_n\text{-Protein} + {}_n\text{ClO}_4^- \rightleftharpoons (\text{CE} \cdot NH_3^+)_n\text{-Protein}(ClO_4^-)_n \text{ (org)}$$

where CE and $(NH_3^+)_n$-Protein denote DCH18C6 and Cyt-c, respectively. Multiple DCH18C6 molecules interact with the protonated amino groups of the lysine residues of Cyt-c. Perchlorate ion acts as a counteranion for the positively charged lysine residues. An aqueous solution containing Cyt-c was observed using CD spectroscopy in both the absence and presence of DCH18C6 and perchlorate ion. The secondary structure was apparently influenced by the addition of DCH18C6 in such a manner that complexation with the crown ether caused transformation to a higher structure for Cyt-c. From the result of analysis using reversed-phase high-performance liquid chromatography (HPLC), Cyt-c complexed with the crown ether became hydrophobic enough to be transferred into the hydrophobic PEG-rich phase.

As DCH18C6 exhibits strong affinity for potassium ion, the extraction of Cyt-c into the PEG-rich phase is influenced by the coexisting potassium ion. The amount of Cyt-c in the PEG-rich phase decreased with increasing concentrations of potassium ion. Therefore, Cyt-c complexed with DCH18C6 in the PEG-rich phase can be recovered into a salt-rich phase using K_2SO_4 by ion exchange of potassium ion and the cationic protein. Figure 4.22 shows the back-extraction profiles of Cyt-c complexed with DCH18C6 from a PEG-rich phase to a salt-rich phase as a function of potassium ion concentration. The amount of Cyt-c in the PEG-rich phase decreased with increasing concentrations of potassium ion; however, most of the protein formed a precipitate in the presence of a relatively small amount of potassium ion. The amount of precipitation decreased in the presence of higher concentrations of potassium ion, and Cyt-c was effectively stripped into the salt-rich phase. Cyt-c was quantitatively recovered from the PEG-rich phase into the salt-rich phase in the presence of 300 mM potassium ion.

The lysine-rich protein Cyt-c was selectively extracted into the PEG-rich phase over a wide pH range, conditions under which the extraction of other cationic proteins Lyso and ribonuclease A was relatively small. At the optimum conditions,

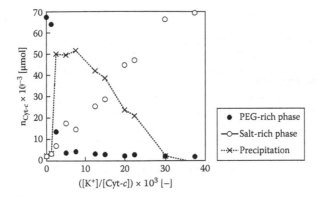

FIGURE 4.22 The effect of K+:Cyt-c molar ratio on the amount of Cyt-c back-extracted from a PEG-rich phase to a salt-rich phase. $n_{Cyt-c, \, ini} = 70 \times 10^{-3}$ μmol, $[K_2SO_4] = 0$–375 mM. (From Oshima, T., Suetsugu, A., and Baba, Y., *Anal. Chim. Acta*, 674, 2010, 211–219.)

Cyt-c could be selectively extracted using DCH18C6 over other cationic proteins. More than 93% of Cyt-c was transferred to the PEG-rich phase via complexation with DCH18C6, 18% of Lyso was transferred to the PEG-rich phase due to self-distribution, and the partition of RibA into the PEG-rich phase was quite small (0.5%).

4.4 PERSPECTIVE

As discussed herein, the macrocyclic compounds calixarenes and crown ethers are found to be effective as complexing reagents for proteins. The complexation results in the solubilization of proteins in organic solutions, which has a potential for applications in protein separation, protein refolding, and enzymatic reactions. In the case of lysine recognition, the degree of modification of the calixarene or crown ether is dominated by the number of lysine residues on the protein surface. The complexation process offers a novel selectivity for proteins based on the difference in the number of protein lysine residues.

At present, more sophisticated ligands are under development using calixarene and other platforms for more precise recognition of proteins.[79–81] Complementary multivalent interactions between ligand and protein functional groups should result in more powerful and specific complexation. Synthetic ligands that form 1:1 supramolecules with particular proteins via multiple interactions should be able to recognize the entire protein structure and modify its functionality. Such ligands would be powerful extractants and exhibit specificity to a targeted protein in a liquid–liquid extraction process.

REFERENCES

1. Walter, H., Johansson, G., and Brooks, D.E., Partitioning in aqueous two-phase systems: recent results, *Anal. Biochem.*, 197, 1–18, 1991.
2. Kula, M.R., Trends and future prospects of aqueous two-phase extraction, *Bioseparation*, 1, 181–189, 1990.

3. Hatti-Kaul, R., Aqueous two-phase systems: a general overview, *Mol. Biotechnol.*, 19, 269–277, 2001.

4. Mazzola, P.G., Lopes, A.M., Hasmann, F.A., Jozala, A.F., Penna, T.C.V., Magalhaes, P.O., Rangel-Yagui, C.O., and Pessoa Jr., A., Liquid–liquid extraction of biomolecules: an overview and update of the main techniques, *J. Chem. Technol. Biotechnol.*, 83, 143–157, 2008.

5. Pires, M. J., Aires-Barros, M. R., and Cabral, J. M. S., Liquid–liquid extraction of proteins with reversed micelles, *Biotechnol. Prog.*, 12, 290–301, 1996.

6. Goklen, K. E., and Hatton, T.A., Extraction of low molecular-weight proteins by selective solubilization in reversed micelles, *Sep. Sci. Technol.*, 22, 831–841, 1985.

7. Woll, J. M., Hatton, T. A., and Yarmush, M. L., Bioaffinity separations using reversed micellar extraction, *Biotechnol. Prog.*, 5, 57–62 1989.

8. Coughlin, R. W., and Baclaski, J. B., N-laurylbiotinamide as affinity surfactant, *Biotechnol. Prog.*, 6, 307–309, 1990.

9. Chen, J.-P., and Jen, J.-T., Extraction of concanavalin A with affinity reversed micellar systems, *Sep. Sci. Technol.*, 29, 1115–1132, 1994.

10. Adachi, M., Harada, M., Shioi, A., Takahashi, H., and Katoh, S., Bioaffinity separation of concanavalin A in reverse micellar system composed of AOT/butanol or nonionic surfactant, *J. Chem. Eng. Jpn.*, 29, 982–989, 1996.

11. Sun, Y., Ichikawa, S., Sugiura, S., and Furusaki, S., Affinity extraction of proteins with a reversed micellar system composed of Cibacron Blue-modified lecithin, *Biotechnol. Bioeng.*, 1998, 58, 58–64, 1998.

12. Zhang, T., Liu, H., and Chen, J., Affinity extraction of BSA by mixed reversed micellar system with unbound triazine dye, *Biochem. Eng. J.*, 4, 17–21, 1999.

13. Adachi, M., Yamazaki, M., Harada, M., Shioi, A., and Katoh, S., Bioaffinity separation of trypsin using trypsin inhibitor immobilized in reverse micelles composed of a nonionic surfactant, *Biotechnol. Bioeng.*, 53, 406–408, 1997.

14. Adachi, M., Shibata, K., Shioi, A., Harada, M., and Katoh, S., Selective separation of trypsin from pancreatin using bioaffinity in reverse micellar system composed of a nonionic surfactant, *Biotechnol. Bioeng.*, 58, 649–653 1998.

15. Oshima, T., Goto, M., and Furusaki, S., Complex formation of cytochrome *c* with a calixarene carboxylic acid derivative: a novel solubilization method for biomolecules in organic media, *Biomacromolecules*, 3, 438–444, 2002.

16. Oshima, T., Baba, Y., Shimojo, K., and Goto, M., Recognition of lysine residues on protein surfaces using calixarenes and its application, *Curr. Drug Discov. Technol.*, 4, 220–228, 2007.

17. Gutsche, C. D., *Monographs in supramolecular chemistry: calixarenes*, vol. 1., Royal Society of Chemistry, Cambridge, 1989.

18. Shinkai, S. Calixarenes—the third generation of supramolecules, *Tetrahedron*, 49, 8933–8968, 1993.

19. Bohmer, V., Calixarenes, macrocycles with (almost) unlimited possibilities, *Angew. Chem. Int. Ed. Engl.*, 34, 713–745, 1995.

20. Ikeda, A., and Shinkai, S., Novel cavity design using calix[*n*]arene skeletons: toward molecular recognition and metal binding, *Chem. Rev.*, 97, 1713–1734, 1997.

21. Asfari, Z., Bohmer, V., Harrowfield, J., Vicens, J., and Saadioui, M., *Calixarenes 2001*, Kluwer Academic, Dordrecht, 2001.

22. Lumetta, G. J., Rogers, R. D., and Gopalan, A. S., ACS Symposium Series, *Calixarene molecules for separations*, American Chemical Society, Washington, DC, 1999.

23. Ludwig, R., Calixarenes in analytical and separation chemistry, *Fresenius J Anal. Chem.*, 367, 103–128, 2000.

24. Haverlock, T.J., Bonnesen, P.V., Sachleben, R.A., and Moyer, B.A., Applicability of a calixarene-crown compound for the removal of cesium from alkaline tank waste, *Radiochim. Acta*, 76, 103–108, 1997.
25. Buschmann, H.-J., Mutihac, L., and Jansen, K., Complexation of some amine compounds by macrocyclic receptors, *J. Incl. Phenom.*, 39, 1–11, 2001.
26. Mutihac, L., Buschmann, H.-J., Mutihac, R.-C., and Schollmeyer, E., Complexation and separation of amines, amino acids, and peptides by functionalized calix[*n*] arenes, *J. Incl. Phenom.*, 51, 1–10, 2005.
27. Chang, S.-K., Jang, M.J., Han, S.Y., Lee, J.H., Kang, Y.S., and No, K.T., Molecular recognition of butylamines by calixarene-based ester ligands, *Chem. Lett.*, 21, 1937–1940, 1992.
28. Mutihac, L., Buschmann, H.-J., and Diacu, E., Calixarene derivatives as carriers in liquid-membrane transport, *Desalination*, 148, 253–256, 2002.
29. Chang, S. K., Son, H. J., Hwang, H. S., and Kang, Y. S., Selective transport of amino acids derivatives through calix[6]arene-based liquid membrane, *Bull. Korean Chem. Soc.*, 11, 364–365 (1990).
30. Chang, S. K., Hwang, H. S., Son, H., Yonk, J., and Kang, Y. S., Selective transport of amino acid esters through a chloroform liquid membrane by a calix[6]arene-based ester carrier, *J. Chem. Soc. Chem. Commun.*, 217–218, 1991.
31. Okada, Y., Kasai, Y., and Nishimura, J., The selective extraction and transport of amino acids by calix[4]arene-derived esters, *Tetrahedron Lett.*, 36, 555–558, 1995.
32. Oshima, T., Goto, M., and Furusaki, S., Extraction behavior of amino acids by calix[6]arene carboxylic acid derivative, *J. Incl. Phenom.*, 43, 77–86, 2002.
33. Oshima, T., Inoue, K., Furusaki, S., and Goto, M., Liquid-membrane transport of amino acids by a calix[6]arene carboxylic acid derivative, *J. Membr. Sci.*, 217, 87–97, 2003.
34. Oshima, T., Inoue, K., Uezu, K., and Goto, M., Dominant factors affecting extraction behavior of amino compounds by a calix[6]arene carboxylic acid derivative, *Anal. Chim. Acta*, 509, 137–144, 2004.
35. Shimojo, K., Oshima, T., and Goto, M. Calix[6]arene acetic acid extraction behavior and specificity with respect to nucleobases. *Anal. Chim. Acta*, 521, 163–171, 2004.
36. Oshima, T., Oishi, K., Ohto, K., and Inoue, K., Extraction behavior of catecholamines by calixarene carboxylic acid derivatives, *J. Incl. Phenom.*, 55, 79–85, 2006.
37. Oshima, T., Higuchi, H., Ohto, K., Inoue, K., and Goto, M., Protein extraction and stripping with calixarene derivatives, *J. Ion Exch.*, 14, 373–376, 2003.
38. Oshima, T., Higuchi, H., Ohto, K., Inoue, K., and Goto, M., Protein extraction and stripping with calixarene derivatives, *J. Ion Exch.*, 14, 373–376, 2003.
39. Elove, G.A., Bhuyan, A.K., and Roder, H., Kinetic mechanism of cytochrome *c* folding: involvement of the heme and its ligands, *Biochemistry*, 33, 6925–6935, 1994.
40. Fedurco, M., Redox reactions of heme-containing metalloproteins: dynamic effects of self-assembled monolayers on thermodynamics and kinetics of cytochrome *c* electron-transfer reactions, *Coord. Chem. Rev.*, 209, 263–331, 2000.
41. An, W.T., Jiao, Y., Sun, X.H., Zhang, X.L., Dong, C., Shuang, S.M., Xia, P.F., and Wong, M.S., Synthesis and binding properties of carboxylphenyl-modified calix[4] arenes and cytochrome c, *Talanta*, 79, 54–61, 2009.
42. Mohsin, M.A., Banica, F.-G., Oshima, T., and Hianik, T., Electrochemical impedance spectroscopy for assessing the recognition of cytochrome *c* by immobilized calixarenes, *Electroanalysis*, 23, 1229–1235, 2011.

43. Oshima, T., Higuchi, H., Ohto, K., Inoue, K., and Goto, M., Selective extraction and recovery of cytochrome c by liquid–liquid extraction using a calix[6]arene carboxylic acid derivative, *Langmuir*, 21, 7280–7284, 2005.

44. Nakashima, K., Oshima, T., and Goto, M., Extraction of amino acids by calixarenes in an aliphatic organic solvent, *Solvent Extr. Res. Devel. Jpn.*, 9, 69–79, 2002.

45. Oshima, T., Suetsugu, A., Baba, Y., Shikaze, Y., Ohto, K., and Inoue, K., Liquid-membrane transport of cytochrome c using a calix[6]arene carboxylic acid derivative as a carrier, *J. Membr. Sci.*, 307, 284–291, 2008.

46. Martinez-Aragon, M., Goetheer, E.L.V., and de Haan, A.B., Host-guest extraction of immunoglobulin G using calix[6]arenas, *Sep. Purif. Technol.*, 65, 73–78, 2009.

47. Shimojo, K., Oshima, T., Naganawa, H., and Goto, M., Calixarene-assisted protein refolding via liquid–liquid extraction, *Biomacromolecules*, 8, 3061–3066, 2007.

48. Zaks, A., and Klibanov, A.M., Enzyme-catalyzed processes in organic solvents, *Proc. Natl. Acad. Sci. USA*, 82, 3192–3196, 1985.

49. Klibanov, A. M., Improving enzymes by using them in organic solvents, *Nature*, 409, 241–246, 2001.

50. Oshima, T., et al., unpublished data.

51. Kobayashi, S., Uyama, H., and Kimura, S., Enzymatic polymerization, *Chem. Rev.*, 101, 3793–3818, 2001.

52. Oshima, T., Sato, M., Shikaze, Y., Ohto, K., Inoue, K., and Baba, Y., Enzymatic polymerization of o-phenylendiamine with cytochrome c activated by a calixarene derivative in organic media, *Biochem. Eng. J.*, 35, 66–70, 2007.

53. Semedo, M.C., Karmali, A., Barata, P.D., and Prata, J.V., Extraction of hemoglobin with calixarenes and biocatalysis in organic media of the complex with pseudoactivity of peroxidase, *J. Mol. Catal. B Enzym.*, 62, 96–103, 2010.

54. Pedersen, C.J., The discovery of crown ethers, *Science*, 241, 536–540, 1988.

55. Cram, D.J., The design of molecular hosts, guests, and their complexes, *Science*, 240, 760–767, 1988.

56. Izatt, R.M., Pawlak, K., Bradshaw, J.S., and Bruening, R.L., Thermodynamic and kinetic data for macrocycle interaction with cations, anions, and neutral molecules, *Chem. Rev.*, 95, 2529–2586, 1995.

57. Buschmann, H.-J., Mutihac, L., and Jansen, K., Complexation of some amine compounds by macrocyclic receptors, *J. Incl. Phenom.*, 39, 1–11, 2001.

58. Izatt, R.M., Lamb, J.D., Izatt, N.E., Rossiter Jr., B.E., Christensen, J.J., and Haymore, B.L., A calorimetric titration study of the reaction of several organic ammonium cations with 18-crown-6 in methanol, *J. Am. Chem. Soc.*, 101, 6273–6276, 1979.

59. Odell, B., and Earlam, G., Dissolution of proteins in organic solvents using macrocyclic polyethers: association constants of a cytochrome c-[1,2-14C2]–18-crown-6 complex in methanol, *J. Chem. Soc. Chem. Commun.*, 359–361, 1985.

60. Bowyer, J.R., and Odell, B., Solubilisation of ferricytochrome C in methanol using a crown ether: absorption, circular dichroism and EPR spectral properties, *Biochem. Biophys. Res. Commun.*, 127, 828–835, 1985.

61. Reinhoudt, D.N., Eendebak, A.M., Nijenhuis, W.F., Verboom, W., Kloosterman, M., and Schoemaker, H.E., The effect of crown ethers on enzyme-catalysed reactions in organic solvents, *J. Chem. Soc. Chem. Commun.*, 399–400, 1989.

62. Unen, D.-J., Engbersen, J. F. J., and Reinhoudt, D. N., Why do crown ethers activate enzymes in organic solvents, *Biotech. Bioeng.*, 77, 248–255, 2002.

63. Yamada, T., Shinoda, S., Kikawa, K., Ichimura, A., Teraoka, J., Takui, T., and Tsukube, H., Supramolecular complex of cytochrome c with lariat ether: solubilization, redox behavior and catalytic activity of cytochrome c in methanol, *Inorg. Chem.*, 39, 3049–3056, 2000.

64. Paul, D., Suzumura, A., Sugimoto, H., Teraoka, J., Shinoda, S., and Tsukube, H., Chemical activation of cytochrome *c* proteins via crown ether complexation: cold-active synzymes for enantiomer-selective sulfoxide oxidation in methanol, *J. Am. Chem. Soc.*, 125, 11478–11479, 2003.
65. Suzumura, A., Paul, D., Sugimoto, H., Shinoda, S., Julian, R.R., Beauchamp, J.L., Teraoka, J., and Tsukube, H., Cytochrome *c*-crown ether complexes as supramolecular catalysts: cold-active synzymes for asymmetric sulfoxide oxidation in methanol, *Inorg. Chem.*, 44, 904–910, 2005.
66. Julian, R.R., and Beauchamp, J.L., Site specific sequestering and stabilization of charge in peptides by supramolecular adduct formation with 18-crown-6 ether by way of electrospray ionization, *Int. J. Mass Spectrom.*, 210–211, 613–623, 2001.
67. Ly, T., and Julian, R.R., Using ESI-MS to probe protein structure by site-specific noncovalent attachment of 18-crown-6, *J. Am. Soc. Mass Spectrom.*, 17, 1209–1215, 2006.
68. Frański, R., Schroeder, G., Kamysz, W., Niedzialkowski, P., and Ossowski, T., Complexes between some lysine-containing peptides and crown ethers—electrospray ionization mass spectrometric study, *J. Mass Spectrom.*, 42, 459–466, 2007.
69. Huddleston, J.G., Willauer, H.D., Swatloski, R.P., Visser, A.E., and Rogers, R.D., Room temperature ionic liquids as novel media for 'clean' liquid–liquid extraction, *Chem. Commun.*, 1765–1766, 1998.
70. Anderson, J.L., Ding, J., Welton, T., and Armstrong, D.W., Characterizing ionic liquids on the basis of multiple solvation interactions, *J. Am. Chem. Soc.*, 124, 14247–14254, 2002.
71. Carda-Broch, S., Berthod, A., and Armstrong, D.W., Solvent properties of the 1-butyl-3-methylimidazolium hexafluorophosphate ionic liquid, *Anal. Bioanal. Chem.*, 375, 191–199, 2003.
72. Smirnova, S.V., Torocheshnikova, I.I., Formanovsky, A.A., and Pletnev, I.V., Solvent extraction of amino acids into a room temperature ionic liquid with dicyclohexano-18-crown-6, *Anal. Bioanal. Chem.*, 378, 1369–1375, 2004.
73. Shimojo, K., Nakashima, K., Kamiya, N., and Goto, M., Crown ether-mediated extraction and functional conversion of cytochrome *c* in ionic liquids, *Biomacromolecules*, 7, 2–5, 2006.
74. Shimojo, K., Kamiya, N., Tani, F., Naganawa, H., Naruta, Y., and Goto, M., Extractive solubilization, structural change, and functional conversion of cytochrome *c* in ionic liquids via crown ether complexation, *Anal. Chem.*, 78, 7735–7742, 2006.
75. Kopperschlager, G., and Birkenmeier, G., Affinity partitioning and extraction of proteins, *Bioseparation*, 1, 235–254, 1990.
76. Huddleston, J.G., and Lyddiatt, A., Aqueous two-phase systems in biochemical recovery: systematic analysis, design, and implementation of practical processes for the recovery of proteins, *Appl. Biochem. Biotechnol.*, 26, 249–279, 1990.
77. Oshima, T., Suetsugu, A., and Baba, Y., Extraction and separation of a lysine-rich protein by formation of supramolecule between crown ether and protein in aqueous two-phase system, *Anal. Chim. Acta*, 674, 211–219, 2010.
78. Kai, C., Oshima, T., and Baba, Y., Enhanced extraction of amino compounds using dicyclohexyl-18-crown-6 as a ligand in an aqueous two-phase system, *Solvent Extr. Res. Devel. Jpn.*, 17, 83–93, 2010.
79. Peczuh, M.W., and Hamilton, A.D., Peptide and protein recognition by designed molecules, *Chem. Rev.*, 100, 2479–2494, 2000.

80. Yin, H., and Hamilton, A.D., Strategies for targeting protein–protein interactions with synthetic agents, *Angew. Chem. Int. Ed. Engl.*, 44, 4130–4163, 2005.

81. Oshima, T., and Baba, Y., Recognition of exterior protein surfaces using artificial ligands based on calixarenes, crown ethers, and tetraphenylporphyrins, *J. Incl. Phenom. Macrocycl. Chem.*, 2013.

5 Interfacial Molecular Aggregation in Solvent Extraction Systems

Hitoshi Watarai
Institute for NanoScience Design,
Osaka University, Osaka, Japan

CONTENTS

5.1 INTRODUCTION

The principle of solvent extraction utilizes the difference in the solubility of molecules and ions in the aqueous phase and organic phase.[1] However, in the case where any solute molecules are hard to be solubilized to both bulk phases, the molecules tend to adsorb at the interface. Such interfacial adsorption can happen very often in the case where the molecules in a two-phase system form supramolecular compounds or molecular aggregates, even though the molecule itself can be well dissolved in either or both phases, since such high-molecular-weight compounds are less soluble in the bulk phases. It has been reported that the partition constants of solutes are primarily determined by the difference in the cavity formation energies for both phases, especially for the aqueous phase, from the calculation according to the scaled particle theory.[2–4]

The liquid–liquid interface is an extremely thin, two-dimensional mixed-liquid state, less than a few nanometers in thickness, in the absence of any adsorbed auxiliary molecules, as suggested by Benjamin.[5,6] In the interface, the properties such as cohesive energy density, electrical potential, dielectric constant, and viscosity are expected to be different from those of the bulk phase. Solute molecules adsorbed at the interface can behave like a solute in a two-dimensional solution or in a nanofilm, depending on the interfacial tension or interfacial pressure. When the concentration of the solute adsorbed at the interface increases, the interface becomes ready to be saturated, since the interfacial saturation concentration can be as low as 10^{-10} mol/cm^2. This is the reason why we can very often observe molecular aggregation at the liquid–liquid interface in solvent extraction systems. Sometimes, the interfacial molecular aggregation exhibits local inhomogeneity in the interface. The molecules in the aggregates are usually orientated, reflecting their hydrophilic and hydrophobic properties. Therefore, various specific chemical phenomena, which are rarely observed in bulk liquid phases, can be observed at liquid–liquid interfaces.[7–9] However, the mechanisms of the aggregate formation at the liquid–liquid interface and the role of the aggregate in the solvent extraction systems are still less understood. These situations are mainly due to the lack of experimental methods required for the characterization of the chemical species and aggregates adsorbed at the interface.[1,10] In the last few decades, various new measurement methods have been invented, which has brought a breakthrough in the study of interfacial reactions.[11]

Solvent extraction systems of metal ions utilize the solubility differences between two bulk phases, and accordingly, the role of the liquid–liquid interface has been thought to be less important in comparison with other factors, such as the solubility, acid dissociation constant, and stability constant, which all determine the distribution ratio of the metal ions. However, in the 1970s, researchers engaged in hydrometallurgy-activated discussions on the role of the interface in solvent extraction kinetics, since the extraction rate in the process industry is a big issue with significant economic consequences. The extractants of metal ions used in hydrometallurgy are generally all interface-active reagents, more or less.[12] The kinetic process of solvent extraction of metal ions depends intrinsically on the mass transfer across the interface. Therefore, the study of the role of the interface is very important to know the real extraction mechanism and to control the extraction rates. However, the traditional techniques used for the study of the interfacial phenomena, which included interfacial tension measurement and the Lewis cell method, were not enough to study the interfacial reaction mechanisms by clarifying the relationship between the interfacial concentration and the extraction rate. In the beginning of the 1980s, the high-speed stirring (HSS) method was invented by Watarai and Freiser.[13,14] Thereafter, various new methods were investigated, and the roles of interfacial reaction and interfacial aggregation in solvent extraction systems have been studied.

The most specific role of liquid–liquid interfaces in the solvent extraction kinetics of metal ions will be the catalytic effect of the interface on the solvent extraction rates.[15] Shaking or stirring of a solvent extraction system generates a wide interfacial area or a high specific interfacial area defined as the interfacial area divided by a bulk phase volume. Almost all extractants and auxiliary ligands used in solvent extraction are more or less interfacially active, since they have both hydrophilic and hydrophobic groups. Interfacial adsorption of the extractant or an intermediate complex at the liquid–liquid interface can accelerate the extraction rate.[7] Another specific role of the interfacial reaction is the formation of aggregates of various organic molecules or metal complexes, which sometimes results in the formation of a third phase in solvent extraction of metal ions. The adsorption of extractants and complexes at the interface also plays a critical role in coalescence. Slow phase disengagement can be just as adverse to process applications as slow kinetics. Since coalescence will be affected by the structure and properties (e.g., interfacial viscosity) of the interfacial layer, interfacial aggregation must have a big effect. Related problems are emulsion formation and entrainment or haze.

In this chapter, studies of the interfacial aggregation of molecules and complexes in model two-phase systems and metal extraction systems are reviewed from the point of view of combining with the development of the experimental methods of the aggregation at liquid–liquid interfaces. This subject is not limited to metal extraction but also relates to biological and other samples.[9]

5.2   FORMATION OF AGGREGATES AND MICELLES IN SOLVENT EXTRACTION SYSTEMS

Solvent extraction systems include not only smaller molecules, but also supramolecules or aggregates, which are produced from the extractant or metal complexes, sometimes forming micelles. Colloidal species have been found in some extraction systems with a range of aggregation numbers. Extensive reviews have been reported, such as that written by A. S. Kertes.[16]

5.2.1   AGGREGATION OF AMINES AND AMMONIUM SALTS

In 1948, Smith and Page reported the extraction of simple acid from the aqueous phase into an organic phase with high molecular amines such as methyldioctylamine, but the mechanism was not clarified at that time.[17] Allen and McDowell found anomalous dependencies of the uranium extraction ratio on tri-n-octylamine sulfate and di-n-decylamine sulfate concentrations in benzene, which were thought to be due to metastable conditions induced by the vigorous agitation customarily employed in separatory-funnel equilibrations. This observation just suggested the effect of an interfacial reaction.[18] Shanker et al. reported that the extraction of Ru(III) with amines from sulfuric acid solutions decreased in the order primary amines > secondary amines ≥ tertiary amines. Within the group of tertiary amines of tri-n-hexylamine, tri-n-octylamine, triisooctylamine, and tri-n-nonylamine, the extraction of Ru(III) decreased with increasing chain length and increased with the increase of branching in the amine. Aggregation of amines was shown to be responsible for unexpected mass-action relationships.[19] Kertes and Markovits found that tridodecylammonium salts formed oligomers in organic solvents by vapor-pressure lowering.[20] Danesi et al. also observed the aggregation of alkylammonium salts in benzene and reported aggregation constants, which depended on the anions in the order $ClO_4^- > NO_3^- > Br^- > Cl^-$.[21] They explained this order as being related to the decreasing interionic separation and corresponding decrease in the dipole moment of the ion pair. An explanation is presented for the extraction of excess nitric acid and water by trilaurylamine. The acceptable description of the excess acid and water extraction is based on the aggregation of the amine nitrate species. By vapor-pressure lowering, average values of ~1.6 in chlorobenzene and 2.3–3.3 in o-xylene have been measured, these values being independent of the aqueous nitric acid concentration between 0.5 and 6 M nitrate in dodecane, xylene, and chlorobenzene.[22] The extraction of hydrochloric, hydrobromic, and hydriodic acids by dilaurylamine into chloroform and the aggregation (dimerization) of the salts has been studied by the two-phase pH titration technique.[23] Aggregation of ligand and complex was observed in the distribution of iron(III) halides with trilaurylammonium (TLA) chloride or bromide (X). The predominant metal complexes were $TLAHFeX_4$ and $TLAHX.TLAHFeX_4$, which turned out actually to be the ion pair $[TLAH...Cl...HTLA][FeX_4]$, in addition to the various homogeneous oligomers of both TLAHX and $TLAHFeX_4$.[24] This follows from Danesi's observations, because the anions will be large, and the dipole moments of the ion pairs will also be large. The values of ΔG° at 25°C were determined for the ionization, dimerization, and trimerization of tetra-n-butylammonium picrate in different organic phases. The various values of ΔG° were related in a simple way to the bulk dielectric constant of the solvents.[25] Associated constants are reported for the aggregation of bis(3,5,5-trimethylhexyl) ammonium sulfate and hydrogensulfate in chloroform solution by vapor phase osmometry.[26]

5.2.2 Aggregation of Acidic Extractants

The aggregation number of Versatic acid (HA) in the n-hexane phase was determined as 2–6, and the principal extracted species in the extraction of sodium and strontium were $NaA\cdot2HA$, $NaA\cdot3HA$, and $SrA_2\cdot3HA$.[27] The aggregation number of dinonylnaphthalene sulfonic acid (DNNSA) was determined to be 7 by the solvent extraction of Ce(III), which suggested micelle formation.[28] The micelles of dinonylnaphthalene sulfonic acid with the aggregation number 7 in the n-heptane phase extracted Eu^{3+}, Sc^{3+}, or Zn^{2+} from the aqueous phase containing 0.50 or

1.00 M perchloric acid.[29] The extraction of magnesium ions from phosphoric acid using DNNSA was studied. The DNNSA easily formed reversed micelles in the organic phase with the aggregation number of 8.[30] The dimer formation and further aggregation of di(2-ethylhexyl)phosphoric acid (HDEHP) have been extensively studied since it was used as an effective extractant for uranium as well as the rare earth metals.[31] The average aggregation number of di(2-ethylhexyl) phosphoric acid (HDEHP) molecules in heptane was obtained as 53.34 from the light scattering measurement with 0.5 M HDEHP and 0.17 M H_2O in heptane at various angles. The result was explained in terms of reversed micelle formation of HDEHP in the heptane–water system.[32] It was shown that Cu(I), reduced from Cu(II), combined in a 1:1 stoichiometric ratio with Cyanex 272, Cyanex 302, and Cyanex 301 ligands to form multinuclear oligomeric complexes in which the ligands bridged between metal centers. Since conventional slope analysis was not possible under these conditions, the complex stoichiometries and geometries were investigated from analysis of the electronic ^{31}P-NMR (nuclear magnetic resonance) and fast atom bombardment–mass spectrometry (FAB–MS) spectra of the complex species.[33] When the Y(III) concentration was increased in the extraction with di(2-ethylhexyl)phosphoric acid (D2EHPA), the system showed a major complexity that was attributed to the formation of aggregates in the organic phase. For the highest metal concentration, a gel of polymeric metal complexes was observed.[34] In the extraction of aluminum(III) and beryllium(II) with mono(2-ethylhexyl) phosphoric acid (M2EHPA) and the mixture of M2EHPA and D2EHPA, the formation of reversed micelles was observed. The aggregation number of M2EHPA was determined in the range of 14–26 at low loading of the metals depending on the M2EHPA concentration and pH.[35] The aggregation of several metal complexes formed during solvent extraction with P,P'-di(2-ethylhexyl) methane diphosphonic acid ($H_2DEH[MDP]$) in deuterated toluene has been investigated by small-angle neutron scattering (SANS). The SANS results showed that the Ca(II)-, Al(III)-, La(III)-, and Nd(III)-$H_2DEH[MDP]$ compounds exhibited little or no tendency to aggregate. Uranyl complexes of $H_2DEH[MDP]$ at high metal loading formed small aggregates with an aggregation number of approximately 14. This behavior was in sharp contrast with the Th(IV)-$H_2DEH[MDP]$ system, which formed large aggregates in the organic phase under high metal-loading conditions.[36] When high concentrations of Nd(III) are extracted into the toluene phase having HDEHP, Cyanex 272, or Cyanex 301, all systems form dinuclear complexes with a 2:6 Nd(III):extractant ratio.[37] On the relationship between the aggregations of metal complexes with the loading ratio of metal ion, a new stoichiometric scheme was proposed in the extraction of zinc(II) and other metals with D2EHPA, which considered that the complex formed on the interface aggregates in the organic phase and released free extractant for increased extraction. The proposed mechanism successfully explained the hitherto unexplained features of metal-D2EHPA extraction systems.[38] The aggregation of 5-dodecylsalicylaldoxime and 5-nonylsalicylaldoxime in toluene and n-heptane was studied by vapor-pressure osmometry, and dimer formation was observed in toluene, but aggregates of three and four molecules were observed in n-heptane.[39]

5.2.3   AGGREGATION OF COORDINATING EXTRACTANTS AND OTHERS

Hydration and aggregation of di(2-ethylhexyl)sulfoxide in a dodecane-water system was studied.[40] Further, equilibrium analysis of aggregation behavior in the extraction of Cu(II) from sulfuric acid by didodecylnaphthalene sulfonic acid was investigated.[41] The organic phases of tri-*n*-butyl phosphate (TBP) in equilibrium with acid solutions are organized in reverse interacting aggregates, and these interactions govern the formation of the third phase. The shape and size of the aggregates were not modified by varying the acid nature or HNO_3 or TBP concentrations.[42] PrX_3-CMPO [oxide octyl(phenyl)-(*N,N*-diisobutylcarbamoyl-methyl)phosphine oxide] (X = NO_3, Cl) complexes in C_6D_6 were found to form large aggregates that have an ellipsoidal shape by small-angle neutron scattering studies.[43] Self-aggregation of Zn chlorophylls was observed in the fluorous phase more than the hydrocarbon phase.[44] In the solvent extraction of Ag^+ and Li^+ with p-*tert*-butyl thiacalix[4]arenes tetra-substituted at the lower rim by tertiary amide groups, the formation of nanosized aggregates was observed in the organic phase with the degree of the extraction more than 67%.[45] The instability leading to the appearance of a third phase during the extraction of nitric acid by malonamide extractants in various diluents had a supramolecular origin: it is a "long range" attraction between polar cores of the reversed micelles formed by the extractants.[46] The aggregation of *N,N'*-dimethyl-*N,N'*-dioctylhexylethoxymalonamide (DMDOHEMA) in *n*-heptane was investigated by vapor-pressure osmometry, and small-angle neutron and X-ray scattering (SANS and SAXS, respectively). Two approaches were taken to model the aggregation of the diamide and the water extraction as a function of the diamide concentration: (1) a single aggregation equilibrium with an average aggregation number of 4.28 ± 0.05, and (2) a competition equilibrium between two types of aggregates, namely, aggregates of the reversed micelle type with 4 diamides per aggregate, and an oligomeric structure composed of about 10 diamide molecules, which appears at high extractant concentration (0.1 mol/L). In both cases, the supramolecular speciation representing the monomers–aggregates distribution was determined. The larger aggregates could extract about five times more water than monomers.[47] The aggregation tendency of *N,N,N',N'*-tetraoctyl diglycolamide (TODGA) in different acids followed the order $HClO_4$ > HNO_3 > HCl, which was similar to the extraction pattern of trivalent actinides/lanthanides in these acids. The presence of Eu(III) ions facilitated the aggregation of TODGA. The aggregation studies revealed that about a 2 nm aggregate size of ligand was the critical size required for efficient extraction of trivalent actinide and lanthanide ions.[48] It was reported that the extracted solute on the aggregation of malonamide extractant as reversed micelles in organic phases led to third-phase formation.[49] Third-phase formation of neodymium(III) and nitric acid with unsymmetrical *N,N*-di-2-ethylhexyl-*N',N'*-dioctyldiglycolamide was observed.[50] The coagulation of reversed micelles of *N,N'*-dimethyl-*N,N'*-dibutyltetradecylmalonamide (DMDBTDMA) into worm-like aggregates in the *n*-dodecane phase was observed by liquid–liquid extraction following contact with acidic and neutral aqueous media containing a Ce(III) ion.

With the SAXS method, the growth of solute architectures was shown to prelude the formation of a third phase. The presence of acid was shown to promote the growth of these micellar chains, and therefore promoted third-phase formation.[51] The formation of a third phase was observed in some conditions in the process with dimethyldibutyltetradecylmalonamide (DMDBTDMA), a potential extractant used in the DIAMEX process. SAXS studies showed that the extractant molecules of DMDBTDMA self-assembled into small aggregates in the organic phase, when the organic phase was contacted with an aqueous phase (acidic or not). These aggregates typically have a polar core of ca. 6–7 Å radius. Within the organic phase, the aggregates were subject to three major interactions: (1) the destabilizing van der Waals interaction, (2) the stabilizing hard-sphere repulsion, and (3) a repulsive steric contribution from the aliphatic chains of the extractant molecules due to the remaining 80% of the volume of the chains. The observable macroscopic effect, namely, the formation of a third phase, was thought to be the macroscopic translation of the effect of these three interactions acting at a microscopic level.[52] Recently, the application of supramolecular solvents, which are surfactant-rich and self-assembled nanostructured liquids, to an analytical extraction and concentration scheme of organic compounds was reviewed.[53] Supramolecular solvents were successfully applied to the extraction of polycyclic aromatic hydrocarbons (PAHs) pesticides, surfactants, bioactive compounds, dyes, endocrine disruptors, etc., and are further promising to offer advantages in extraction efficiency and rapidity, which are suited for miniaturization of the extraction system.

5.3 FORMATION OF MICROEMULSIONS AND MICRODROPLETS IN SOLVENT EXTRACTION SYSTEMS

The cloud point extraction (CPE) method was invented by Watanabe et al. in 1976[54]; it was thought to be a rapid and simple separation and extraction technique as an alternative to organic solvents. An extension of the cloud point method was reported with the name of cold-induced aggregation microextraction (CIAME), which utilized the ionic liquid as an extracting medium, and it was applied to preconcentrate cobalt(II) ions from water samples as a step prior to its spectrophotometric determination with 1-(2-pyridylazo)-2-naphthol (PAN). Very small amounts of 1-hexyl-3-methylimidazolium hexafluorophosphate ([Hmim][PF$_6$]) and 1-hexyl-3-methylimidazolium bis(trifluoromethylsulfonyl) imide ([Hmim] [Tf$_2$N]) as hydrophobic ionic liquids and extractant solvents were dissolved in the sample solution containing Triton X-114 (antisticking agent). Then, the solution was cooled in the ice bath and a cloudy solution was formed. After centrifuging, the fine droplets of the extractant phase had settled to the bottom of the conical-bottom glass centrifuge tube, and were then collected for analysis.[55] In the CIAME spectrophotometric analysis of mercury with Michler thioketone (TMK), a limit of detection of 0.3 ng mL^{-1} was attained.[56] In the analysis of gold, sodium hexafluorophosphate (NaPF$_6$) was added to the sample solution containing a Au-TMK complex and a very small amount of 1-hexyl-3-methyl-imidazolium tetrafluoroborate [Hmim][BF$_4$]. Under the optimum conditions, the limit of detection was 0.7 ng mL^{-1} and the relative standard deviation was 1.65% for 50 ng mL^{-1} gold.[57]

Aggregation and cloud point (CP) behavior, as well as CP extraction of lanthanide ions, has been studied for novel nonionic cyclophanic surfactants with the varied length of polyoxyethylene and hydrophobic moieties (CnEm) based on the calix[4]arene platform in their mixtures with Triton X100 (TX100). The cyclophanic structure of CnEm was determined as the key reason of the formation of large lamellar-like aggregates with TX100, exhibiting the unusual CP behavior and CP extraction efficiency.[58] A novel method, termed ionic liquid cold-induced aggregation dispersive liquid–liquid microextraction, combined with high-performance liquid chromatography (HPLC), was developed for the determination of three phthalate esters in water samples. Good relative recoveries for phthalate esters in tap, bottled mineral, and river water samples were obtained in the ranges of 91.5–98.1%, 92.4–99.2%, and 90.1–96.8%, respectively.[59]

When concentrated ammonia (>2 M) in water was used for neutralization of a water(NH$_3$)–Versatic 10–isooctane system, a one-phase microemulsion was observed at the phase inversion pH. The one-phase region where maximal solubilization took place was at 50% degrees of neutralization.[60] Aggregation, reversed micelles, and microemulsions in liquid–liquid extraction systems were reviewed by Osseo-Asare, who focused on the tri-n-butylphosphate diluent-water-electrolyte systems.[61]

5.4 MOLECULAR AGGREGATION AT LIQUID–LIQUID INTERFACES

5.4.1 STUDIES BY CONVENTIONAL METHODS

The experimental methods applicable to the measurements of interfacial aggregation reaction are not numerous. The most traditional method is the interfacial tension measurement. It is not simply proportional to the interfacial concentration of the adsorbed molecules, but it should be analyzed by the Gibbs adsorption isotherm equation to obtain the interfacial concentration, termed the interfacial excess, since the interfacial tension is defined as the Gibbs free energy per unit area. Nevertheless, it is a useful technique to measure the extent of interfacial adsorption and determine the aggregation or micelle formation at the interface and in the bulk phases. The equation of the Gibbs adsorption isotherm is represented by

$$\Gamma = -\frac{1}{RT}\frac{\partial\gamma}{\partial\ln C} \tag{5.1}$$

where Γ is the surface (or interface) excess, which can be thought of as the interfacial concentration (mol/cm^2), γ is the interfacial tension (dyne/cm or mN/m), and C is the equilibrium concentration of the adsorbing species in

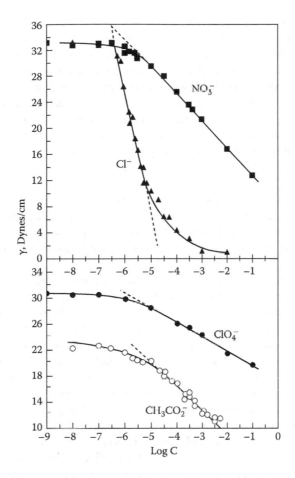

FIGURE 5.1 Gibbs adsorption isotherm plot of interfacial tension vs. log concentration for o-xylene solutions of tridodecylmethylammonium salts against 2 M aqueous solutions of their corresponding acid.

a bulk phase (mol/L). Horwitz et al. measured Gibbs adsorption isotherms for tridodecylmethylammonium and trioctylmethylammonium salts in an o-xylene/2 M acid aqueous solution phase at 25°C.

The effect on interfacial activity from changing dodecyl to octyl groups was experimentally imperceptible, but salts with different anions gave substantially different Gibbs curves. As shown in Figure 5.1, the interfacial concentration of tridodecylmethylammonium ion measured from the slopes was found to increase in the order $ClO_4^- < CH_3CO_2^- < NO_3^- < Cl^-$.[62] In the Cl^- system, an aggregation of the ammonium ion is observed as the deviation from the linearity. Surface aggregation of calcium di(2-ethylhexyl)phosphate was investigated by the air–water surface tension method and by electron microscopy observation. These aggregates ranged in diameter from 12 to 25 nm and, on the average, were about

3 nm in height.[63] The extraction selectivity of HDEHP for the different metal ions investigated could be arranged in decreasing order as $Co^{2+} > Mg^{2+} > Ni^{2+}$, $Ca^{2+} > Sr^{2+} > Ba^{2+}$, $Cu^{2+} > Na^+$. Interfacial tension results suggested the formation of reversed micelles in HDEHP–n-hexane–aqueous metal chloride systems under certain conditions. Reversed micelles appear to play an important role in enhancing the extraction of metal ions from an aqueous phase to a nonpolar organic phase through substrate solubilization in the micellar phase.[64] The association microstructures, which form in the hydrometallurgical extraction of nickel from nitrate media of ionic strength 0.25 M by di(2-ethylhexyl)phosphoric acid (HDEHP) in n-heptane, were characterized under constant high loading conditions. The metal–extractant aggregates were cylindrical reversed micelles containing between five and six water molecules per HDEHP monomer. Furthermore, it appeared that the metal–extractant aggregates were mixed reversed micelles consisting of predominantly nickel-HDEHP and to a lesser extent NaDEHP. The water in the metal–extractant reversed micelles was found to exist mainly in three different states: free water and bound water in the aqueous core of the aggregates, and trapped water in the interfacial layer among the hydrocarbon chains of the extractant molecules near the micellar core.[65] The aggregation of amine-type extractants was studied in sulfuric acid media and applied to systems containing molybdenum, tungsten, and rhenium. Diisododecylamine (DIDA) and trioctylamine (TOA) form reversed micelles in nonpolar and low-polar diluents when a certain concentration of water and acid is available in the organic phase. The radius of the micelle can be as large as 2 nm, depending on the experimental conditions. The aggregation number was found to be as large as 126. Acid ions and metal anions are situated in the water pool of aggregates, whereby metal anions are coordinated to the polar amine head groups.[66] The aggregation state of a variety of fluorinated and nonfluorinated alkylphenoxy alcohols and 1-decanol in n-heptane was determined by osmometry at 25°C. These alcohols were found to form aggregates containing between three and five (most frequently four) molecules.[67] A description of ion extraction equilibria was rationalized using the adsorption isotherm approach. All ions present in the two-phase systems equilibrate between their hydrated state and "adsorbed" states. Considering extraction equilibrium as a sum of adsorption isotherms corresponding to the different states of aggregation of the extractant molecule in the solvent, the resulting constant was derived as representative of both the efficiency of the extraction and the structure of the solution.[68] Goto et al. have found that a calix[6]areneacetic acid derivative forms a supramolecular complex with urea-denatured cytochrome c at the oil–water interface, which enables quantitative transfer of the protein from an 8 M urea aqueous solution into an organic phase through a proton-exchange mechanism.[69]

5.4.2 STUDIES BY NEWLY DEVELOPED EXPERIMENTAL METHODS

Many methods have been used to investigate the chemical reaction and kinetics in the liquid–liquid systems, which include a Lewis cell, a single-drop method, and a rotating-disc method.[70] All of these methods, however, could not simultaneously measure both the extraction rate and interfacial concentration of extractant until the high-speed stirring method succeeded in determining the interfacial amount of extractant under the vigorous stirring.[13,14] After that time, various experimental methods were developed for the measurements of the interfacial reaction rate, the interfacial species, the interfacial concentration, and the interfacial chirality.

5.4.2.1 High-Speed Stirring (HSS) Method

The HSS method made it possible to measure both an interfacial concentration and an extraction rate for the first time by a simple principle (Figure 5.2). When a two-phase system is highly stirred or agitated in a vessel, the interfacial area is extremely extended, increasing the interfacial amount of the adsorbed molecule, and therefore the bulk concentration in the organic phase is decreased. From the spectrophotometric measurement of the change of the organic phase concentration of solute molecule, the interfacial amount of the lost solute from the organic phase was determined.[13] The interfacial area in the stirred system is about 500 times larger than that in a standing condition. Therefore, the specific interfacial area, defined by an interfacial area divided by a volume of a bulk phase, in a HSS condition attains as high as 400 cm^{-1}. The saturated interfacial concentration of an ordinary compound is the order of 10^{-10} mol/cm^2. Therefore, when a 50 mL organic phase–50 mL aqueous phase system is vigorously stirred and the interface is saturated by solute molecules, the total amount of the interfacial molecule corresponds to the bulk phase concentration change of 10^{-4} M, which is readily detectable by conventional spectrometry. Using HSS spectroscopy, the species adsorbed from the organic phase to the interface can be identified from the spectrum, and the consumption rate of the extractant and the formation rate of the extracted complex can also be measured simultaneously. The first-order rate

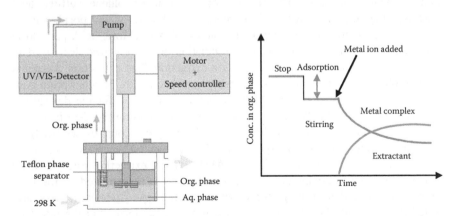

FIGURE 5.2 Schematic drawing of the high-speed stirring (HSS) apparatus. The increase in the interfacial area by the stirring can increase the amount of the interfacial molecules, which results in a decrease in the organic-phase concentration. The diagram on the right shows a schematic presentation of an observed concentration change.

equation for the extraction rate of Ni(II) with 2'-hydroxy-5'-nonyl-acetophenone oxime (LIX65N) in n-heptane and the interfacial amount of LIX65N was confirmed experimentally.[71] The measurements were carried out by employing 50 mL for each phase at the stirring speed of 5000 rpm in a high-speed stirring condition and 200 rpm in a static condition. The flow rate circulating the organic phase through a polytetrafluoroethylene (PTFE) phase separator and a flow cell in a diode-array spectrophotometer was 10–20 mL/min. Though the HSS method cannot measure the interface directly, it is very useful for the determination of the adsorptivity of organic phase solutes in dilute concentration and for the study of extraction kinetics, because of its high sensitivity and selectivity with the use of spectrophotometry. The interfacial adsorptivity of seven β-diketones, including five trifluoromethyl derivatives, was measured in heptane–water systems by the HSS method. Preferential interfacial adsorption of the corresponding enolate ions was observed in all the systems. The interfacial concentration of the enolate form was governed by the Langmuir isotherm between the interfacial and the aqueous concentrations of the enolate form. Therefore, the interfacial concentration was also governed by the partition constant and the acid dissociation constant of the β-diketones.[72] The kinetic synergistic effect of 1,10-phenanthroline (phen) on the extraction rate of Ni(II) with dithizone (HDz) into chloroform[73] was studied by the HSS method combined with photodiode-array spectrophotometry. The initial extraction rate of the adduct complex $NiDz_2phen$ depended upon the concentrations of both HDz and phen, suggesting the formation of $NiDzphen^+$ as the rate-controlling step. When [HDz] < [phen], the initial extraction of $NiDz_2phen$ competed with the formation of an intermediate complex, which was adsorbed at the interface and assigned most probably to $NiDzphen_2^+$. The intermediate complex was gradually converted to $NiDz_2phen$, the extracted species.[74] Ion-association adsorption of diprotonated tetraphenylporphyrin and monoprotonated octaethylporphyrin was observed in the toluene–sulfuric acid systems, suggesting the interfacial aggregation of the protonated porphyrins in the higher concentration. In other acid systems of hydrochloric acid, perchloric acid, and trifluoroacetic acid, the ion-association adsorption was decreased in this order, accompanied by ion-association extraction. The extraction constants of the protonated porphyrins were inversely correlated with the adsorption constants of the protonated species.[75] The structure isomers of diazine derivatives were recognized by the interfacial reaction with a palladium(II)-2-(5-bromo-2-pyridylazo)-5-diethylaminophenol(5-BrPADAP)-chloride complex (PdLCl) at the toluene–water interface applying the HSS method.[76] The catalytic effect of N,N-dimethyl-4-(2-pyridylazo)aniline (PADA) on the extraction rate of Ni(II) with 1-(2-pyridylazo)2-naphthol (Hpan) in toluene was investigated by the HSS method. The rate-determining step of the extraction rate of $Ni(pan)_2$ was suggested to be the interfacial reaction between $Ni(pada)_2^+$ formed at the interface and Hpan from the organic phase, followed by a successive fast ligand substitution reaction with another Hpan.[77] The interfacial adsorption behavior of 2-hydroxy-5-nonylbenzophenone oxime (LIX65N) at the heptane–water system was examined by the HSS method, changing the pH of the aqueous phase over a wide range.[78]

A reversible decrease of the organic-phase concentration was observed due to the stirring in the acidic condition at the pH of 3, suggesting the interfacial adsorption of the neutral species of LIX65N. However, at a high pH of 12, the organic-phase concentration due to the stirring exhibited a slow and large decrease. When the stirring was stopped, the recovery of the interfacial compound to the organic phase was very slow, suggesting a metastable interfacial aggregation of the acid dissociation form of LIX65N. Interfacial adsorption equilibrium constants of a μ-oxo dimer of (5,10,15,20-tetraphenylporphyrinato)iron(III) and those of other metal–porphyrinato complexes were determined by the HSS method.[79,80]

5.4.2.2 Centrifugal Liquid–Membrane (CLM) Method

Both requirements of a high specific interfacial area and a direct spectroscopic observation of the interface were attained by the centrifugal liquid–membrane (CLM) method shown in Figure 5.3.[81] About 100 μL volumes of organic and aqueous phases are introduced into a cylindrical glass cell with a diameter of 19 mm. When the cell was rotated at the speed of 7000–10,000 rpm, a stacked two-liquid–membrane, each with a thickness of 50–100 μm, was produced inside the cell wall, which attained the specific interfacial area over 100 cm⁻¹. UV-vis spectrophotometry was used for the measurement of the interfacial species as well as those in the bulk phases. This method can measure the interfacial reaction rate as fast as

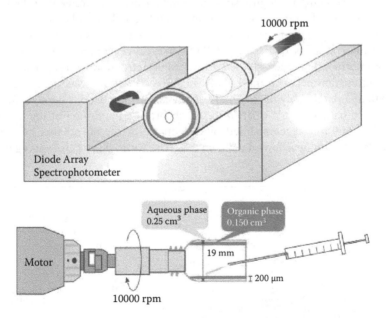

FIGURE 5.3 Schematic illustration of the centrifugal liquid–membrane (CLM) apparatus (top). The glass cylinder cell (i.d. = 19 mm, bottom figure) is rotated at a speed of 7000–10,000 rpm. Introduction of 50–200 μL of the organic and the aqueous phase generated two-liquid layers with a thickness of 50–100 μm for each phase. Reagents can be injected from a small hole at the end side of the cell.

on the order of seconds. In the first paper of CLM method,[81] the equilibrium and
kinetics of the protonation of 5,10,15,20-tetraphenylporphyrin (H_2TPP) and the
demetalation of 5,10,15,20-tetraphenylporphyrinatozinc(II) (ZnTPP) at the dodecane–aqueous acid interface were investigated. The consumption of H_2TPP in the
bulk dodecane phase and the production of the diprotonated aggregate, $(H_4TPP^{2+})_n$,
adsorbed at the liquid–liquid interface, were directly measured from the spectral
change. The observed rate constant of the demetalation of ZnTPP depended upon
the first order of the acidity function, suggesting that the rate-determining step
is the formation of the monoprotonated intermediate, $[ZnTPPH]^+$, at the interface. The demetalation rate constant of ZnTPP was determined as $(8.6 \pm 1.3) \times
10^{-5}$ dm^3 mol^{-1} s^{-1} at 298 K. The aggregation rate of H_4TPP^{2+} was governed by
the molecular diffusion of H_2TPP in the bulk dodecane phase. The aggregation
rate of the diprotonated species, H_4TPP^{2+}, at the dodecane–sulfuric acid solution
interface was measured by a two-phase stopped flow method.[82] From the kinetic
measurements, it was found that the stagnant layer of 1.4 µm still existed in the
dodecane phase side of the droplet interface, even under the highly dispersed
system. In the CLM method, the liquid–membrane phase of 50–100 µm thickness behaved as a stagnant layer where the H_2TPP molecule migrated according
to its self-diffusion rate, followed by a rapid protonation and aggregation at the
interface.

The CLM method is highly advantageous for the measurement of the formation of interfacial aggregates of dyes and metal complexes when the aggregates
are formed only at the interface and exhibit characteristic spectral changes. The
aggregation kinetics of the nickel(II)-5-Br-PADAP complex at the heptane–
water interface has been studied by CLM spectrophotometry.[83] The interfacial aggregation between the ligands in NiL_2 complexes was observed as the
growth of a remarkably intense and narrow absorption band (J band) at 588 nm,
showing a bathochromic shift from the absorption maximum of the monomer
complex (569 nm), accompanied by a decrease of the absorbance of the free
ligand at 452 nm, as shown in Figure 5.4. The formation of the aggregate was

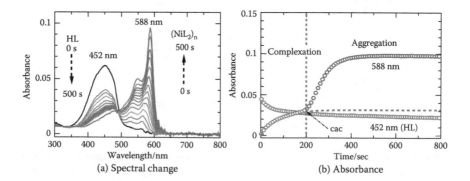

FIGURE 5.4   (a) Spectral change of the interfacial aggregation of Ni(II)-5-Br-PADAP.
(b) Critical aggregation concentration (cac) observed from the absorbance change at 588 nm.

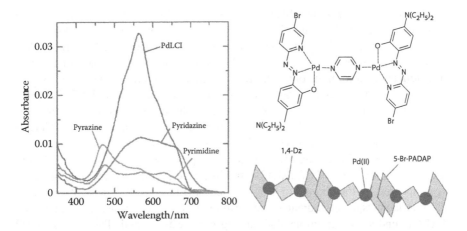

FIGURE 5.5   Absorption spectra of PdLCl and the interfacial aggregates of ternary PdL-diazine complexes measured by the CLM method. $[PdLCl]_{init} = 5.6 \times 10^{-5}$ M, [pyridazine] $= 2.0 \times 10^{-4}$ M, [pyrimidine] $= 8.0 \times 10^{-3}$ M, [pyrazine] $= 8.0 \times 10^{-5}$ M, $[ClO_4^-] = 0.1$ M, $[Cl^-] = 0$ M, pH 2.0. The right illustrations show a probable unit and structure of the membrane-like interfacial aggregate formed from the pyrazine complexes.

initiated at the time when the interfacial concentration of the monomer complex attained the critical aggregation concentration (cac) of 2.4×10^{-10} mol cm^{-2}. The initial formation rate of the interfacial aggregate was proportional to the concentrations of both free Ni(II) and free ligand after attaining the cac. The observed rate constant for the aggregation, 4.1×10^1 M^{-1} s^{-1}, was smaller than the formation rate constant of the NiL$_2$ complex at the interface. Interfacial aggregate formation was also observed for the extraction of Fe(II) and Fe(III)-5-Br-PADAP complexes.[84]

The formation of the interfacial aggregates of PdL (L = 5-Br-PADAP) and diazine isomer complexes was investigated by CLM–UV-vis spectroscopy, CLM–resonance Raman spectroscopy, and optical microscopic images. The high molar absorptivity of the PdLCl complex (4.33×10^4 M^{-1} cm^{-1} at 564 nm) was drastically reduced by the formation of the interfacial aggregates with a diazine isomer, giving different-shaped spectra to each other (Figure 5.5). The ability for the formation of the interfacial aggregate of the PdL-diazine isomer complexes increased in the order 1,3-diazine < 1,2-diazine < 1,4-diazine.[85] It is interesting that the interfacial aggregate can discriminate the small difference in the molecular structure of isomers. This will give us an idea that the interfacial aggregate can be used as a new supramolecular analytical reagent.

CLM-UV-vis spectrometry can measure directly the spectra of the interfacially formed complexes, but for the determination of the structure of the unknown interfacial complex, other information is required. To complement this drawback of CLM spectrometry, a mass spectrometry technique was applied to the liquid–liquid systems. A two-phase microflow system was directly introduced into an electrospray ionization (ESI) MS instrument, and the molecular weight of the complexes formed in the interface of the two-phase flow system was observed.[86] The inside silica capillary, having a tip with 7 μm i.d., generated small droplets of the toluene phase containing 5-Br-PADAP (HL) into the aqueous phase flow (including Cu^{2+} ion) in the outer stainless capillary with 0.2 mm i.d. The ESI-MS spectra showed clearly the formation of a 1:1 CuL complex and 1:2 CuL(HL) complex, as well as a Cu_2L_2 complex. The two-phase sheath flow technique was applied to the analysis of the interfacial aggregation mechanism of the Cu(II)-5-(octadecyloxy)-2-(2-thiazolylazo)phenol (TARC18) complex at the heptane–water interface.[87] When the pH of an aqueous phase was increased from 4 to 6, the 1:1 complex of Cu(II)-TARC18 at the interface further formed an aggregate, accompanied by the change of UV-vis spectra (Figure 5.6). The MS spectra of the interfacial species indicated the formation of a 2:3 complex for Cu(II) and TARC18 under the conditions that the aggregate was formed. Then, the spectral change due to the interfacial aggregation was analyzed by the reaction scheme including the 2:3 complex. Furthermore, the addition of a purine base of adenine or guanine into the system resulted in the disruption of the aggregate by the formation of a new three-element complex of 1:1:1 for Cu(II), TARC18, and the base, showing a bathochromic shift in the spectra. Thus, the interfacial molecular recognition ability of the Cu(II)-TARC18 aggregate for the hydrophobic bases was found.

The CLM method can also be combined with various kinds of spectroscopic methods. The fluorescence lifetime of the interfacially adsorbed zinc-tetraphenylporphyrin complex quenched by methylviologen at the dodecane–water interface was observed by a nanosecond time-resolved laser-induced fluorescence method.[88]

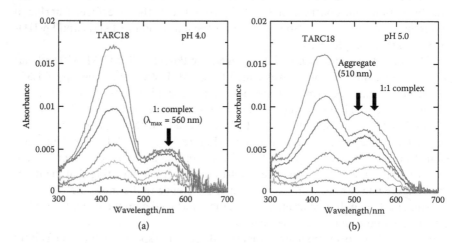

FIGURE 5.6   (a) Interfacial formation of Cu(II)-TARC18 complex at pH 4.0 and (b) the interfacial aggregation of Cu(II)–TARC18 complex at pH 5.0 observed by the CLM method. Heptane phase: 150 mL, $[TARC18]_T = 3.3 \times 10^{-6} - 4.0 \times 10^{-5}$ M. Aqueous phase: 250 mL, $[Cu(II)]_T = 1.0 \times 10^{-3}$ M, $[ClO_4^-] = 0.1$ M, and $[MES] = 1.0 \times 10^{-3}$ M.

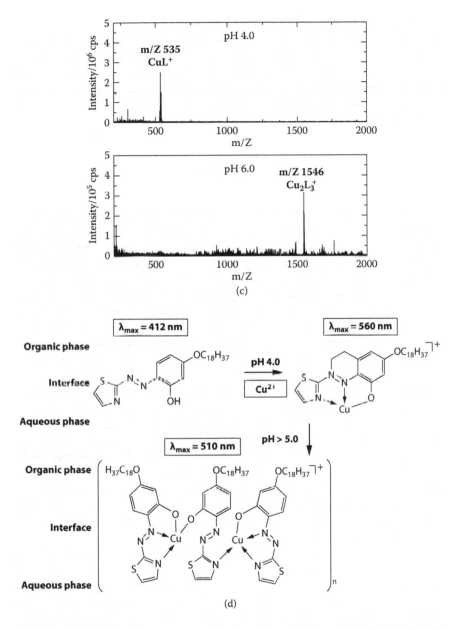

FIGURE 5.6 (*Continued*)   (c) Direct MS observation of the interfacial Cu(II)–TARC18 complex by a two-phase micro-flow API-MS method. The monomer complex was observed at pH 4.0, but the aggregate was detected at pH 6.0. Heptane phase: $[TARC18]_T = 5.0 \times 10^{-4}$ M; flow rate 0.015 mL/h. Aqueous phase: $[Cu(II)]_T = 1.25 \times 10^{-4}$ M; $[CH_3COO^-] = 1.0 \times 10^{-3}$ M; flow rate 0.2 mL/h. (d) Expected reaction scheme of interfacial aggregation.

5.4.2.3   Total Interfacial Reflection Spectroscopy

The refractive index of the organic phase is usually higher than that of the aqueous phase. Therefore, an incident light beam irradiated from the organic phase side to the interface can be totally reflected at the interface. In the total internal reflection (TIR) condition at the liquid–liquid interface, one can observe an interfacial reaction in the evanescent layer in the aqueous phase side with a thickness of ca. 100 nm calculated by

$$d = \frac{\lambda}{4\pi\sqrt{(n_1^2 \sin^2 \theta_1 - n_2^2)}} \qquad (5.2)$$

where d is the penetration depth, λ the wavelength of the beam irradiated to the interface from phase 1 to phase 2, n_1 and n_2 the refractive indexes of phase 1 and phase 2 $(n_1 > n_2)$, and θ_1 the incident angle, which is set larger than the critical angle θ_c $(= \sin^{-1}(n_2/n_1))$. Fluorometry is the most sensitive method to detect the interfacial species[89] and its dynamics.[90] Time-resolved laser spectrofluorometry is a powerful tool for the elucidation of rapid dynamic phenomena at the interface.[91] The aggregation of the Pd(II)-tetrapyridylporphine (tpyp) complex at the toluene–water interface was observed by the fluorescence decay time measurement of tpyp, since the fluorescence of the tpyp ligand is quenched by coordination with the Pd(II) ion. A toluene solution of tpyp was contacted with a $PdCl_2$ aqueous solution under an acidic condition. Pd(II) is bound to the nitrogen atoms of the pyridyl group, not to the central pyrrole nitrogens of tpyp. Depending on the concentration ratio of Pd(II)/tpyp, two types of aggregates were observed: an aggregate (AS1) observed in a lower tpyp concentration had a 3:1 composition in a Pd/tpyp ratio with a shorter fluorescence lifetime of 0.15 ns, and another one (AS2), observed in a higher tpyp concentration, had a 1:1 composition with a longer fluorescence lifetime of 1.1 ns. The 1:1 aggregate (λ_{max} = 668 nm) showed a red shift of 12 nm from the 3:1 aggregate (λ_{max} = 656 nm), and suggested a π stacking interaction in the 1:1 aggregate (Figure 5.7).[92] By considering the result obtained by a rotational relaxation measurement, the structures of AS1 and AS2 were postulated as shown in Figure 5.7.

Time-resolved total internal reflection fluorometry was extended to probing the single molecule at the interface. The first observation of the single-molecule detection at the liquid–liquid interface has been accomplished by means of total internal reflection fluorescence microscopy.[93,94] The apparatus consisted of an inverted microscope, an oil immersion objective (60×, NA 1.4), a cw-Nd:YAG laser (532 nm), and an avalanche photodiode detector (APD). A pinhole, 50 μm in diameter, was attached just in front of the photodiode to restrict the observation area down to a diameter of 830 nm (d_{obs}). p-Polarized laser light was irradiated to the interface in a flat two-phase microcell at an angle of incidence of 73° from the vertical line, which was larger than the critical angle. Fluorescence emitted by interfacial molecules was collected by the objective lens and focused on the pinhole after passing through a band-pass filter (path range, 587.5–612.5 nm).

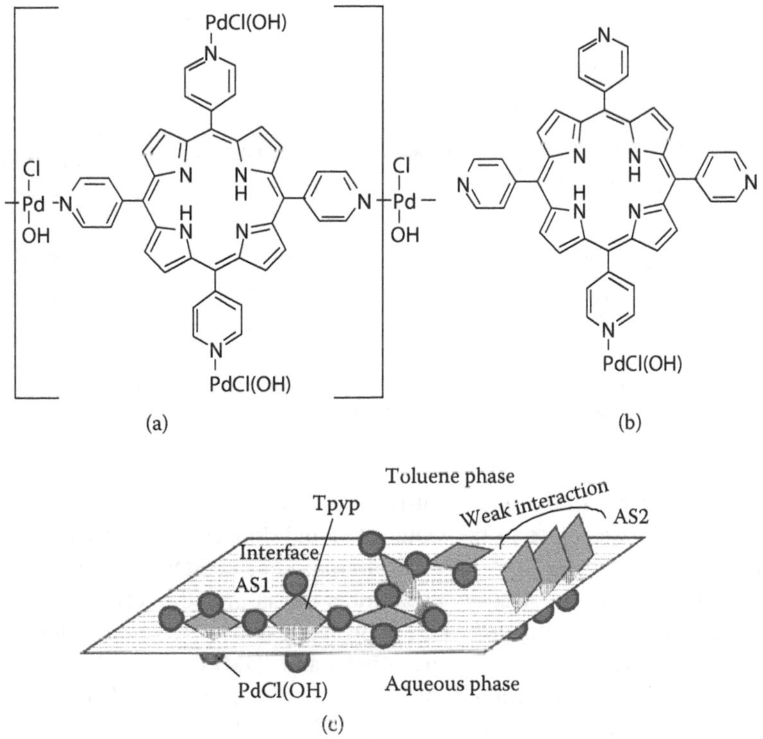

FIGURE 5.7 Possible structures of the aggregates of Pd(II)-tpyp complex suggested from the quenching effect of Pd(II) on the fluorescence life time of tpyp: (a) A unit of AS1, (b) a unit of AS2, and (c) a schematic illustration of AS1 and AS2 aggregates at the interface.

Time-resolved photon counting was carried out using a multichannel scalar. The overall detection efficiency of fluorescence at the interface was calculated to be 3.3%. The single-molecule detection at the liquid–liquid interface having a supramolecular monolayer of a surfactant was carried out by using 1,1′-diocta decyl-3,3,3′,3′-tetramethylindocarbocyanine (DiI) as a fluorescent probe, which is a monovalent cation with two C_{18} alkyl chains. The influence of two kinds of surfactants, sodium dodecyl sulfate (SDS) and dimyristoyl phosphatidylcholine (DMPC), was examined to evaluate the effect on the lateral diffusion mobility of single molecules at the dodecane–water interface. All DiI molecules adsorbed at the interface at a lower concentration than 1×10^{-7} M. In the presence of a few DiI molecules (0.02 molecules on average) in the observation area, an intermittent fluorescence photon bundle was observed during the measurement time of 16 ms. The maximum duration of the photon burst (t_{max}) meant the period in which the single DiI molecule migrated through the diameter of the observation area. From the value of t_{max}, the lateral diffusion coefficient of DiI was calculated. By using the Einstein–Stokes equation, the apparent viscosity of the surfactant aggregate

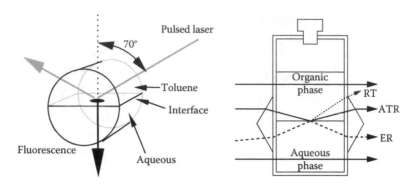

FIGURE 5.8 Total internal reflection cell used for the interfacial fluorescence lifetime measurement (left) and the external reflection absorption spectrometry (right).

layer was evaluated. In the surfactant-free system, the viscosity of 1.4 mPa s was obtained, which was close to the value of liquid dodecane (1.4 mPa s) and was higher than that of water (0.89 mPa s). The presence of SDS at the interface had no effect on the apparent viscosity even under the saturated interfacial concentration of SDS, 2.5×10^{-10} mol/cm^2. This is probably due to the same molecular size between SDS and dodecane. On the other hand, DMPC showed a remarkable retardation effect. The maximum interfacial viscosity in the DMPC system was 0.170 Pa s, which was 100 times higher than that of the surfactant-free interface. The DMPC molecule has two alkyl chains fixed to the polar group. This may be the cause of the high viscosity of the aggregated monolayer of DMPC molecules. These results demonstrated that a single molecule can probe the hydrodynamic properties of the interfacial molecular aggregate and monolayer.

In the case that an organic phase contains light-absorbing compounds, an external reflection (ER) absorption spectrometry is more useful than a total internal reflection spectrometry.[95,96] Another advantage of the ER method is its higher sensitivity than the TIR method, especially using s-polarized light. Typical optical cells used for total internal reflection spectrometry and external reflection absorption spectrometry are shown in Figure 5.8.

5.4.2.4 Raman Spectroscopy of Liquid–Liquid Interface

Raman scattering spectrometry has some advantages over infrared (IR) spectrometry in the sensitivity, the space resolution, and the applicability to aqueous solution. It can attain high sensitivity, when it is performed under the resonance Raman condition, or under the surface-enhanced condition with silver or gold nanoparticles as a source of plasmon. These techniques have been applied successfully for the measurement of extremely small amounts of molecules adsorbed at the liquid–liquid interface.

The combination of resonance Raman microscope spectrometry and the centrifugal liquid–membrane (CLM) method is a powerful technique to observe directly the molecules and molecular aggregates formed at the liquid–liquid

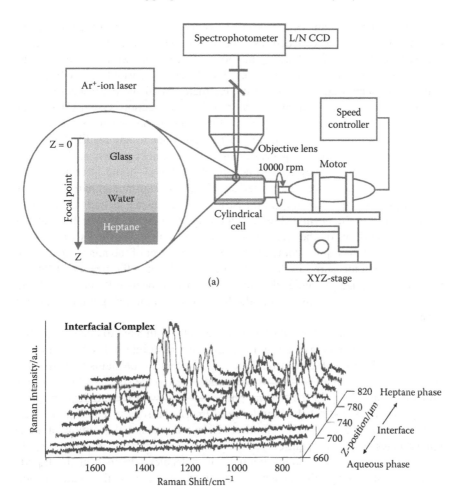

FIGURE 5.9   (a) Schematic drawing of the apparatus of the centrifugal liquid–membrane (CLM)-resonance Raman microprobe spectroscopy. Organic phase, 250 μL, and aqueous phase, 250 μL. (b) 3D CLM/Raman spectra depicting the formation of interfacial complex of PdLCl.

interface, as well as those in the bulk phases by traveling the focal point of an objective lens. A schematic diagram of the CLM–Raman microscope spectrometry system is shown in Figure 5.9. It was used to measure the complex formation rate between Pd(II) and 5-Br-PADAP (HL) at the heptane–water interface.[97] Figure 5.9 shows 3D spectra depicting the formation of an interfacial complex of PdLCl. Raman peaks at 1599, 1408, and 1303 cm^{-1} were assigned to PdLCl formed at the interface. The resonance Raman spectra of the interface can provide the information on the nano-environment of the complex, that is, the solvent effect of the interface and the extent of the aggregation at the interface. The solvent effect on the resonance Raman spectra of PdLCl reflected a change in the ratio of the azo and imine resonance structures. The azo/imine intensity ratios of I_{1482}/I_{1463} at 1482 and 1463 cm^{-1} and I_{1307}/I_{1284} at 1307 and 1284 cm^{-1} were used to evaluate the dielectric constants of the interfaces. The dielectric constants of the heptane–water and toluene–water interfaces were estimated as 40 and 62, respectively.[98]

Formation of metal nanoparticles in solvent extraction systems has been reviewed by Osseo-Asare.[99] Nanoparticle formation in the solvent extraction systems should become more of a matter of interest, since it may readily happen in solvent extraction systems. In general, nanoparticles can be present as dispersions in the bulk phases, so they tend to adsorb at the liquid–liquid interface. Silver and gold nanoparticles are widely used as substrates of surface-enhanced Raman scattering (SERS) spectroscopy. In the SERS measurements, its high signal intensity compared to the normal Raman scattering, by a factor of over 10^5, enables the characterization of the trace amount of molecules adsorbed to the metal nanoparticles at the liquid–liquid interface. A monolayer of the nonresonant molecules has been measured by the SERS method.[100] When the irradiated light is absorbed by the analyte molecules, the overlapped resonance Raman effect will further enhance the signals, so that the Raman spectra of a single molecule can be observed.[101]

SERS from oleate-stabilized silver nanoparticles with 5 nm diameter adsorbed at the toluene–water interface was measured in a total internal reflection (TIR) condition.[102] Based on the electromagnetic enhancement mechanism of SERS, it was concluded that an oleate ion would be adsorbed to the silver surface through the carboxylate group in the toluene phase side and with the ethylene group in the aqueous phase side at the liquid–liquid interfacial region. The liquid–liquid interface is advantageous to the creation of a two-dimensional assembly of metal nanoparticles, which is required for the SERS method. Thus, it will be applicable to detect extremely small amounts of molecules adsorbed at the liquid–liquid interface, though the molecules are adsorbed finally on the metal surface. Dodecanethiol (DT) in the cyclohexane phase reacted with citrate-reduced silver nanoparticles in the aqueous phase at the interface. The interfacial SERS spectra showed that the relative intensity ratios of the ν(C-S) band of gauche to trans conformers decreased with the increase of the initial concentration of DT, suggesting the change from the liquid-like structure to the solid-like structure of the DT on the nanoparticles. The residue of the negative charges on the nanoparticles at the interface was detected by the resonance SERS peaks of the adsorbed cationic porphyrin.[103]

5.4.2.5 Optical Chirality Measurement

Optical chirality of molecules and molecular aggregates is the characteristic information to study the stereochemical property of biological, pharmaceutical, and metal coordination compounds. Chiral structures of amino acids, proteins, DNAs, drugs, and natural compounds in solutions have been determined from the measurement of circular dichroism (CD). Molecular aggregates and supramolecules have exhibited optical chirality, which was induced by a vortex force by stirring or rotating of the solution depending on the stirring directions, clockwise or counterclockwise.[104,105] The CD spectra of molecular aggregates formed at the liquid–liquid interfaces have been measured for the first time by applying the centrifugal liquid–membrane (CLM) method.[106] The CLM cell was set horizontally in the sample chamber of the circular dichroism spectropolarimeter in almost the same manner as the CLM–UV-vis spectrometry shown in Figure 5.3. By this technique, the CD spectra of the interfacial aggregates of tetraphenylporphyrin diacid (H_4TPP^{2+}) were observed. H_2TPP monomer in toluene shows the Soret band at 419 nm and Q band at 515 nm. Formation of the interfacial J-aggregate of the diprotonated form (($H_4TPP^{2+})_n$) can be recognized from the red shift of the Soret band and Q band to 473 and 720 nm, respectively. Therefore, the observed bisignate CD spectrum centered at 473 nm was assigned to a characteristic CD signal of the exciton coupling, corresponding to the helical conformation of the J-aggregate at the interface.[107] Furthermore, the interfacial aggregates of diprotonated tetraphenylporphyrin were successfully employed as the chiral recognition reagent of 2-alkylalcohol at the interface, as shown in Figure 5.10.[108] The J-aggregate of the diprotonated tetraphenylporphyrin formed at the dodecane–water interface showed circular dichroism spectra

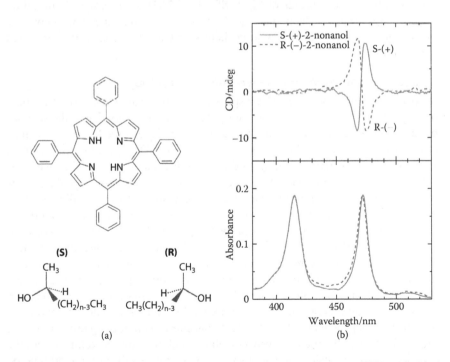

FIGURE 5.10 (a) Molecular structures of tetraphenylporphyrin (H_2TPP) and chiral 2-alkyl alcohols. (b) CLM-CD spectra (upper) and CLM absorbance spectra (lower) of H_4TPP^{2+} aggregates in the presence of S-(+)-2-nonanol and R-(-)-2-nonanol. These spectra were measured 1 minute after introducing the H_2TPP dodecane solution. $[H_2TPP]_{org} = 2.8 \times 10^{-5}$ M, $[2\text{-nonanol}]_{org} = 1.0 \times 10^2$ M, $[H_2SO_4]_{aq} = 4$ M.

around 473 nm, corresponding to the chirality of 2-alkylalcohol, which was added in the dodecane phase. The CD intensity was stronger in a longer alkyl-alcohol than 2-butanol. The chiral recognition ability of the J-aggregate was highly sensitive even for such a low concentration of 1.0×10^{-6} M 2-nonanol. The phenomenon suggested the preferential interaction between the nanosized J-aggregates of H_4TPP^{2+} and the chiral alcohols at the interface. This result demonstrated a potential use of the J-nanoaggregate as a chiral recognition reagent at the interface, where the surface-active analyte molecule in the bulk phase can be preferentially concentrated.

A chiral thioether-substituted phthalocyanine ((2,3,9,10,16,17,23,24-octakis-1-phenylethylthiophthalocyaninato)-magnesium(II) [MgPc(SEtPh)$_8$]) formed aggregates at the toluene-water interface with soft-metal ions such as palladium(II) ion. The CLM-CD method and matrix-assisted laser desorption ionization time-of-flight mass spectrometry (MALDI-TOF/MS) and scanning electron microscope (SEM) were used to characterize the interfacial chiral aggregate. The toluene solution of MgPc(SEtPh)$_8$ showed no optical chirality. However, when the system included PdCl$_2$, a helical J-aggregate (head to-tail fashion) of MgPc(SEtPh)$_8$ bound by PdCl$_2$ was formed in the toluene–water interface. One J-aggregate unit contained five 1:2 complexes of MgPc(SEtPh)$_8$-2Pd(II) on average.[109] Formation of helical aggregates at the liquid–liquid interface is a promising subject not only in analytical chemistry and separation chemistry, but also in material science and nanotechnology.

5.4.2.6 Second Harmonic Generation Spectroscopy

Second harmonic generation (SHG) spectroscopy is advantageous to other spectroscopic techniques in its inherent high interfacial selectivity due to the second-order nonlinear susceptibility, and in its high sensitivity afforded by the resonance effect of the photon-absorbing molecule. Therefore, SHG spectroscopy can provide specific information relating to the electronic state of molecules, which is exhibited only at the interface. SHG spectroscopy has been applied to the studies of molecular orientation at the interface[110,111] and solvation at the alkane–water interface,[112] and aggregation of dye molecules at the electrochemical liquid–liquid interfaces.[113] The interfacial properties of the N,N-di(2-ethylhexyl)isobutyramide (DEHiBA) and N,N'-dimethyl-N,N'-dibutyl-2-tetradecylmalonamide (DMDBTDMA) extractant molecules used in the nuclear industry were examined by the SHG method to obtain a better understanding of ion transfer across the water–oil interface. Real-time SHG experiments and titration were carried out to follow the kinetics of nitric acid extraction. The SHG intensity evolution was strongly dependent on the extractant concentration in the organic phase, and the SHG intensity fluctuations were correlated with the nitric acid flux across the interface. These intensity fluctuations were the signature of a strong modification of extractant concentration at the interface that was close to a critical aggregation concentration of extractant in the organic phase.[114]

The most selective technique to measure the optical chirality at the surface is second harmonic generation–circular dichroism (SHG-CD) spectroscopy, as demonstrated by the pioneering work of Hicks and coworkers.[115,116] However, the nonlinear optical activity of the liquid–liquid interface was attempted for the first time in our work.[117] A combination of SHG-CD and magneto-optical effect at the liquid–liquid interface will be a future subject.

5.4.2.7 Magnetophoretic Measurement of Interfacial Aggregates

The effect of a magnetic field on the interfacial aggregation has rarely been studied, though the application of magnetic force to separation and characterization of microparticles in liquids has been suggested by Giddings, who invented field flow fractionation.[118] Recently, the magnetoanalysis of micro- and nanoparticles has been reviewed.[119] The magnetic force on molecules is very weak in comparison with the electrostatic force. However, when the molecules form micrometer-sized aggregates, the magnetic force effect is not negligible, even if they are diamagnetic aggregates. The migration of nano- and microparticles due to the magnetic force is called magnetophoresis, analogous to electrophoresis, and magnetophoretic velocimetry is proposed as a new analytical method to measure the magnetic susceptibility and the amount of some composites in individual microparticles.[120] By measuring the size and the migration velocity of a microparticle with a charge coupled device (CCD) microscope, the magnetic susceptibility of the particle was obtained. By using a high magnetic field gradient generated with a superconducting magnet, this technique allowed us to measure a volume magnetic susceptibility at the 10^{-6} level with enough sensitivity to detect the attomole level of Mn(II) ion extracted in a single organic droplet in an aqueous phase.[121] The experimental setup of the magnetophoresis is shown in Figure 5.11(a), where a cryogen-free superconducting magnet of 10 T is used to give a magnetic field gradient of 4.7×10^4 T²/m. A pair of permanent magnets and a magnetic circuit can also be used for the magnetophoresis of molecular aggregate and particles.[122] The magnetophoretic velocity v_m of a droplet with radius r in a medium is represented by the equation

$$v_x = \frac{2}{9} \frac{(\chi_p - \chi_m)}{\mu_0 \eta} r^2 B \left(\frac{dB}{dx} \right) \tag{5.3}$$

where χ_p is the magnetic susceptibility of the droplet, χ_m is the magnetic susceptibility of the medium, η is the viscosity of the medium, μ_0 is the permeability of a vacuum, and B is the magnetic flux density. Figure 5.11(b) shows the magnetophoretic migration of a single pure 2-fluorotoluene droplet ($\chi_p = -7.2 \times 10^{-6}$) in water ($\chi_m = -9.1 \times 10^{-6}$). The less diamagnetic 2-fluorotoluene droplet was attracted toward the gap of the pole pieces like a paramagnetic particle in water. The migration velocity was maximized near the edge of the pole pieces, where the value of $B(dB/dx)$ becomes largest. When the organic droplets contained

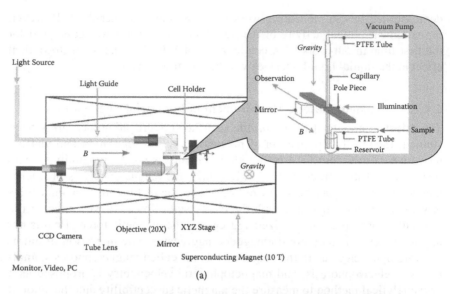

(a)

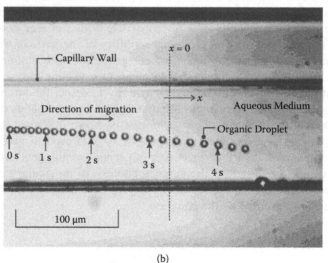

(b)

FIGURE 5.11 (a) Schematic drawing of the magnetophoretic velocimetry apparatus installed in the bore of a superconducting magnet. (b) The magnetophoretic behavior of a single microdroplet of 2-fluorotoluene in water. This microphotograph was made by superimposing the images captured at 1/5 second intervals. The dashed line indicates the position of the edge of pole pieces ($x = 0$).

lauric acid and were dispersed in a dilute Dy(III) aqueous solution, the observed χ_p of the single droplet depended on the radius by Equation (5.4),[123]

$$\chi_p = \frac{3\chi_{Dy}^M C_{int}}{r} + \chi_{2FT}^V \qquad (5.4)$$

where C_{int} is the interfacial concentration of dysprosium(III) (mol m^{-2}), the super-scripts M and V refer to the molar and volume magnetic susceptibilities (m^3 mol^{-1}), respectively, and the subscripts Dy and 2FT indicate dysprosium(III) and 2-fluorotoluene, respectively. This equation means that from the size dependence of the observed magnetic susceptibility of single droplets, one can obtain the interfacial concentration of a paramagnetic ion such as Dy(III). For example, under the condition of 1.0×10^{-2} M lauric acid in 2-fluorotoluene and dispersed in 5.0×10^{-4} M Dy(III) aqueous solution, the interfacial concentration of Dy(III) as the 1:1 complex with laurate was determined as 2.6×10^{-10} mol cm^{-2}, as shown in Figure 5.12.[123] The measurement of the magnetophoretic velocity under different pH and Dy(III)

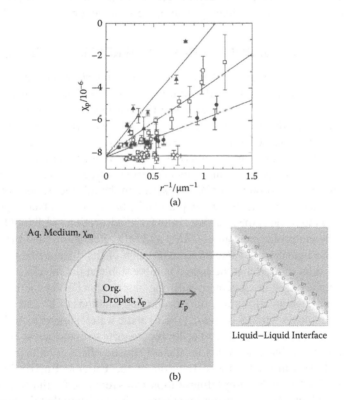

(a)

(b)

FIGURE 5.12   (a) Contribution of interfacial adsorption of Dy(III) complex on the magnetic susceptibility of droplets of 2-fluorotoluene. The organic droplet contained capric acid (filled circle), lauric acid (open square), or stearic acid (triangle). The open circles show the absence of dysprosium(III) and/or carboxylic acid. (b) Schematic illustration of the interfacial adsorption of Dy(III) carboxylate complex.

concentration gave the interfacial adsorption equilibrium of the Dy(III)-laurate complex, which was represented by the Langmuir equation.[124] This technique is an absolute analytical method because it requires only magnetic susceptibility to determine the interfacial concentration. Therefore, the magnetophoretic velocimetry is a universal method to determine the interfacial amount of adsorbed species, provided that the magnetic susceptibility of the adsorbed species is different from those of the organic and aqueous phases.

As for the aggregate formed in the solvent extraction system, the magnetic force effect was observed in the synergic extraction rate of Eu(III) with 2-thenoyltrifluoroacetone (Htta) and oxalate (ox⁻) at the dodecane–water interface.[125] The effect of the magnetic field was thought to be the magnetophoretic migration of paramagnetic microaggregate of the Eu(III)-tta-ox complex formed in the aqueous phase close to the interface. The microaggregate was migrated to the interface by the magnetic force and reacted with Htta molecules in the organic phase at the interface to be extracted after forming $Eu(tta)_3$. The magnetic force-assisted acceleration of the extraction rate of metal ions will be useful for the other extraction systems, including paramagnetic metal ions.

5.5 COMPUTER SIMULATION OF INTERFACIAL AGGREGATION

Some fundamental properties of the liquid–liquid interface have been reported using the molecular dynamics (MD) simulations. A remarkable attenuating effect on the rotational dynamics of molecules at the interface was found as the surface tension is reduced due to a local density gap. This gave rise to a local effective viscosity, which increased as the surface tension was decreased.[126] The first report that used MD simulation for examination of the solvent extraction rate dealt with the solvent effect of heptane and toluene on the extraction rate of Ni(II) with 5-Br-PADAP.[127] The slower interfacial reaction in the toluene–water system than in the n-heptane–water system was explained by the stronger solvation of toluene to the ligand molecule at the interface. Interfacial adsorption of LIX65N was simulated by the molecular mechanics (MM) and MD calculations, and the solvation energy of the ligand was evaluated as a function of the distance from the interface. The results showed that the ligand is more stable at the interface than in the bulk n-heptane phase in agreement with the observation.[128] Vayssière and Wipff reported an MD simulation study of concentrated solutions of K^+ and Sr^{2+} cations and their complexes with 18-crown-6 (18C6) at the water–supercritical CO_2 interface, and compared picrate versus perfluorooctanoate counterions at 305 and 350 K.[129] A molecular dynamics study on the synergistic effect of chlorinated cobalt bis(dicarbollide) anions for the extraction of the cationic Eu(III)-2,6-bis(5,6-isopropyl-1,2,4-triazin-3-yl)-pyridine complex was reported for the octanol–water systems[130] and nitrobenzene–water interfaces.[131] It was found that the ion association reactions between the cation and the anion at the interface were important processes for the synergistic extraction. A molecular modeling study of the aggregation behavior of nickel(II), cobalt(II), lead(II), and zinc(II) bis(2-ethylhexyl) phosphate complexes was done by Ibrahim and Neuman.[132] The cobalt-extractant species formed rod-like reversed micelles, but did not show the formation of any open-water channel. The zinc- and lead-extractant species formed ellipsoidal (or deformed spherical) reversed micelles with fewer water molecules located at the core of the micelles, which is in accord with the conventional view of reversed micelles. The molecular modeling of the reversed micellar system of NaDEHP in n-heptane clearly showed that the main forces between the NaDEHP molecules were van der Waals and electrostatics forces. Water molecules were in some cases acting as an "antimicellization" agent and had no gluing effect on the formation of NaDEHP reversed micelles. In fact, large NaDEHP reversed micelles formed in "dry" conditions.[133] The computer simulation technique is a powerful tool for the discussion and understanding of interfacial phenomena with molecular pictures. The progress of computing technology of molecular modeling will bring more rapid and reliable information on the interfacial aggregation of molecules.

5.6 SUMMARY

Studies of the mechanism of molecular aggregation and third-phase formation in solvent extraction systems are important subjects to improve the extraction efficiency, selectivity, and extraction rate in process metallurgy. To characterize the species contained in the aggregates, modern measurement methods developed in the field of interfacial science can be used. However, the solvent extraction systems are more complicated than the subjects treated by physical and colloid chemists. Therefore, researchers studying solvent extraction chemistry and technology sometimes have to invent desirable methods or make modifications to the conventional methods by their own effort.

Finally, I summarize some future prospects:

1. Molecular aggregation is an inevitable phenomenon in solvent extraction processes under high loading conditions. Therefore, it should be more positively taken into consideration for the design of extraction schemes and utilized for the improvement of selectivity and extractability of metal ions. It is interesting that aggregation, micelle formation, and phase separation are induced by rather small amounts of heat energy or consumption of additives.
2. Molecular aggregates can be manipulated more easily than molecules by external fields or external forces. Therefore, a combination of the aggregate-forming extraction and external force, such as magnetic force, dielectric force, or acoustic force, will open new possibilities.
3. Direct reduction or oxidation of metal ions from the aggregate in solvent extraction systems will be used for the production of nanoparticles. Fundamental research of formation and separation of nanoparticles in solvent extraction systems will bring interesting and useful technologies.
4. The computational simulation method will be more conventionally employed in the near future for understanding and modeling the structure of the molecular aggregate and the design of supramolecular extractants.

188 Ion Exchange and Solvent Extraction

But now, the real aggregation reaction is turning out to be more complicated than expected. However, the author believes that there is a beautiful landscape of possibility beyond the steep mountains of science of molecular aggregations and supramolecular extraction systems.

ACKNOWLEDGMENTS

This work was supported by a Grant-in-Aid for Scientific Research (A) (no. 2450220) of the Ministry of Education, Culture, Sports, Science and Technology of Japan.

REFERENCES

1. Rydberg, J., Musikas, C., and Choppin, G. R. (eds.). 1992. *Principles and practices of solvent extraction*. Marcel Dekker, New York.
2. Pierotti, R. 1976. A scaled particle theory of aqueous and nonaqueous solutions. *Chem. Rev.* 76:717–726.
3. Watarai, H. Tanaka, M., and Suzuki, N. 1982. Determination of partition coefficients of halobenzenes in heptane/water and 1-octanal/water systems and comparison with the scaled particle calculation. *Anal. Chem.* 54:702–705.
4. Watarai, H., Oshima, H., and Suzuki, N. 1984. Regularities of the partition coefficients of bis, tris, and tetrakis(acetylacetonato)metal (II, III and IV) complexes. *Quant. Struct.-Act. Relat.* 3:17–22.
5. Benjamin, I. 1991. Molecular dynamics study of the free energy functions for electron-transfer reactions at the liquid–liquid interface. *J. Phys. Chem.* 95:6675–6683.
6. Nelson, K. V., and Benjamin, I. 2010. A molecular dynamics-empirical valence bond study of an SN_2 reaction at the water/chloroform interface. *J. Phys. Chem. C.* 114:1154–1163.
7. Watarai, H., Teramae, N., and Sawada, T. (eds.). 2005. *Interfacial nanochemistry*. Kluwer Academic/Plenum Publishers, New York.
8. Volkov, A. G., and Deamer, D. W. (eds.). 1996. *Liquid–liquid interfaces, theory and methods*. CRC Press, New York.
9. Volkov, A. G., Deamer, D. W., Tanelian, D. L., and Markin, V. S. (eds.). *Liquid interfaces in chemistry and biology*. John Wiley & Sons, New York, 1998.
10. Watarai, H. 1993. What's happening at the liquid–liquid interface in solvent extraction chemistry? *Trends Anal. Chem.* 12:313–318.
11. Watarai, H. 2007. Measurement of complex formation and aggregation at the liquid–liquid interface. In *Advanced chemistry of monolayers at interfaces 14*, ed. IMAE Toyoko. Elsevier, Amsterdam, pp. 277–308.
12. Flett, D. S. 1977. Chemical kinetics and mechanisms in solvent extraction of copper chelates. *Acc. Chem. Res.* 10:99–104.
13. Watarai, H., and Freiser, H. 1983. Role of the interface in the extraction kinetics of zinc and nickel ions with alkyl-substituted dithizones. *J. Am. Chem. Soc.* 105: 189–190.
14. Watarai, H., and Freiser, H. 1983. Effect of stirring on the distribution equilibria of n-alkyl-substituted dithizones. *J. Am. Chem. Soc.* 105:191–194.
15. Watarai, H., Tsukahara, S., Nagatani, H., and Ohashi, A. 2003. Interfacial nanochemistry in liquid–liquid extraction systems. *Bull. Chem. Soc. Jpn.* 76:1471–1492.

16. Kertes, A. S. 1977. Critical evaluation of some equilibrium constants involving alkylammonium extractants. International Union of Pure and Applied Chemistry. Commission on Equilibrium Data. Oxford, Pergamon Press.
17. Smith, E. L., and Page, J. E. 1948. The acid-binding properties of long-chain aliphatic amines. *J. Soc. Chem. Ind. Lond.* 67:48–51.
18. Allen, K. A., and McDowell, W. J. 1960. Anomalous solvent extraction equilibria due to violence of agitation. *J. Phys. Chem.* 64(7):877–880.
19. Shanker, R., Venkateswarlu, K. S., and Shanker, J. 1968. Extraction of ruthenium(III) by long-chain amines from sulphuric acid solutions. *J. Less-Common Metals* 15:75–88.
20. Kertes, A. S., and Markovits, G. 1968. Activity coefficients, aggregation, and thermodynamics of tridodecylammonium salts in nonpolar solvents. *J. Phys. Chem.* 72:4202–4210.
21. Danesi, P. R., Orlandinio, F., and Scibona, G. 1968. Computer equilibria evaluation of the aggregation of some alkylammonium salts in benzene. *J. Inorg. Nucl. Chem.* 30:2513–2519.
22. Bac, R. 1970. The extraction of excess nitric acid and water by tri-(long-chain)-alkylamine nitrates in organic diluents. *J. Inorg. Nucl. Chem.* 32:3655–3666.
23. Casey, A. T., Cattrall, R. W., and Davey, D. E. 1971. The aggregation of the chloride, bromide and iodide salts of dilaurylamine in chloroform. *J. Inorg. Nucl. Chem.* 33:535–542.
24. Levy, O., Markovits, G., and Kertes, A. S. 1971. Aggregation equilibria in solvent extraction of iron(III) halides. *J. Inorg. Nucl. Chem.* 33:551–557.
25. Oldmaann, S., and Cave, G. C. B. 1971. Ions and aggregates of tetra-n-butylammonium picrate in nonaqueous solvents at 25°C: a solvent extraction study. *Can. J. Chem.* 49:1726–1735.
26. Cattrall, R. W., and Slater, S. J. E. 1974. A study of the aggregation of the sulphate and bisulphate salts of bis(3,5,5-trimethylhexyl)amine in chloroform solution by vapour phase osmometry. *J. Inorg. Nucl. Chem.* 36:841–845.
27. McDowell, W. J., and Harmon, H. D. 1969. Sodium and strontium extraction from sodium nitrate solutions by n-octane solutions of a branched aliphatic monocarboxylic acid: mechanism and equilibria. *J. Inorg, Nucl. Chem.* 31:1473–1485.
28. Van Dalen, A., Gerritsma, K. W., and Wijkstra, J. 1974. Determination of the aggregation number of dinonyl naphthalene sulfonic acid by metal ion extraction. *J. Colloid Interf. Sci.* 48:122–126.
29. Morris, D. F. C. 1975. Aggregation of a liquid cation-exchanger. *J. Colloid Interf. Sci.* 51:52–57.
30. Yu, J., and Liu, D. 2010. Extraction of magnesium from phosphoric acid using dinonylnaphthalene sulfonic acid. *Chem. Eng. Res. Des.* 88:712–717.
31. Sekine, T., and Hasegawa, Y. 1977. *Solvent extraction chemistry*. Dekker, New York, p. 451.
32. Bhattacharyya, S. N., and (Nandi) Ganguly, B. 1987. Study of the aggregation behavior of di(2-ethyl hexyl)phosphoric acid in heptane in the presence of water. *J. Colloid Interf. Sci.* 118:15–19.
33. Sole, K. C., and Hiskey, J. B. 1995. Solvent extraction of copper by Cyanex 272, Cyanex 302 and Cyanex 301. *Hydrometallurgy* 37:129–147.
34. Anticó, E., Masanaa, A., Hidalgo, M., Salvadb, V., Iglesias, M., and Valiente, M. 1996. Solvent extraction of yttrium from chloride media by di(2-ethylhexyl)phosphoric acid in kerosene. Speciation studies and gel formation. *Anal. Chim. Acta* 327:267–276.

35. Sato, H., Kubokawa, K., and Komasawa, I. 1998. Equilibrium study on the extraction of aluminum and beryllium by a mixture of mono(2-ethylhexyl)phosphoric acid and bis(2-ethylhexyl)phosphoric acid. *Solvent Extr. Res. Dev. Jpn.* 5:51–59.

36. Chiarizia, R., Urbani, V., Thiyagarajan, P., and Herlinger, A. W. 1999. Aggregation of complexes formed in the extraction of selected metal cations by P,P'-di(2-ethylhexyl) methanediphosphonic acid. *Solvent Extr. Ion Exch.* 17:113–132.

37. Jensen, M. P., Chiarizia, R., Urban, V., and Nash, K. L. 2000. Aggregation of the neodymium complexes of HDEHP, Cyanex 272, Cyanex 302, and Cyanex 301 in toluene. *JAERI-Conference 2002-004 (Proceedings of the International Symposium NUCEF 2001)*, 281–288.

38. Kumar, S., and Tulasi, G. L. 2005. Aggregation vs. breakup of the organic phase complex. *Hydrometallurgy* 78:79–91.

39. Rúa, M. S., Almela, A., and Elizalde, M. P. 2006. Aggregation equilibria of the components of the commercial extractants LIX 622 and LIX 622N in toluene and *n*-heptane. *Fluid Phase Equilibria* 244:111–116.

40. Moyer, B. A., Caley, C. E., and Baes, C. F., Jr. 1988. Hydration and aggregation of monofunctional and other neutral oxygen-donor extractants: the di(2-ethylhexyl) sulfoxide, dodecane, water system. *Solvent Extr. Ion Exch.* 6:785–817.

41. Moyer, B. A., Baes, C. F., Jr., Case, G. N., Lumetta, G. J., and Wilson, N. M. 1993. Equilibrium analysis of aggregation behavior in the extraction of Cu(II) from sulfuric acid by didodecylnaphthalene sulfonic acid. *Sep. Sci. Technol.* 28:81–113.

42. Nave, S., Mandin, C., Martinet, L., Berthon, L., Testard, F., Madicc, C., and Zemb, Th. 2004. Supramolecular organisation of tri-n-butyl phosphate in organic diluent on approaching third phase transition. *Phys. Chem. Chem. Phys.* 6:799–808.

43. Thiyagarajan, P., Diamond, H., and Horwitz, E. E. 1988. Small-angle neutron scattering studies of the aggregation of $Pr(NO_3)_3$-CMPO and $PrCl_3$-CMPO complexes in organic solvents. *J. Appl. Cryst.* 21:848–852.

44. Tamiaki, H., Nishiyama, T., and Shibata, R. 2007. Self-aggregation of zinc chlorophylls possessing perfluoroalkyl chains in fluorous solvents: selective extraction of the self-aggregates with fluorous phase and accelerated formation of the ordered supramolecules in this phase. *Bioorg. Med. Chem. Lett.* 17:1920–1923.

45. Stoikov, I. I., Yushkova, E. A. Yu., Zhukov, A., Zharov, I., Antipin, I. S., and Konovalov, A. I. 2008. Solvent extraction and self-assembly of nanosized aggregates of p-tert-butylthiacalix[4]arenes tetrasubstituted at the lower rim by tertiary amide groups and monocharged metal cations in the organic phase. *Tetrahedron* 64:7489–7497.

46. Berthon, L., Testard, F., Martinet, L., Zemb, T., and Madic, C. 2010. Influence of the extracted solute on the aggregation of malonamide extractant in organic phases: consequences for phase stability. *C. R. Chimie* 13:1326–1334.

47. Meridiano, Y., Berthon, L., Crozes, X., Sorel, C., Dannus, P., Antonio, M. R., Chiarizia, R., and Zemb, T. 2009. Aggregation in organic solutions of malonamides: consequences for water extraction. *Solvent Extr. Ion Exch.* 27:607–637.

48. Pathak, P. N., Ansari, S. A., Kumar, S., Tomar, B. S., and Manchanda, V. K. 2010. Dynamic light scattering study on the aggregation behaviour of N,N,N',N'-tetraoctyl diglycolamide (TODGA) and its correlation with the extraction behavior of metal ions. *J. Colloid Interf. Sci.* 342:114–118.

49. Berthon, L., Martinet, L., Testard, F., Madic, C., and Zemb, Th. 2007. Solvent penetration and sterical stabilization of reverse aggregates based on the DIAMEX process extracting molecules: consequences for the third-phase formation. *Solvent Extr. Ion Exch.* 25:545–576.

50. Ravi, J., Prathibha, T., Venkatesan, K. A., Antony, M. P., Srinivasan, T. G., and Rao, P. R. V. 2012. Third phase formation of neodymium (III) and nitric acid in unsymmetrical N,N-di-2-ethylhexyl-N',N'-dioctyldiglycolamide. *Sep. Purif. Technol.* 85:96–100.

51. Ellis, R. J., and Antonio, M. R. 2012. Coordination structures and supramolecular architectures in a cerium(III)–malonamide solvent extraction system. *Langmuir* 28:5987–5998.

52. Erlinger, C., Gazeau, D., Zemb, T., Madic, C., Lefran, L., Hebrant, M., and Tondre, C. 1998. Effect of nitric acid extraction on phase behavior, microstructure and interactions between primary aggregates in the system dimethyldibutyltetradecylmalonamide (DMDBTDMA)/n-dodecane/water: a phase analysis and small angle X-ray scattering (SAXS) characterization study. *Solvent Extr. Ion Exch.* 16:707–738.

53. Gomez, A. B., Sicilia, M. D., and Rubio, S. 2010. Supramolecular solvents in the extraction of organic compounds. A review. *Anal. Chim. Acta* 677:108–130.

54. Miura, J., Ishii, H., and Watanabe, H. 1976. Extraction and separation of nickel chelate of 1-(2-thiazolylazo)-2-naphthol in nonionic surfactant solution. *Bunseki Kagaku* 25:808–809.

55. Gharehbaghi, M., Shemirani, F., and Farahani, M. D. 2009. Cold-induced aggregation microextraction based on ionic liquids and fiber optic-linear array detection spectrophotometry of cobalt in water samples. *J. Hazard. Mater.* 165:1049–1055.

56. Baghdadi, M., and Shemirani, F. 2008. Cold-induced aggregation microextraction: a novel sample preparation technique based on ionic liquids. *Anal. Chim. Acta* 613:56–63.

57. Mahpishanian, S., and Shemirani, F. 2010. Ionic liquid-based modified cold-induced aggregation microextraction (M-CIAME) as a novel solvent extraction method for determination of gold in saline solutions. *Miner. Eng.* 23:823–825.

58. Mustafina, A., Zakharova, L., Elistratova, J., Kudryashova, J., Soloveva, S., Garusov, A., Antipin, I., and Konovalov, A. 2010. Solution behavior of mixed systems based on novel amphiphilic cyclophanes and Triton X100: aggregation, cloud point phenomenon and cloud point extraction of lanthanide ions. *J. Colloid Interf. Sci.* 346:405–413.

59. Zhang, H., Chen, X., and Jiang, X. 2011. Determination of phthalate esters in water samples by ionic liquid cold-induced aggregation dispersive liquid–liquid microextraction coupled with high-performance liquid chromatography. *Anal. Chim. Acta* 689:137–142.

60. Jääskeläinen, E., and Paatero, E. 1999. Properties of the ammonium form of Versatic 10 in a liquid–liquid extraction system. *Hydrometallurgy* 51:47–71.

61. Osseo-Asare, K. 1991. Aggregation, reversed micelles, and microemulsions in liquid–liquid extraction: the tri-*n*-butylphosphate diluent-water-electrolyte systems. *Adv. Colloid Interf. Sci.* 37:123–173.

62. Vandegrift, G. F., Lewey, S. M., Dyrkacz, G. R., and Horwitz, E. P. 1980. Interfacial activity of liquid/liquid extraction reagents. II. Quaternary ammonium salts. *J. Inorg. Nucl. Chem.* 42:127–130.

63. Gaonkar, A. G., Fereshtehkhou, S., and Neuman, R. D. 1986. Study of surface aggregation of calcium di(2-ethylhexyl)phosphate by electron microscopy. *J. Colloid Interf. Sci.* 112:298–301.

64. Gaonkar, A. G., and Neuman, R. D. 1987. Interfacial activity, extractant selectivity, and reversed micellization in hydrometallurgical liquid/liquid extraction systems. *J. Colloid Interf. Sci.* 119:251–261.

65. Neuman, R. D., and Park, S. J. 1992. Characterization of association microstructures in hydrometallurgical nickel extraction by di (2-ethylhexyl) phosphoric acid. *J. Colloid Interf. Sci.* 152:41–53.

66. Gerhardt, N. I., Palant, A. A., and Dungan, S. R. 2000. Extraction of tungsten (VI), molybdenum (VI) and rhenium (VII) by diisododecylamine. *Hydrometallurgy* 55:1–15.
67. Delmau, L. H., Bonnesen, P. V., Herlinger, A., and Chiarizia, W. R. 2005. Aggregation behaviour of solvent modifiers for the extraction of cesium from caustic media. *Solvent Extr. Ion Exch.* 23:145–169.
68. Testard, F., Berthon, L., and Zemb, T. 2007. Liquid–liquid extraction: an adsorption isotherm at divided interface? *C. R. Chimie* 10:1034–1041.
69. Shimojo, K., Oshima, T., Naganawa, H., and Goto, M. 2007. Calixarene-assisted protein refolding via liquid–liquid extraction. *Biomacromolecules* 8:3061–3066.
70. Hanna, G. J., and Noble, R. D. 1985. Measurement of liquid–liquid interfacial kinetics. *Chem. Rev.* 85:583.
71. Watarai, H., Takahashi, M., and Shibata, K. 1986. Interfacial phenomena in the extraction kinetics of nickel(II) with 2'-hydroxy-5'-nonyl- acetophenone oxime. *Bull. Chem. Soc. Jpn.* 59:3469–3473.
72. Watarai, H., Kamada, K., and Yokoyama, S. 1989. Interfacial adsorption of β-diketones in vigorously stirred heptane/aqueous phase systems. *Solvent Extr. Ion Exch.* 7:361–376.
73. Freiser, B. S., and Freiser, H. 1970. Analytical applications of mixed ligand extraction equilibria nickel-dithizone-phenanthroline complex. *Talanta* 17:540–543.
74. Watarai, H., Sasaki, K., Takahashi, K., and Murakami, J. 1995. Interfacial reaction in the synergistic extraction rate of Ni(II) with dithizone and 1,10-phenanthroline. *Talanta* 42:1691–1700.
75. Chida, Y., and Watarai, H. 1996. Interfacial adsorption and ion-association extraction of protonated tetraphenylporphyrin and octaethylporphyrin. *Bull. Chem. Soc. Jpn.* 69:341–347.
76. Ohashi, A., Tsukahara, S., and Watarai, H. 1999. Isomer recognizing adsorption of palladium(II)-2-(5-bromo-2-pyridylazo)-5-diethylaminophenol with diazine derivatives at the toluene-water interface. *Anal. Chim. Acta* 394:23–31.
77. Onoe, Y., Tsukahara, S., and Watarai, H. 1998. Catalytic effect of N,N-dimethyl-4-(2-pyridylazo)aniline on the extraction rate of Ni(II) with 1-(2-pyridylazo)-2-naphthol: ligand-substitution mechanism at the liquid–liquid interface. *Bull. Chem. Soc. Jpn.* 71:603–608.
78. Watarai, H., and Sasabuchi, K. 1985. Interfacial adsorption of 2-hydroxy-5-nonylbenzophenone oxime in static and vigorously stirred distribution systems. *Solvent Extr. Ion Exch.* 3:881–893.
79. Nagatani, H., and Watarai, H. 1998. Formation and interfacial adsorption of the μ-oxo dimer of (5,10,15,20-tetraphenylporphyrinato)iron(III) in dodecane/aqueous acid systems. *J. Chem. Soc. Faraday Trans.* 94:247–252.
80. Nagatani, H., and Watarai, H. 1997. Specific adsorption of metal complexes of tetraphenylporphyrin at dodecane-water interface. *Chem. Lett.* 167–168.
81. Nagatani, H., and Watarai, H. 1998. Direct spectrophotometric measurement of demetalation kinetics of 5,10,15,20-tetraphenylporphyrinatozinc(II) at the liquid–liquid interface by a centrifugal liquid membrane method. *Anal. Chem.* 70:2860–2865.
82. Nagatani, H., and Watarai, H. 1996. Two-phase stopped-flow measurement of the protonation of tetraphenylporphyrin at the liquid–liquid interface. *Anal. Chem.* 68:1250–1253.
83. Yulizar, Y., and Watarai, H. 2003. In situ measurement of aggregate formation kinetics of nickel(II)-pyridylazoaminophenol complex at the heptane-water interface by centrifugal liquid membrane spectrophotometry. *Bull. Chem. Soc. Jpn.* 76:1379–1386.
84. Yulizar, Y., Monjushiro, H., and Watarai, H. 2004. Interfacial aggregate growth process of Fe(II) and Fe(III) complexes with pyridylazophenol in solvent extraction system. *J. Colloid Interf. Sci.* 275:560–569.

85. Ohashi, A., Tsukahara, S., and Watarai, H. 2003. Molecular recognition of diazine isomers and purine bases by the aggregation of palladium(II)-pyridylazo complex at the toluene/water interface. *Langmuir* 19:4645–4651.
86. Watarai, H., Matsumoto, A., and Fukumoto, T. 2002. Direct electrospray ionization mass spectroscopic measurement of micro-flow oil/water system. *Anal. Sci.* 18:367–368.
87. Watarai, H., and Oyama, H. 2008. In situ measurements of aggregation and disaggregation of Cu(II) complex at liquid/liquid interface. *Anal. Chem.* 80:8348–8352.
88. Nagatani, H., and Watarai, H. 1999. Heterogeneous fluorescence quenching reaction between (5,10,15,20-tetraphenylporphyrinato)zinc(II) and methylviologen at dodecane-water interface. *Chem. Lett.* 701–702.
89. Watarai, H., and Funaki, F. 1996. Total internal reflection fluorescence measurements of protonation equilibria of rhodamine B and octadecylrhodamine B at a toluene/water interface. *Langmuir* 12:6717–6720.
90. Tsukahara, S., Yamada, Y., and Watarai, H. 2000. Effect of surfactants on in-plane and out-of-plane rotational dynamics of octadecylrhodamine B at toluene-water interface. *Langmuir* 16:6787–6794.
91. Fujiwara, M., Tsukahara, S., and Watarai, H. 1999. Time-resolved total internal reflection fluorometry of ternary europium(III) complexes formed at the liquid/liquid interface. *Phys. Chem. Chem. Phys.* 1:2949–2951.
92. Fujiwara, N., Tsukahara, S., and Watarai, H. 2001. In situ fluorescence imaging and time-resolved total internal reflection fluorometry of palladium(II)-tetrapyridylporphine complex assembled at the toluene-water interface. *Langmuir* 17:5337–5342.
93. Hashimoto, F., Tsukahara, S., and Watarai, H. 2001. Single molecule detection of cyanine dye at the dodecane-water interface. *Anal. Sci.* 17:181–183.
94. Hashimoto, F., Tsukahara, S., and Watarai, H. 2003. Lateral diffusion dynamics for single molecules of fluorescent cyanine dye at the free and surfactant-modified dodecane-water interface. *Langmuir* 19:4197–4204.
95. Fujiwara, K., and Watarai, H. 2001. Axial hydration and adsorption of chloro(5,10,15,20-tetraphenylporphyrinato) manganese(III) at the toluene/water interface, studied by external reflection spectrophotometry. *Bull. Chem. Soc. Jpn.* 74:1885–1890.
96. Moriya, Y., Nakata, S., Morimoto, H., and Ogawa, N. 2004. Interfacial adsorption state of protonated lipophilic porphyrins in a liquid–liquid system by using reflection spectrometry. *Anal. Sci.* 20:1533–1536.
97. Ohashi, A., and Watarai, H. 2001. Resonance Raman spectroscopic detection of pyridylazo complex formed at liquid–liquid interface in centrifugal liquid membrane system. *Chem. Lett.* 1238–1239.
98. Ohashi, A., and Watarai, H. 2002. Azo-imine resonance in palladium(II)-pyridylazo complex adsorbed at liquid–liquid interfaces studied by centrifugal liquid membrane-resonance Raman microprobe spectroscopy. *Langmuir* 18:10292–10297.
99. Osseo-Asare, K. 2010. Nanoscience in aqueous processing. XXV International Mineral Processing Congress (IMPC) 2010 Proceedings/Brisbane, QLD, Australia, September 6–10.
100. Tarabara, V. V., Nabiev, I. R., and Feofanov, A. V. 1998. Surface-enhanced Raman scattering (SERS) study of mercaptoethanol monolayer assemblies on silver citrate hydrosol. Preparation and characterization of modified hydrosol as a SERS-active substrate. *Langmuir* 14:1092–1098.
101. Otto, A. 2002. What is observed in single molecule SERS, and why? *J. Raman Spectrosc.* 33:593–598.

102. Yamamoto, S., Fujiwara, K., and Watarai, H. 2004. Surface-enhanced Raman scattering from oleate-stabilized silver colloids at a liquid/liquid interface. *Anal. Sci.* 20:1347–1352.

103. Yamamoto, S., and Watarai, H. 2006. Surface-enhanced Raman spectroscopy of dodecanethiol-bound silver nanoparticles at the liquid/liquid interface. *Langmuir* 22:6562–6569.

104. Ribo, J. M., Crusats, J., Sagues, F., Claret, J., and Rubires, R. 2001. Chiral sign induction by vortices during the formation of mesophases in stirred solutions. *Science* 292(5524):2021–2022.

105. Tsuda, A., Alam, Md. A., Harada, T., Yamaguchi, T. Ishii, N., and Aida, T. 2007. Spectroscopic visualization of vortex flows using dye-containing nanofibers. *Angew. Chem.* 119:8346–8350.

106. Wada, S., Fujiwara, K., Monjushiro, H., and Watarai, H. 2004. Measurement of circular dichroism spectra of liquid/liquid interface by centrifugal liquid membrane method. *Anal. Sci.* 20:1489–1491.

107. Wada, S., Fujiwara, K., Monjushiro, H., and Watarai, H. 2007. Optical chirality of protonated tetraphenylporphyrin J-aggregate formed at the liquid/liquid interface in a centrifugal liquid membrane cell. *J. Phys. Condens. Matter* 19:375105.

108. Watarai, H., Mitani, K., Morooka, N., and Takechi, H. 2012. Chiral recognition of 2-alkyl alcohols with porphyrin J-nanoaggregates at liquid–liquid interface. *Analyst* 137(14):3238–3241.

109. Adachi, K., Chayama, K., and Watarai, H. 2006. Formation of helical J-aggregate of chiral thioether-derivatized phthalocyanine bound by palladium(II) at the toluene/water interface. *Langmuir* 22:1630–1639.

110. Corn, R. M., and Higgins, D. A. 1994. Optical second harmonic generation as a probe of surface chemistry. *Chem. Rev.* 94:107–125.

111. Eisenthal, K. B. 1996. Liquid interfaces probed by second-harmonic and sum-frequency spectroscopy. *Chem. Rev.* 96:1343–1360.

112. Steel, W. H., and Walker, R. A. 2003. Solvent polarity at an aqueous/alkane interface: the effect of solute identity. *J. Am. Chem. Soc.* 125:1132–1133.

113. Nagatani, H., Samec, Z., Brevet, P.-F., Fermin, D. J., and Girault, H. H. 2003. Adsorption and aggregation of *meso*-tetrakis(4-carboxyphenyl)porphyrinato zinc(II) at the polarized water-1,2-dichloroethane interface. *J. Phys. Chem. B* 107:786–790.

114. Martin-Gassina, G., Gassina, P. M., Coustonb, L., Diata, O., Benichouc, E., and Brevet, P. F. 2012. Nitric acid extraction with monoamide and diamide monitored by second harmonic generation at the water/dodecane interface. *Colloids Surf. A Physicochem. Eng. Aspects* 413(5):130–135.

115. Petralli-Mallow, T., Wong, M. T., Byers, J. D., Yee, H. I., and Hicks, J. M. 1993. Circular dichroism spectroscopy at interfaces: a surface second harmonic generation study. *J. Phys. Chem.* 97:1383–1388.

116. Byers, J. D., Yee, H. I., Petralli-Mallow, T., and Hicks, J. M. 1994. Second-harmonic generation circular-dichroism spectroscopy from chiral monolayers. *Phys. Rev. B Cond. Matter Mater. Phys.* 49:14643–14647.

117. Fujiwara, K., Monjushiro, H., and Watarai, H. 2004. Non-linear optical activity of porphyrin aggregate at the liquid/liquid interface. *Chem. Phys. Lett.* 394:349–353.

118. Giddings, J. C. 1991. *Unified separation science*. Wiley, New York.

119. Suwa, M., and Watarai, H. 2011. Magnetoanalysis of micro/nanoparticles: a review. *Anal. Chim. Acta* 690:137–147.

120. Suwa, M., and Watarai, H. 2001. Magnetophoretic velocimetry of manganese(II) in a single emulsion droplet at the femtomole level. *Anal. Chem.* 73:5214–5219.

121. Suwa, M., and Watarai, H. 2002. Magnetophoretic velocimetry of manganese(II) in a single microdroplet in a flow system under a high gradient magnetic field generated with a superconducting magnet. *Anal. Chem.* 74:5027–5032.

122. Watarai, H., Suwa, M., and Iiguni, Y. 2004. Magnetophoresis and electromagnetophoresis of microparticles in liquids. *Anal. Bioanal. Chem.* 387:1693–1699,

123. Suwa, M., and Watarai, H. 2003. Magnetophoretic velocity of microorganic droplets adsorbed by dysprosium(III) laurate in water. *J. Chromatogr. A* 1013:3–8.

124. Suwa, M., and Watarai, H. 2008. Magnetophoretic evaluation of interfacial adsorption of dysprosium(III) on a single microdroplet. *Anal. Sci.* 24:133–137.

125. Tsukahara, S., Takata, A., and Watarai, H. 2004. Magnetic field enhanced microextraction rate of europium(III) with 2-thenoyltrifluoroacetone and oxalate at dodecane-water interface. *Anal. Sci.* 20:1515–1521.

126. Hill, A. W., and Benjamin, I. 2004. Influence of surface tension on adsorbate molecular rotation at liquid/liquid interfaces. *J. Phys. Chem. B* 108:15443–15445.

127. Watarai, H., Gotoh, M., and Gotoh, N. 1997. Interfacial mechanism in the extraction kinetics of Ni(II) with 2-(5-bromo-2-pyridylazo)-5-diethylaminophenol and molecular dynamics simulation of interfacial reactivity of the ligand. *Bull. Chem. Soc. Jpn.* 70:957–964.

128. Watarai, H., and Onoe, Y. 2001. Molecular dynamics simulation of interfacial adsorption of 2-hydroxy oxime at heptane/water interface. *Solvent Extr. Ion Exch.* 19:155–166.

129. Vayssière, P., and Wipff, G. 2003. Importance of counter-ions in alkali and alkaline-earth cation extraction by 18-crown-6: molecular dynamics studies at the water/sc-CO$_2$ interface. *Phys. Chem. Chem. Phys.* 5:2842–2850.

130. Chevrot, G., Schurhammer, R., and Wipff, G. 2007. Synergistic effect of dicarbollide anions in liquid–liquid extraction: a molecular dynamics study at the octanol–water interface. *Phys. Chem. Chem. Phys.* 9:1991–2003.

131. Chevrot, G., Schurhammer, R., and Wipff, G. 2007. Molecular dynamics study of dicarbollide anions in nitrobenzene solution and at its aqueous interface. Synergistic effect in the Eu(III) assisted extraction. *Phys. Chem. Chem. Phys.* 9:5928–5938.

132. Ibrahim, T. H., and Neuman, R. D. 2006. Molecular modeling study of the aggregation behavior of nickel(II), cobalt(II), lead(II) and zinc(II) bis(2-ethylhexyl) phosphate complexes. *J. Colloid Interf. Sci.* 29:4321–4327.

133. Ibrahim, T. H. 2010. Role of water molecules in sodium bis(2-ethylhexyl) phosphate reversed micelles in n-heptane. *J. Franklin Inst.* 347:875–881.

6 Supramolecular Aspects of Stable Water-in-Oil Microemulsions in Chemical Separations

Mikael Nilsson
Department of Chemical Engineering and Materials Science, University of California, Irvine, California

Peter R. Zalupski
Aqueous Separations and Radiochemistry Department, Idaho National Laboratory, Idaho Falls, Idaho

Mark R. Antonio
Chemical Sciences and Engineering Division, Argonne National Laboratory, Argonne, Illinois

CONTENTS

6.1   INTRODUCTION

The separation and purification of elements and chemicals on a commercial scale is of fundamental importance for our everyday life. Ranging from cosmetics to jet engines, the applications that require pure metals or pure chemicals are numerous. One of the most widespread separation techniques for nonvolatile chemicals is liquid–liquid extraction, or solvent extraction (SX). In the field of hydrometallurgy, these two-phase processes have been used for half a century on an industrial level,[1] and operating knowledge for most processes is good. The current solvent extraction industry gained momentum during the 1940s from the studies of plutonium purification, where solvent extraction played a key role and new solvating and chelating extraction reagents were developed. Some of these reagents are still used today in commercial metal refineries. Solvent extraction still is the benchmark technique for separation of elements in used nuclear fuel, and since the 1940s, several advanced processes have been developed. While some are now well established, fundamental knowledge of the chemical interactions is still incomplete. Both historical and contemporary research shows that neutral and acidic extracting reagents used to extract and purify metal ions quite often display surface-active tendencies, form self-assembled aggregate structures, and yield organized, supramolecular nonaqueous environments. Some of the most prevalent supramolecular entities are of the reverse-micelle type, where amphiphilic molecules shield the polar heads through the formation of a hydrophilic core and by orienting their lipophilic tails into the diluent. At certain conditions, depending on several factors, including temperature, organic diluent, ionic media, and ionic strength, to name just four, the reverse micelles undergo further attractive associations resulting in the splitting of the organic phase to form a phase that resides between the aqueous and organic phases. This undesirable phenomenon is known as third-phase formation. A half-century worth of other research has indicated that acidic extracting reagents are prone to form hydrogen-bonded aggregates in the organic phase as well as stable water-in-oil microemulsions. These supramolecular assemblies can dramatically alter the efficiency of the extraction process. The influence of the effects depends upon the system studied and has been reported as both favorable (synergistic) and unfavorable (antagonistic) in terms of impact on the extraction efficiency and rate of phase transfer. In order to successfully model liquid–liquid extraction processes, the presence of microemulsions and their influence on the mechanism of extraction must be taken into account to refine predictions based solely on metal ion coordination chemistry models. Stable water-in-oil microemulsions have been observed for many extraction reagents developed for the separation of elements in used nuclear fuel (including lanthanides (Ln), actinides (An), and numerous fission products of the s-, p-, and d-block elements), as well as conventional metal refining. Knowledge about the formation and role of microemulsions in separation processes provides a new window through which to view the performance and reliability of industrial and analytical chemical operations in an original manner. This review collocates aspects of fundamental and practical research to present insights into solvent-mediated interactions between solutes governing the formation, structure, and effect of aggregate formation in organic solutions employed for chemical separations.

A solvent extraction process for metal ions relies on the electrostatic interaction of one (or more) extraction reagent(s) in an oil phase with metal ions in an aqueous phase, forming lipophilic molecular metal–extractant complexes that are transferred across a water–oil interface. Numerous studies support the formation of metal–extractant complexes, and there are well-established techniques for analyzing extraction data, for example, slope analysis, Job plots, etc.[2] Conventional slope analysis often interprets the extraction data as one or more monomers of the reagent (commonly referred to as an amphiphilic ligand) interacting with the metal ion. In addition to the ligand interacting with metal ions (the extracted solute), there are numerous examples of ligand–ligand interactions, as well as ligand interactions with other solutes, such as water and acids, in the nonaqueous phase. The result of the solvent-mediated solute–solute, solute–ligand, and ligand–ligand interactions is the formation of dimers, trimers, and higher-order aggregates. The exact nature of these interactions arises through a combination of hydrogen bonds between the polar groups of the amphiphilic ligands as well as myriad electrostatic-based interactions and attractive van der Waals forces. For certain ligands, there is evidence of metal ion complexation through partial disruption of the aggregate structure resulting in a chelate ring formation, such as when a dimer structure of two O–H–O rings opens up and one hydrogen may be replaced by a metal ion. Some extracting reagents are able to extract water and mineral acids. Tri-*n*-butyl phosphate (TBP) is a prime example,[3] where the extracted water and acid may support the hydrogen bond formation to build higher-order TBP aggregates. It has been shown that as the aggregates grow in the nonaqueous phase, they have a tendency to arrange themselves into spherical forms, where the lipophilic groups interact with the bulk organic phase and the polar groups interact with hydrogen-bonded water and acid as well as with polar groups of the other ligands in the aggregate, effectively forming a reverse micelle. A conceptual illustration of a spherical reverse micelle (RM) is provided in Figure 6.1.

Once the reverse micelles have formed, studies have shown that as the acidity of the aqueous phase in contact with the organic phase containing the reverse micelles increases, the reverse micelles can start to accommodate water and acid inside the polar core, forming an aqueous droplet. As the amount of water

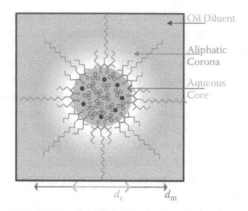

FIGURE 6.1 Conceptual illustration of a spherical reverse micelle of diameter d_m and its two constituent parts: (1) The inner core of diameter d_c contains water, acid (triangular arrangements of open circles represent NO_3^-), and metal cation solutes (filled circles), as well as the hydrophilic, polar portions (O donor atoms) of the amphiphile extractant molecules. (2) The outer corona of thickness $d_s = d_m - d_c$ consists of the oleophilic, nonpolar portions (*n*-alkyl group chains) of the extractant that extend out into the bulk, paraffinic diluent.

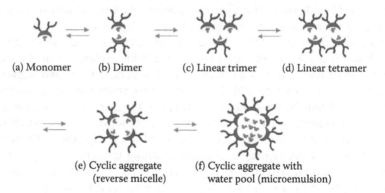

FIGURE 6.2 The progression of monomers (a) of surface-active extracting reagents to form dimers (b), trimers (c), polymers (c and d), reverse micelles (e), and microemulsions (f). (Adapted from Osseo-Asare, K., *Adv. Colloid Interface Sci.*, 37(1–2), 1991, 123–173.)[4]

increases in the reverse-micelle core, the droplet will swell, and the organic phase changes into a thermodynamically stable microemulsion. The progression of monomers to dimers to spherical aggregates to microemulsions is shown in Figure 6.2.

In general, microemulsions can be formed when surface-active molecules organize themselves in clusters to produce micelles or reverse micelles as described above. An organic diluent and an aqueous phase are required, and sometimes a co-surfactant is added to facilitate the formation of the microemulsion. Micelles have an outer corona of polar groups and an inner core of hydrophobic hydrocarbon groups (e.g., long-chained aliphatic groups), an environment that is suited for the solubilization of hydrophobic molecules. Reverse micelles have instead the hydrocarbon chains on the outer corona, and the polar groups are oriented to form the core architecture (as shown in Figure 6.1), which is suited for the solubilization of aquated metal complexes. Israelachvili et al.[5] studied the shape of different aggregates and related the structures forming to the ratio of the volume (per molecule) of the hydrocarbon tail of the amphiphile, v, and the optimal surface area of the amphiphile, a_0 (i.e., the area each molecule contributes to the total surface area of the aggregate at the minimum free energy per amphiphile in a micelle), and the length of the hydrocarbon tail, l (approximately equal to 90% of the full length of the extended carbon chain). They outlined boundaries in terms of values of the $v/(a_0\, l)$ ratio where various structures are formed (see Figure 6.3).[6] For $v/(a_0\, l) \leq 1/3$, normal micelles form. For values of $1/3 < v/(a_0\, l) < 1/2$, rodlike micelles are formed, and for values of 1/2 to 1, lamellar structures are formed, and finally, for values above 1, reverse micelles form. This means that for reverse micelles to form, the hydrocarbon volume must be relatively large or the surface area requirement must be small. Most extracting reagents may be characterized by a relatively high $v/(a_0\, l)$ parameter, never forming normal micelles.

Stable microemulsions have been observed for a long time as a separate liquid phase from pure aqueous or pure organic phases. Winsor[7,8] classified four types of equilibrated emulsified systems, as described in Figure 6.4. Type I is an organic phase in contact with a phase consisting of an oil-in-water microemulsion,

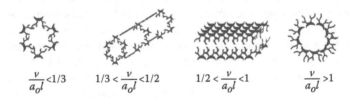

$$\frac{v}{a_0 l} < 1/3 \qquad 1/3 < \frac{v}{a_0 l} < 1/2 \qquad 1/2 < \frac{v}{a_0 l} < 1 \qquad \frac{v}{a_0 l} > 1$$

FIGURE 6.3 Various structures of aggregates forming at different ratios of the hydrocarbon volume, v, to the surface area, a_0, and the length of the hydrocarbon chain, l. (From Paatero, E., and Sjöblom, J., *Hydrometallurgy*, 25(2), 1990, 231–256.)[6]

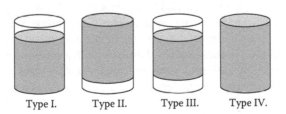

Type I.   Type II.   Type III.   Type IV.

FIGURE 6.4 Types of Winsor microemulsions.[8] Gray areas denote microemulsion; the oil phase is assumed to be less dense than the water phase.

comprised of micelles. Type II is an aqueous phase in contact with a water-in-oil microemulsion—aggregates formed by reverse micelles are the principal systems of interest here. Type III is sometimes called third-phase formation and consists of an aqueous and an organic phase with a microemulsion layer between them. Type IV is one single phase consisting of micelles or reverse micelles where no single organic or aqueous phase is present.

Emulsified systems have been shown to feature some physical properties of interest to industrial applications in separation science and technology.[9–17] The observed enhancement in the interfacial area may improve reaction kinetics, and the formation of micellar structures may benefit solubility in nonaqueous media. In addition, significant water content in the microemulsion phase renders it relatively inexpensive when compared to a molecular organic solvent.[18] Systems consisting of multiple surface-active amphiphiles are especially prone to the formation of aggregates and microemulsions. Coincidentally, systems with such multicomponent chemistry are commonly used in SX processes as described above.[4,19–31]

The phenomenon of clustering of solutes in organic solvents and the formation of microemulsions impacts a wide variety of issues of both basic and practical interest on a broad scale from oil field chemistry to pharmaceutical research to separation science and technology.[32] Aggregation behaviors that result from self-assembly and directed-organization processes in bulk solutions stand at the very core of supramolecular chemistry.[33] However, the forces that drive and limit bulk ordering, in terms of kinetics, thermodynamics, colligative, and spectroscopic properties, are not well understood. A primary reason for this deficiency is that studies are usually approached from a solute, namely, metal-centered, standpoint using models and procedures appropriated from coordination chemistry. For example, SX studies of metal ions—wherein selected metals are preferentially moved out of an aqueous cocktail and into an organic phase via complexation with an extractant molecule—are typically performed under the ideal conditions of low concentrations of metals and extractants, thereby promoting the existence of discrete, mononuclear metal-extractant complexes in the organic phase. The complexes identified in this way, however, are unlikely to exist under the conditions met in practice, where the extractant concentration is generally high and the organic phase often approaches saturation with respect to the extracted species. Under these more realistic conditions, the discrete metal–extractant entities familiar to coordination chemistry undergo self-assembly, leading to the formation of much larger species, and to phase separation phenomena. These phenomena have been understood by complementing the metal-centered approach traditionally followed in coordination chemistry with a solvent-centered approach in which the major emphasis is on supramolecular chemistry, vis-à-vis aggregation and micellization of extractant molecules (e.g., surfactants) in nonpolar media.[34,35] Selected aspects of aggregate formation in an assortment of solvent extraction systems are reviewed here. It has been generally accepted that third-phase formation is attributed to aggregate formation, and this topic has already received an extensive review in a previous volume of this book series.[36] Although some overlap with this work is unavoidable, insofar as possible, we have attempted to focus on aggregates forming in the absence of third-phase formation.

6.2 EXPERIMENTAL AND ANALYTICAL CONSIDERATIONS FOR SX SYSTEMS WITH EXTENSIVE AGGREGATIVE BEHAVIOR

With more than a century of research on classical surfactant micelles as a guide, it is clear that the coordination of observations from multidisciplinary investigations—in a bottom-up, progressive zoom manner, spanning atomic, molecular, and supramolecular length scales—of extractant molecules and their complexes with solutes will elucidate the processes in liquid–liquid extraction in a clear and definitive fashion. The science of microemulsions, in general, and micelles, in particular, has advanced through use of an armamentarium of experimental methods,[37,38] including vapor-pressure osmometry (VPO)[39,40]; ultracentrifugation[41]; viscosimetry[42]; freezing- and cloud-point measurements; nuclear magnetic resonance (NMR)[43–48]; electron spin resonance[49]; electronic, optical, vibrational, and photon correlation[50–52]; fluorescence[53] and X-ray spectroscopies; light[42,54,55]; X-ray[56–61] and neutron[62–66] scattering; dielectric measurement[67–69]; electro-optical Kerr effect[70,71]; and positron annihilation.[72]

The methods in liquid–liquid extraction using commercial instrumentation (e.g., distribution, slope, and continuous variation analyses, mass spectrometry, etc., as mentioned elsewhere in this review) provide information regarding the composition and possible stoichiometry of extraction systems. Methods commonly associated with large-scale user facilities for X-ray absorption spectroscopy (XAS),[73–83] large-angle[84] and small-angle scattering techniques (both small-angle X-ray scattering (SAXS) and small-angle neutron scattering (SANS)[85–89]) provide metrical information about the core and corona of the reverse micelle (as illustrated in Figure 6.1 and as described below in this review). In addition to the above methods, the electroanalytical method of voltammetry—the electrochemical equivalent of spectroscopy—has found recent application. Whereas both cyclic voltammetry and differential pulse voltammetry techniques have been used beforehand to study Faradaic couples, such as the ferro-ferricyanide $[Fe(CN)_6]^{4-}/[Fe(CN)_6]^{3-}$ response in reverse micelles,[90–96] and non-Faradaic ones, such as with Aerosol OT (AOT) (1,4-bis-2-ethylhexylsulfosuccinate, the most extensively studied anionic surfactant) at an oil–water (O/W) interface in studies of incipient micellization,[97] their application to the study of organized media with regard to liquid–liquid extraction of cerium has only recently received attention.[98] In view of the limited conductivities of the organic phases in SX,[99,100] the aforementioned electrochemical experiments were approached in a manner different from conventional, two-phase (consisting of a solid working electrode surface (phase 1) immersed into a single solution (phase 2) containing the analyte and electrolyte) applications in highly conducting media. Instead, a three-phase voltammetry method was applied, wherein the solid working electrode (phase 1) has a thin-film coating of the water-immiscible, analyte-containing organic phase (phase 2) that is immersed into an aqueous electrolyte (phase 3).[101] It is surprising that the technique has not been applied to liquid–liquid extraction before now, because the very nature of the process, involving the contact of two immiscible liquids—namely, an aqueous phase and an aliphatic hydrocarbon one—is precisely tailored to the use of the three-phase methodology. Moreover, in view of intersecting interests in the use of the coordination chemistry of trivalent Ln ions to probe the confined aqua environments of reverse-micelle cores[76,102–106] and for use in templating the preparation of ultrafine particles,[107–112] including quantum dots,[113,114] as well as geopolitical pressures pushing for the development of more efficient rare-earth refining technologies,[115] new approaches to the understanding of old problems will drive the field to high-performance systems.

The morphologies of architectures formed in interacting solution systems employed in solvent extraction can oftentimes be challenging to discern from small-angle scattering data. This is because interactions between reverse-micelle units influence the scattering in the low to medium region of photon momentum transfer (denoted by the symbol Q), which is the very region that holds information on the extended structure of aggregates.[116] The micellar model of SX as pioneered by Erlinger et al.[99,117] and Chiarizia et al.[74,118–123] was based upon the interpretation of small-angle scattering data using the Baxter model for sticky hard spheres.[124,125] The micelles, they proposed, behave like small adhesive spheres that, upon incorporation of metal ions or acid, attract each other. When the strength of the interaction exceeds $2k_BT$, where k_B corresponds to the Boltzmann constant and T is temperature, the reverse micelles "condense" and form a discrete "third" phase, which limits processing capacity and can be dangerous in nuclear fuel reprocessing due to criticality issues.[126] Although helpful in understanding the physical reasons behind third-phase formation, the Baxter model is a simplification of the dynamic, fluid properties of solvent extraction supramolecular structures. Going beyond Baxter, in efforts to understand the supramolecular structures of the higher-ordered solution architectures, the scattering contribution from the interactions must be separated from the scattering produced by aggregates. This has recently been achieved using the generalized indirect Fourier transformation (GIFT) method, which uses a model to approximate the interactions between micellar cores and simultaneously converts the scattering contribution from the structure of the mesoscopic architectures into real-space functions known as pair distance distribution functions (PDDFs). As shown in Figure 6.5, a PDDF describes the scattering entity by a distribution of distances between points within the assembly, with r being the distance and $p(r)$ the relative number of particular distances that occur; this number becomes zero at the maximum linear extent of the scattering particle. As such, the PDDF provides specific metrical parameters about the aggregates. The short-axis diameter of the core is obtained from the inflection point after the initial peak in the same way as that for a cylinder (determined from the second derivative of the function).[127,128] An example of the type of information obtained by use of GIFT is provided in the display of Figure 6.5, showing the PDDF plots for organic solutions of the diamide DMDOHEMA (dimethyl-dioctyl-hexaethoxy-malonamide) after contact with aqueous solutions of lithium nitrate and nitric acid. For the

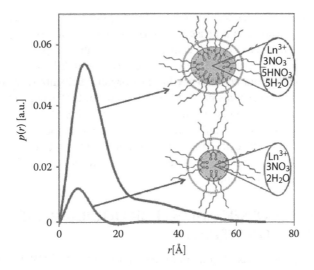

FIGURE 6.5   PDDFs from the SAXS data of organic phases: 0.5 M DMDOHEMA in *n*-heptane after contact with 3 M HNO₃ containing no metal ion (top curve) and after contact with 3 M LiNO₃ containing no metal ion (bottom curve). Both of these functions display a prominent positive peak at low *r*, attributable to the polar core size, which is significantly larger for the acidic system than for the neutral one. The peak in the neutral system is followed by a small negative region and a smaller positive peak, which are typical for inhomogeneous ellipsoids with a core-shell structure.[128] The peak in the acidic system is followed by a tail-like structure, which is indicative of a transition from globular to rod-shaped micelles[129] that would occur as reverse micelles assemble into chains.[130] The saddle point between the initial peak and the tail suggests that the inhomogeneity is preserved through this transition.[128] The cross sections of reverse-micellar unit structures are illustrated for the neutral (bottom curve) and acid (top curve) systems based on the GIFT PDDFs and Baxter modeling results. (From Ellis, R. J. et al., *Chemistry—A European Journal*, DOI: 10.1002/chem.201202880, 2013.)[131]

neutral system (bottom curve), this approach reveals a short ellipsoid axis for the polar core of 11.0 Å, a value that is comparable with the Baxter model that assumes aggregates are spherical in shape with a core diameter of 11.6 Å. For the acidic system (top curve), the short axis of the core cross section is found to be 14.8 Å, a value that is equivalent to that from the Baxter treatment.

Studies of malonamide acid[130–132] and TRUEX acid[133] (transuranic extraction, where a mixture of CMPO (carbamoyl-methyl-phosphine oxide) and TBP is utilized to co-extract trivalent An and Ln) systems alike using the GIFT method show that globule-to-rod transitions driven by acid extraction underpin organic-phase behaviors. This perspective contrasts with the accepted Baxter model as it concerns third-phase formation. Regarding soft-matter science, Israelachvili et al.[5] predicted sphere-to-rod transitions as a function of amphiphile concentration in aqueous solutions and established a basic equation linking chemical potential to curvature variation. Furthermore, Bellini et al.[134] linked critical phenomena in an ionic amphiphile solution to the one-dimensional growth of micellar rods induced by increasing concentrations of amphiphile. The rods become flexible and entangled, leading to a phase transition at the critical concentration. Other soft-matter studies have linked changes in curvature to phase boundaries,[135] some involving the addition of polar solutes to amphiphile systems.[136] The SX of inorganic acids from aqueous media into organic solvents with malonamide, CMPO, and TBP extractants produces structured fluids sharing similarities with nonionic surfactant-in-oil systems that show the assembly of reverse micelles into cylindrical chains whose lengths are controllable by temperature or surfactant concentration.[137–142] The extraction of mineral acids into organic phases of practical relevance to separation systems behaves like soft matter, with phase splitting traceable to mesoscopic roots.

6.3 GENERAL OBSERVATIONS FOR LIQUID–LIQUID EXTRACTION SYSTEMS

The emerging consensus of contemporary studies[4,74,75,99,117–120,123,136,143–161] is that the organic phases used in many liquid–liquid extraction systems are not simple molecular solutions of extractants but rather more complex solutions of extractant aggregates, including dimers, tetramers, etc., and even larger, supramolecular assemblages—particularly those of the reverse-micelle type, as illustrated by a progressive architectural buildup in Figure 6.2. This micellization behavior and the formation of microemulsions are attributable to the self-organization of amphiphilic extractants.[28,162–165] The structures for a number of phosphorus- and nitrogen-based amphiphiles employed in nuclear and hydrometallurgical separation technologies are shown in Table 6.1. The very presence of aggregates in aliphatic, nonpolar diluents, such as n-dodecane, is of prime importance for the understanding of the bulk solution behaviors—in terms of efficiency and selectivity to name just two—of the extraction systems.

The understanding of selected SX processes in terms of aggregation and micellization phenomena in solution phases containing metal ion complexes with the extractants illustrated in Table 6.1 is reviewed. Toward this end, focus is placed upon the various conditions and effects that impact the nucleation of such molecular-scale ordering and the organization of extractants in bulk organic diluents. Such structural information is a critical ingredient to understanding the kinetics[166] and mechanisms involved in driving an ion across a water–oil interface and the corresponding organization/reorganization of the extractant-solute complex in its transformation from an interface-adsorbed species to a solubilized organic-phase complex. Such macromolecular system insights offer new perspectives into chemical separations, wherein the transport of target solutes across an interface between an aqueous and an organic solvent may be facilitated by the self-organization, micellization, and aggregation behaviors of extractants in nonpolar diluents. Examples of increasing structure hierarchy can be the transition from molecules to micelles; from contact ion pairs to correlated fluids of highly loaded diluents (third phases); from electrostatic (Born) to dipole–dipole (van der Waals) forces. Despite decades of international research

TABLE 6.1
Structures and Names of Ligands Referred to in This Review

Name	Structure	Abbreviation/Trademark

Neutral Phosphorous Reagents

Name	Structure	Abbreviation/Trademark
Tributyl phosphate	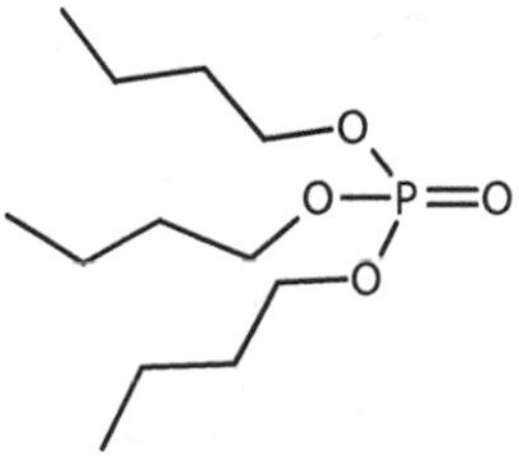	TBP
Trioctyl phosphine oxide	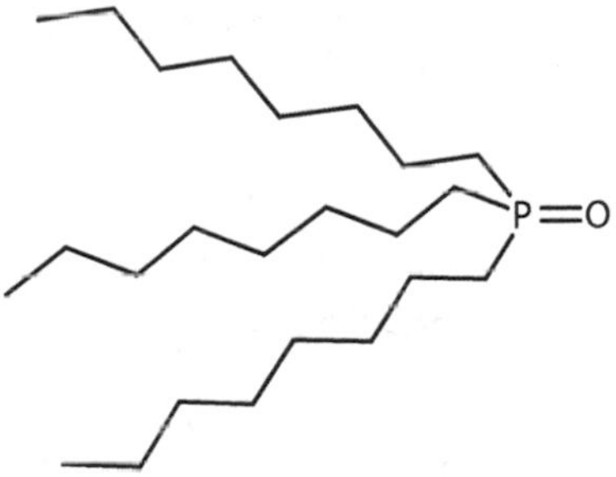	TOPO
Trialkyl phosphine oxide	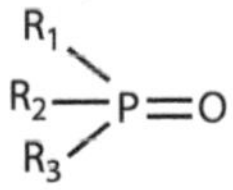 $R_2-\overset{R_1}{\underset{R_3}{P}}=O$	TRPO, Cyanex 923 R1 = R2 = R3 = C_8H_{18}, TOPO R1 = R2 = R3 = C_4H_{10}, TBPO

Quaternary Ammonium Salts

Name	Structure	Abbreviation/Trademark
Trioctyl methylammonium cation		TOMA, Aliquat 336 (mixture of C8 and C10 chloride salts)

(Continued)

TABLE 6.1 (*Continued*)

Structures and Names of Ligands Referred to in This Review

Name	Structure	Abbreviation/Trademark
	Acidic Phosphoric Reagents	
Dibutyl phosphoric acid or dibutyl phosphate monoacid	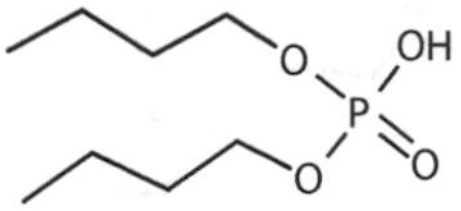	DBP or HDBP
Di-2-ethylhexyl phosphoric acid	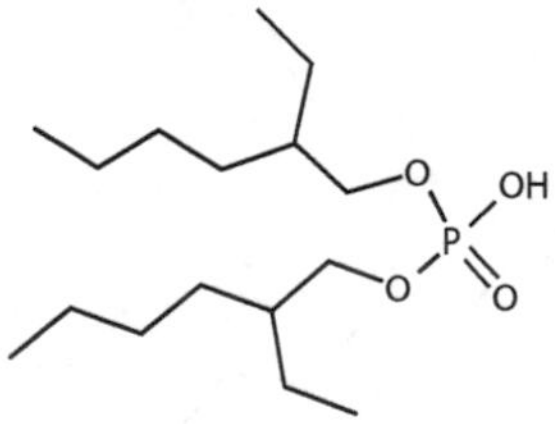	HDEHP or D2EHPA or DEHPA
Di-2-ethylhexyl phosphonic acid	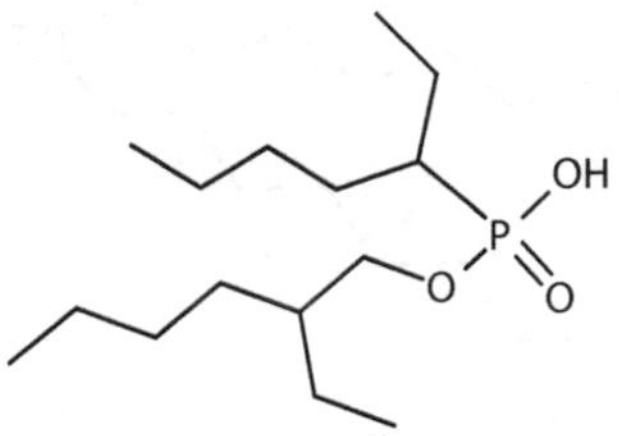	PC88A
2,4,4-trimethyl-pentyl phosphinic acid	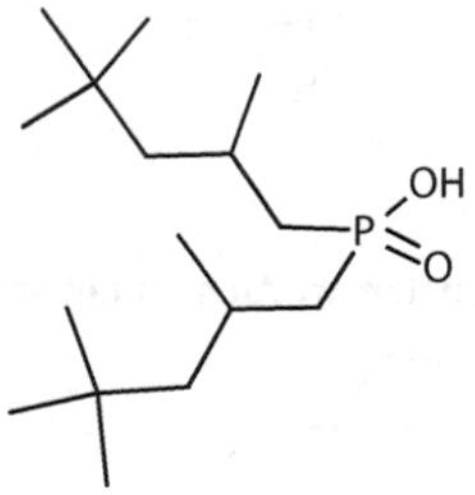	Cyanex 272
P,P′-di(2-ethylhexyl)- alkyl-diphosphonic acid *n* is the alkyl group of different lengths (methylene, ethylene, butylenes)	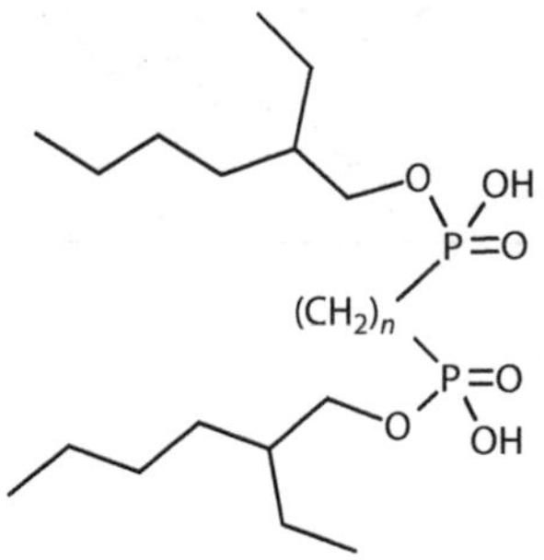	For $n = 1$, $H_2DEH[MDP]$ For $n = 2$, $H_2DEH[EDP]$ For $n = 3$, $H_2DEH[BuDP]$

TABLE 6.1 (*Continued*)
Structures and Names of Ligands Referred to in This Review

Name	Structure	Abbreviation/Trademark
P,P′-di(2-ethylhexyl) benzene-1,2-diphosphonic acid	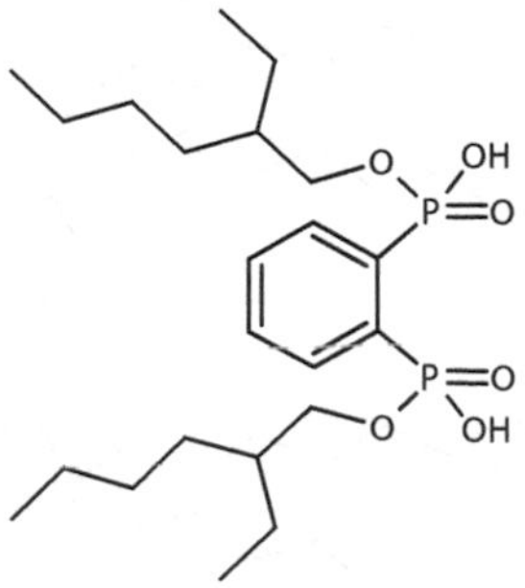	$H_2DEH[1,2\text{-}BzDP]$

Oximes and Quinolines (chelating agents)

7-(1-ethyl-3,3,5,5-tetramethylhexyl)-8-quinolinol	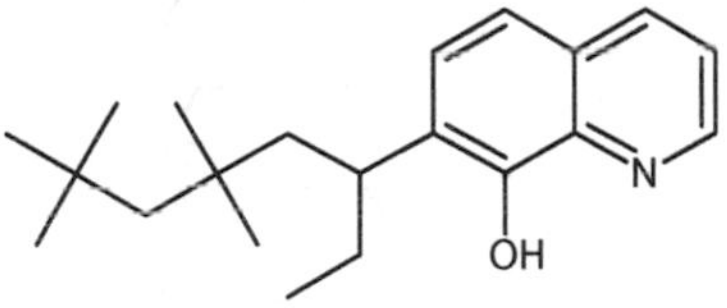	Kelex 100, HQ
5,8-diethyl 7 hydroxy-6-dodecanone-oxime	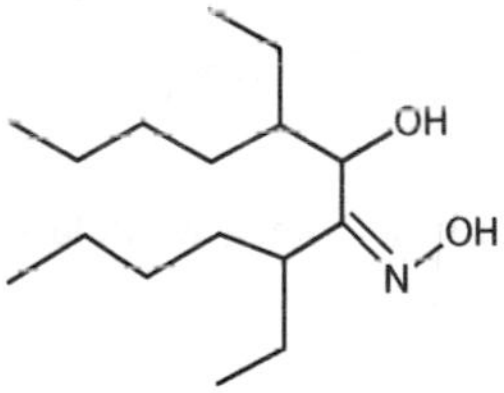	Hydroxyoxime, LIX 63

Neutral Oxygen Donor (amides, glycolamides)

Di-methyl-di-butyl-tetradecyl-malonamide	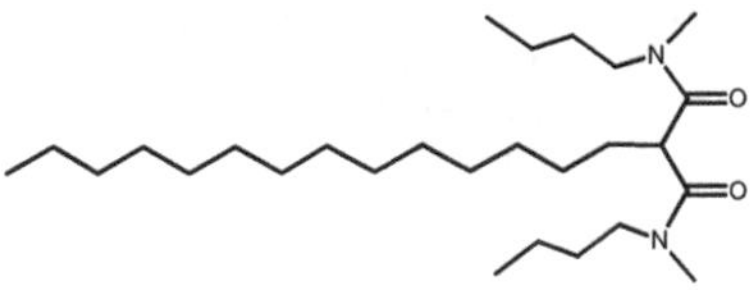	DMDBTDMA
Dimethyl-dioctyl-hexaethoxy-malonamide	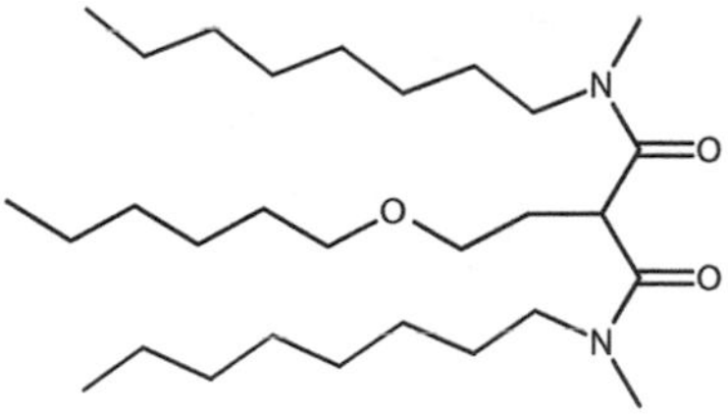	DMDOHEMA

(Continued)

TABLE 6.1 (*Continued*)
Structures and Names of Ligands Referred to in This Review

Name	Structure	Abbreviation/Trademark
Tetraoctyl-diglycolamide	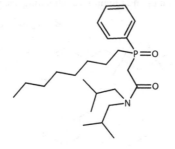	TODGA
Carbamoyl-methyl- phosphine oxide		CMPO

Sulfuric Acid Reagents (thio acids)

Name	Structure	Abbreviation/Trademark
Bis(2,4,4-trimethyl- pentyl)dithio-phosphinic acid	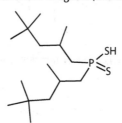	Cyanex 301
Bis(2,4,4-trimethyl- pentyl)monothio- phosphinic acid	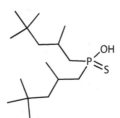	Cyanex 302

TABLE 6.1 (*Continued*)
Structures and Names of Ligands Referred to in This Review

Name	Structure	Abbreviation/Trademark
	Sulfonic Acids (surfactants)	

Di(2-ethylhexyl)sodium sulfosuccinate

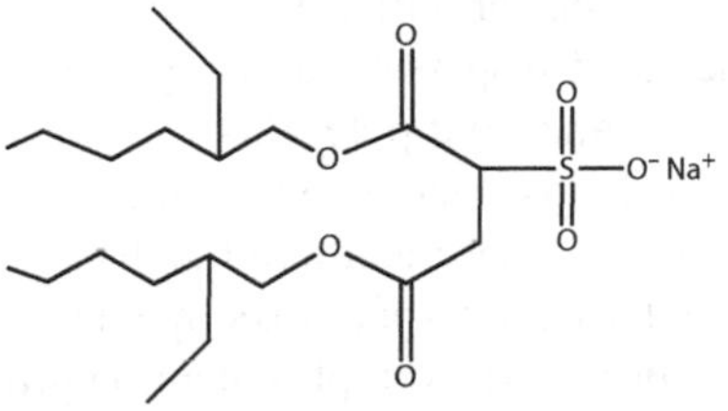

Aerosol OT or AOT

Dinonyl-naphthalene sulfonic acid

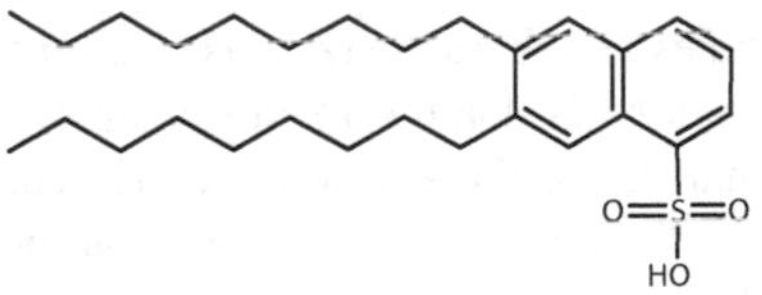

HDNNS HDDNS (if dodecyl chains replaces the nonyl chains)

in SX, the effects of these transitions on selectivity and control, synergism, and antagonism in systems of relevance to the separation science of actinides, lanthanides, and transition metal ions are yet to be fully exploited, as they are in analytical extractions of organic compounds in biological, environmental, and agrifood fields.[167,168] There is sufficient research to suggest that coulombic associations and their correlation effects, as exploited in electrically switched ion exchange (ESIX) for Cs-selective ion[169–171] and rare-earth metal[172,173] separations, and van der Waals forces between polar cores of self-assembled reverse micelles can be tuned to transition from aggregates of solutes to structured fluids in ways different from third-phase formation. The impact and significance of realizing such rational and deliberate controls will provide an innovative entry into a comprehensive and predictive description of the mechanisms and energetics of SX (and ion exchange (IX)) as well as third-phase formation—a topic covered extensively elsewhere.[36,74,89,119–121,126,143,145,150,152,155,174–202] The accumulation of knowledge about the role of associating behaviors and metrical aspects of interactions between extractants, solvents, and solutes (both neutral and charged ones, e.g., H_2O, HNO_3, cations, and anions) at liquid–liquid interfaces offers the prospect to manipulate micelle formation and, in so doing, to preferentially tailor process performance.

6.3.1 Solvent-Centered versus Metal-Centered Extraction

For isolated ions and molecules in organic solutions to self-assemble in aggregates of various sizes and shapes, the single ions or molecules must interact with one another in manners that are driven by the net balance of energetics arising from three different associations: (1) solute–solvent; (2) solute–solute, and (3) solvent–solvent.[203] The solvation of neutral, cationic, and anionic solutes has a significant influence on the chemistry and phase stability of the solution, especially with regard to ion-transfer processes. Despite long-standing practical, fundamental, and theoretical interests, there is no generally agreed-upon model of solution chemistry in solvent extraction for which it is envisioned that three fundamental types of solvent-mediated dispersion and repulsive forces come into play: (1) ion–ion, (2) ion–dipole, and (3) dipole–dipole.[204] Structural aspects of these correlations in bulk, polar fluids, such as water and acetonitrile, have long been examined through analyses of experimental radial distribution functions obtained by neutron and X-ray scattering and spectroscopy methods.[104,205–209] These types of fluids (water and acetonitrile) are examples of amphiprotic and aprotic solvents, respectively, with high dielectric constants and in which the interactions between ions and permanent dipoles are characterized as coulombic. In comparison, current knowledge of the nature of solutes in nonpolar solvents, especially paraffinic ones, is far less mature. In these low-dielectric-constant diluents, such as n-octane and n-dodecane, the most important forces at the origin of the attraction between solute molecules are the various types of van der Waals forces between permanent-permanent, permanent-induced, and induced-induced (dispersion) dipoles. As such, investigations of self-assembly phenomena and third-phase formation in solvent extraction are almost wholly centered on understanding the role of van der Waals forces that are present in the organic phase.

Historically, the emphasis on a metal-centered interpretation of solute transfer in liquid–liquid systems has been a dominating approach in solvent extraction science.[210] Focused on a coordination environment of a metal ion, this approach typically invokes simplified mixtures where discrete metal complexes of uniform stoichiometry are assumed to exist in a solvent continuum. Figure 6.6 depicts one literature example of such a metal-centered understanding of the solute partitioning in solvent extraction, where stoichiometric relationships are clearly defined.[211] In contrast, very limited regard to aggregation behaviors and structural aspects in the nonaqueous environment is given, with the organic phase typically considered as bulk medium.

The scientific consensus from research of the past decade appears to add to the momentum carried by another representation of liquid–liquid systems, where the aggregation behaviors of extractants and metal–extractant complexes in SX affect the distribution of target solutes between liquid phases. In particular, there is a dearth of new information about kinetic and thermodynamic effects, metal–extractant stoichiometries, speciation, and structures especially in process systems employing combinations of extractants, e.g., DIAMEX-SANEX (diamide extraction-selective actinide extraction), which employs the combination of DMDOHEMA and HDHP as well as novel multidentate nitrogen-containing ligands, bis-triazine-pyridines and bis-triazine-bis-pyridines (BTPs and BTBPs), and TRUEX, which employs the combination of CMPO and TBP. A solvent-centered approach seeks relationships between the solution structure and solute

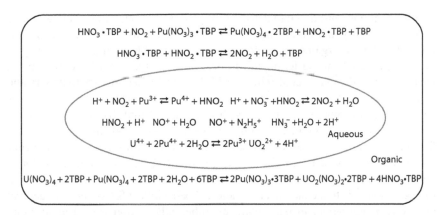

FIGURE 6.6 Metal-centered representation of a liquid–liquid system, where possible stoichiometric relationships are identified to quantify mass transfer. The figure visualizes a lack of structural consideration of the fluid phases in a metal-centered approach. (From Miles, J. H., *Science and Technology of Tributyl Phosphate*, Vol. 3, CRC Press, Boca Raton, FL, 1990.)[211]

partitioning in liquid–liquid systems. Solvent-mediated interactions between solutes, whether it is something as simple as electrostatic ion pairing, hydrogen bonding, or van der Waals interactions, or more complex like micellization, orchestrate solution properties of fundamental and practical importance in both natural and engineered settings, including electron transfer and ion transport, chemical reactivity, crystallization, and precipitation.

It has been observed that separation factors generally, in the absence of oversaturation and steric interferences, improve by preorganizing chelating groups of ligands onto molecular (macrocyclic) platforms, such as calixarenes, cavitands, cryptands, tripodands, trityls, dendrimers, and nanoparticles (creating a polydentate extractant).[212,213] However, the preorganization of extractants in nonpolar diluents mediates the distribution of target solutes between liquid phases in ways that are not as fully understood. The consequences of extractant aggregation (e.g., micellar pseudophase morphology)[214] on solute (e.g., M^{n+}, H_2O, HNO_3, etc.) extraction and the solvent effects driving and limiting the formation of supramolecular aggregates of extractants and their complexes with solutes are only slowly coming to light.[6,30] The physical reality of whether or not there can be more than one metal ion, namely, multinuclear species, per aggregate remains to be demonstrated. With the advent of synchrotron radiation, the application of techniques like high-energy X-ray scattering (HEXS), with its demonstrated sensitivity to elucidate long, nonbonding metal–metal interactions in polynuclear solution species,[207,209,215,216] will advance knowledge in the field in ways that are otherwise not possible. Although the solvent extraction of metal ions is coordination chemistry at the most fundamental level,[217–220] higher-order effects on the process still stand to be exploited vis-à-vis the manipulation of chemical and physical factors that drive and limit self- and directed-assembly of microemulsions and, ultimately, to prevent their condensation into dense, third phases. Although the window of ideal process chemistry in SX is narrower (in terms of maximum loading concentrations for metals before organic-phase collapse) than available with pyrometallurgical methods, which do not suffer the third-phase scourge, the scientific motivations to push the limiting organic concentrations higher and higher are easy to understand.

Advances in the fundamental understanding of supramolecular chemistry in metal ion separation processes through use of multifaceted experimental approaches hold the promise to simplify SX processes and improve separation factors by combining process steps, solvents, and extractants. Sometimes the combinations work (as in DIAMEX-SANEX and TRUEX), and sometimes, for no obvious reason, they do not.[221] This is just one example of a more generic problem that is attributable to the fact that the chemical behaviors of SX systems, in general, and An/Ln separation, in particular, are inherently complicated. Liquid–liquid extraction results (e.g., distribution ratios as well as slope analyses of water, acid, extractant, and metal ion dependencies) are oftentimes not easily interpreted due to the dearth of information about the stoichiometries and structures of the molecular An and Ln complexes solubilized by amphiphilic extractants and their aggregation and micellization in aliphatic diluents. In this regard, the roles that aggregation phenomena play in the basic physics and chemistry of ion transport in SX are underestimated; the dipole–dipole, dipole–ion, and ion–ion interactions at the core of the aggregation of solute species in bulk, nonpolar organic phases employed for SX have significant functions in the macroscopic behaviors of SX systems. It is not hyperbole to state that the consequences of aggregation on metal ion distribution ratios and separation factors are yet to be disentangled. Indeed, with over 60 years of practice by U.S. Department of Energy (DOE)-sponsored researchers, who in 1953 at Oak Ridge National Laboratory performed the first kilogram-scale separation of gadolinium by SX,[222] it can be argued that SX works despite the myriad chemical and physical consequences of aggregation. In this intricate issue, we posit that there are three aspects to consider. Metal distribution and phase behaviors can be tied to the three components that constitute the reverse micelle depicted in Figure 6.1 as follows: (1) The internal core, which contains free and bound water, bound anions, acids, and metal ions as well as the polar portion of the extractant molecule itself, is envisioned to affect the selectivity and efficiency of the extraction. (2) The hydrophobic corona is thought to be responsible for the morphology of the micelles. (3) The micelle–micelle interactions are thought to drive bulk phase behaviors and colligative properties. A process-by-process knowledge of these three issues will drive the field forward in a deliberate and logical manner through studies of various systems, including dialkyl phosphoric acids, trialkyl phosphates, phosphine oxides, diamides, diglycolamides, etc., with variations of hydrophobic alkyl groups and hydrophilic polar head groups.

6.3.2 THERMODYNAMIC OBSERVATIONS IN EXTRACTION SYSTEMS

Thorough thermodynamic discussion of structural rearrangements, aggregation, and self-assembly in nonaqueous mixtures encountered in solvent extraction must procede through systems where energetic demands on solvation are progressively growing. To realize the thermodynamic strains imposed on solvation, it is appropriate to evaluate forces responsible for the solvation of an amphiphile in nonaqueous mixtures. Depending on the structural features of a solute, the London forces (attraction due to dispersion forces), dipole–dipole forces (attractive or repulsive arrangement of electric dipoles), and hydrogen bonding may all contribute to the minimization of the standard molar Gibbs energy of solvation. These forces determine all physical properties of a solvent and, when solute arrives, react to accommodate its presence.

A solute molecule entering a liquid disrupts the long-range order of the solvent, attempting to regain its thermodynamic stability through multiple chemical interactions. The characteristic "dual" hydrophobic-hydrophilic nature of an amphiphile allows the solute to stabilize its chemical potential in both the aqueous and nonaqueous environments. In aqueous solutions, the hydrophobic tails of a surfactant induce a large, unfavorable energy contribution to the free energy of the system due to structural reorganization of the water network to accommodate an amphiphile. As such, the hydrophobic interactions between the hydrocarbon tails are highly disposed to arrange a layer of polar head groups to shield those away from the bulk water. This pronounced effect of the amphiphilic solute on the solvent (water) results in a thermodynamic push to arrange the surfactant into micellar architectures. A much different scenario takes place in nonaqueous media. Singleterry, in his pioneering review of amphiphilic structures in organic solvents, outlined the evidence for aggregative behavior.[223] He argued that an overall favorable gain in the interfacial free energy may be acquired if the amphiphilic monomers arrange their polar head groups together and expose their hydrocarbon tails into the organic solvent. However, the thermodynamic push toward the formation of such solute aggregates, sometimes referred to as "inverted micelles" (equivalent in our terminology to reverse micelles), is not large compared to those present in the aqueous solutions. The thermodynamic demands on solvation, which the amphiphile imposes on the organic solvent, are comparable for the monomeric or the aggregated surfactant.[224] For this reason, the aggregation phenomena in nonaqueous environments tend to be much less dramatic. Instead of an abrupt critical micellar concentration, solute aggregation tends to gradually progress through multiple aggregation equilibria as the solvation capacity of the organic diluent is challenged. Kertes, and later Ruckenstein, illustrated that most binary surfactant–solvent systems do not support the existence of critical micelle concentration, where equilibrium drastically shifts to the micellar stage.[162,224]

In a wet system of two immiscible liquids—aqueous and organic—surfactant molecules adsorb at the liquid–liquid interface and lower the surface tension. Cote and Szymanowski, in an excellent study on the interfacial properties of acidic organophosphorus extractants, illustrated that the migration of dialkyl acid esters from bulk organic solvent to the interface is characterized by a large, favorable free energy of adsorption.[225] The presence of an amphiphilic structure at the interface in general enables various mechanisms for the partitioning of polar solutes (metal, acid, water) into the nonaqueous environment. As increasing amounts of solutes travel into the organic phase, the solubility limits are challenged, resulting in supramolecular ordering. As the liquid phases diversify in composition, so do the structural architectures, progressing from the stoichiometric arrangements on to supramolecular aggregates, such as polymeric structures, reverse micelles, Winsor type II water-in-oil (w/o) microemulsions, or splitting of phases (third-phase formation). When considering the thermodynamic influence of aggregative rearrangements in the nonaqueous environments on the liquid–liquid distribution of those polar solutes, it is appropriate to progress in such order of structural arrangements, beginning with simple binary extractant–diluent mixtures at low metal loading conditions, and ending with ordered microemulsion phases.

Thermodynamic connection between the growing metal-to-ligand ratio and the phase-transfer equilibria follows a simplified qualitative rule. In general, the consumption of the free ligand carries a penalty in the efficiency of the metal ion partitioning. The increasing metal-to-ligand ratio challenges the solvation thermodynamics in nonaqueous environments and triggers progressively expanding structural rearrangements. The search to regain the thermodynamic stability is more difficult in nonaqueous mixtures, relative to aqueous environments. An organic diluent contains 3–10 moles of molecules in the same volume, which would contain 55 moles of water.[226] Therefore, thermodynamics impose more pronounced effects on the order of the nonaqueous phase to dissipate the excess chemical potential, resulting from various nonideal terms such as electrostatic or dipole interactions, solvation effects, mixing terms, etc.

Growing solvation demands to accommodate the increasing amounts of polar solute in the nonaqueous environment typically correspond with the less favorable metal partitioning into the organic phase. As the free ligand becomes scarce, the thermodynamic drive to cross the liquid–liquid boundary is depleted. Such determination easily applies based on the stoichiometric treatment. It is more difficult, however, to quantitatively connect supramolecular rearrangement and the liquid–liquid distribution equilibrium for the neutral solvent extraction systems. The extraction of uranyl ion by TBP, for example, is suppressed by the extraction of acid (as discussed below). The existence of this competitive equilibrium interferes with quantitative assessment of the impact that the micellization of TBP has on the efficiency of extraction. Collectively, in the highly metal-loaded, weakly hydrated liquid–liquid distribution systems, the consumption of free ligand available for metal ion coordination governs sequential rearrangements in the nonaqueous mixtures. Systems supporting large quantities of water introduce yet another realm of supramolecular structuring in solutions.

6.3.3 METAL DISTRIBUTION OBSERVATIONS IN MICROEMULSION SYSTEMS

Throughout the last few decades, the use of water-in-oil microemulsions has received growing attention in the field of solvent extraction, and a review of the role of microemulsions for separation purposes was published in 1997 by Watarai.[14] Osseo-Asare and Keeney[227] noted that the addition of long-chained sulfonate salts (AOT, HDNNS (dinonylnaphthalene sulfonic acid), and HDDNS (didodecylnaphthalene sulfonic acid)) (see Table 6.1) resulted in unexpected synergy with the extraction reagent anti-5,8-diethyl-7-hydroxy-6-dodecanone oxime (LIX 63) for extraction of nickel. Extraction experiments showed that both the rate of extraction and the efficiency of nickel extraction increased with the addition of the sulfonic acids. This synergy was attributed to the formation of reverse micelles that could assist in the complexation between nickel and the oxime. Interfacial tension measurements were made in this study, and the interpretation suggested the formation of reverse micelles. A typical trend observed for surfactants is a decrease in surface tension between the aqueous and the water-in-oil microemulsion as the concentration of surfactant is increased (see further discussion below). A break in the decreasing response indicates that the critical micelle concentration (cmc) has been reached, a value that is diagnostic of the onset of formation of large aggregates that consist of a core containing water and the oriented polar groups (NOH and OH; see Table 6.1) of the hydroxyoxime with which the Ni ions are solubilized. The alkyl chains of the oxime constitute the corona of the reverse micelle (see Figure 6.1).

Later, Osseo-Asare[29] suggested an extraction model and mechanisms for metal extraction in a microemulsion based on a combination of surfactant and extractant, both molecules being able to coordinate to the metal ion. This can be compared to conventional solvent extraction for synergistic systems (see Figure 6.7).

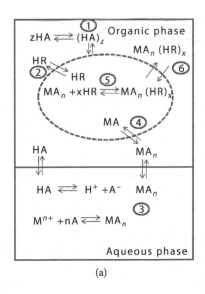

(a)

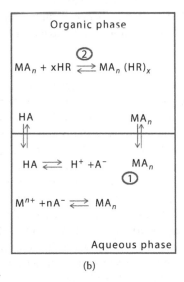

(b)

FIGURE 6.7 Graphical comparison of microemulsion extraction (a) and conventional solvent extraction (b) systems. HA is either the surfactant or the acidic extraction agent. HR is the (solvating) extraction agent.

The mechanism of extraction in the microemulsion systems is explained as a six-step process:

1. The surfactant, HA, forms aggregates, $(HA)_z$, in the organic phase.
2. The extracting reagent, HR, is distributed between the bulk organic phase and the reverse micelles.
3. Surfactant monomers bind to metal ions, M^{n+}, in the form of stoichiometric, charge-neutral complexes, MA_n, for transfer into the bulk organic phase. (Here, the surfactants are assumed to be anionic.)
4. The surfactant–metal complex partitions between the bulk organic phase and the reverse micelles.
5. The surfactant–metal complex reacts with the extracting reagent inside the reverse micelle. The exact stoichiometry of this complex may vary; the extractant could replace the surfactant or simply make a mixed complex, $MA_n(HR)_x$.
6. The final complex is transported into the bulk organic phase.

In contrast, the mechanism of metal extraction using conventional solvent extraction for a synergistic system containing an acidic chelating reagent and a solvating reagent may be explained in only two steps:

1. Acidic chelating reagents, HA, bind to the metal, M^{n+}, thereby neutralizing the charge of the metal ion and transferring it into the organic phase as a neutral complex, MA_n.
2. The metal–chelate complex reacts with the solvating extractant, HR, forming a mixed complex, $MA_n(HR)_x$, thereby completely dehydrating the metal ion.

The two extraction mechanisms are schematically represented by Figure 6.7.

Regardless of the path for extraction, the final result will be the same. That is, the thermodynamic equilibrium is an equation of state. The fact that the microemulsion has some water present in the reverse micelles may make the dehydration of the metal ion less important than models would otherwise suggest.

In a compilation of extraction studies by microemulsions, Osseo-Asare[228] elaborated upon the increase in kinetics when a microemulsion system was formed; the author called this "microemulsion-mediated solvent extraction." The example used was one where a synergistic system of an acidic reagent (carboxylic or sulfonic acid), denoted HA, combined with a hydroxyoxime or Oxine, denoted HR, was used to extract a divalent metal. This is directly comparable to what was used by Osseo-Asare and Keeny in their previous studies,[227] that is, nickel extraction by HDNNS and LIX 63. The classic explanation[229] was given where the synergist (HR) removes the excess water (i.e., completely dehydrates the metal ion) after the metal is extracted, or the synergist facilitates phase transfer and the other reagent removes the excess water.

$$2HA_{(org)} + M(H_2O)_6^{2+} \rightleftharpoons M(H_2O)_2A_{2(org)} + 2H^+ + 4H_2O$$

$$M(H_2O)_2A_{2(org)} + 2HR_{(org)} \rightleftharpoons M(HR)_2A_{2(org)} + 2H_2O$$

or

$$2HR_{(org)} + M(H_2O)_6^{2+} \rightleftharpoons M(H_2O)_2R_{2(org)} + 2H^+ + 4H_2O$$

$$M(H_2O)_2R_{2(org)} + 2HA \rightleftharpoons M(HA)_2R_{2(org)} + 2H_2O$$

The improved kinetics for the microemulsion system are explained in the following way:

1. The HDNNS molecules (HA) form reverse micelles in the organic phase; these micelles solubilize the LIX 63 (HR) molecules, resulting in a high local concentration of the LIX 63 inside the micelles as illustrated by infrared (IR) spectroscopy.
2. The HDNNS attracts a metal ion to the bulk aqueous-organic interface, and two DNNS⁻ ions bind to the metal forming a MA_2 complex.
3. This complex is transferred to the HDNNS microemulsion; this is rapid because there is no need to completely dehydrate the metal ion since there is an aqueous environment inside the reverse micelle.
4. When the metal complex is in the micelle, the high local concentration of LIX 63 results in a rapid complexation/dehydration of the metal complex and a mixed extracted complex is formed.
5. The mixed complex is transferred out of the micelle and into the bulk organic solvent.
6. The micelle may solubilize new, free LIX 63 molecules.

Cote[230] reported on a similar mechanism for the extraction of gallium(III) from leach liquor via a microemulsion-assisted extraction. In his study, he proposed a microemulsion formed by a combination of surfactant, sodium octanoate, and the extraction reagent Kelex 100 (see Table 6.1). Kelex 100 alone extracts Ga(III) very slowly,[231] reaching equilibrium only after several hours, whereas it was observed that the addition of surfactant increases the kinetics, and equilibrium is reached within a few minutes.[232] Cote proposed that the microemulsion droplets formed in the organic phase would solubilize water, and that when the droplets would be in close vicinity to the aqueous–organic interface, the Ga(III)(OH)₄⁻ would rapidly transfer to the "dispersed" aqueous phase. These microemulsion droplets may then disperse in the bulk of the organic phase (see Figure 6.8).

Cote then postulated that the Kelex 100 molecules would be adsorbed on the lipophilic shell of the microemulsion droplets, where it would react with the gallium–hydroxide complex replacing a hydroxide molecule forming a $Ga(OH)_3(Kelex100)^-$ complex. This intermediate complex would react with a second and subsequently a third Kelex 100 molecule in the bulk organic phase

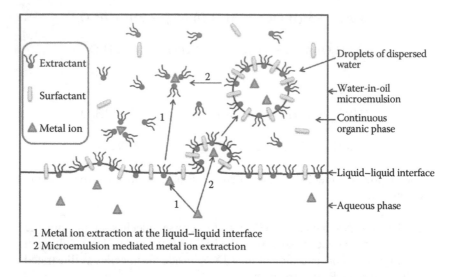

FIGURE 6.8 Schematic of micelle-assisted extraction of metal ion. (From Cote, G., *Radiochim. Acta.*, 91(11), 2003 639–643.)[230]

forming the final Ga(Kelex100)$_3$ complex. New Kelex 100 molecules could then adsorb on the microemulsion droplet to continue extracting and solubilize gallium. The increased kinetic rate is due to the fact that the reaction takes place at the microscopic interface of the reverse micelles and the collected surface area available for this reaction is larger than what would be available for a conventional solvent extraction system.

The work by Osseo-Asare paved the way for other studies where conventional solvent extraction was compared and enhanced by the addition of surface-active reagents. Paatero et al.[30] investigated the extraction of copper in a system consisting of AOT combined with different acidic extracting reagents (di(2-ethylhexyl)dithiophosphoric acid (DEHDTPA), anti-2-hydroxy-5-nonylbenzophenone oxime (HNBPO), or Versatic 10 (a mixture of tri-alkyl carboxylic acids)) under the formation of water-in-oil microemulsions. In this study, the characteristics of the microemulsion were investigated and connected to the extraction efficiency of copper. The water content in the organic (microemulsion) phase at low pH (2) was measured under different conditions using Karl-Fischer titrations. The amount of water solubilized in the organic phase may be used as a direct indication of the extent of reverse-micelle formation, and it was clear that an increase in surfactant, AOT, increased the amount of water present in the organic phase. Different extraction reagents increase or decrease the amount of water compared to AOT alone, following the trend DEHDTPA > AOT alone > Versatic 10 >> HNBPO. It was suggested that the aromatic character of HNBPO explained the lower water uptake. The fact that HNBPO and Versatic 10 exist in their acid form at pH = 2, and thus have little interfacial character, while DEHDTPA is a stronger acid, may be another explanation.

Several studies[21,23,233] have been carried out using microemulsion extraction directed to the industrial application of extraction of gallium from Bayer liquor, an alkaline waste stream from the Bayer process of aluminum recovery from bauxite ore. Two of these studies[21,23] distinguish themselves from the ones previously discussed by the fact that they utilize the extractant 7-(1-ethyl-3,3,5,5-tetramethylhexyl)-8-quinolinol, HQ (which is the active ingredient in Kelex 100), together with a carboxylic acid or alcohol to produce a microemulsion. Thus, no surfactant was added as in other studies; instead, the molecules exhibit surface-active tendencies, and by combining them a microemulsion is formed. By investigating different combinations of carboxylic acids or alcohols, some trends could be seen on the effect they had on the uptake of water in the water-in-oil microemulsion.[23] It was observed that longer aliphatic chains decreased the amount of water in the reverse micelles, perhaps by giving the microemulsion a more hydrophobic character or decreasing the number and size of reverse micelles. One of the reasons the extractant can be included in the shell of the reverse micelle is that the Bayer liquor is alkaline, and at high pH the HQ functions as an amphiphile.

Neutral solvating amphiphiles, such as TBP or TODGA (tetra-octyl-diglycolamide), are hydrogen bond acceptors and coordinate acid and water molecules. Such solvent extraction reagents are efficient, but not very surface active relative to typical micellar agents or ionized phase-transfer reagents, such as sodium salts of monoacidic dialkyl organophosphorus acids. Figure 6.9 visualizes differences in the interfacial activity for several phase-transfer reagents discussed here.

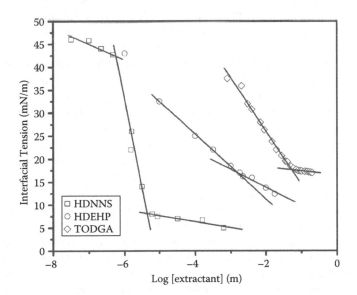

FIGURE 6.9 Relative comparison of interfacial tension for growing concentrations of micellar, ion exchanging, and solvating solvent extraction reagents. (a) HDNNS in hexane/0.5 M KNO$_3$, pH = 2.5, 25°C.[234] (b) HDEHP in hexane/0.05 M CoCl$_2$, pH = 4.0, 25°C.[235] (c) TODGA in n-dodecane/2 M HNO$_3$, 25°C.[153]

Clearly, micellar reagents influence the shape of the interface to the greatest extent, and HDNNS has been shown to form reverse micelles at micromolar concentrations.[234] The interfacial tension of partially saponified HDEHP (di-2-ethylhexyl phosphoric acid) is greatly affected by the presence of metal ions in aliphatic diluents. The characteristic "break" in the interfacial tension curve, though less pronounced than for HDNNS, does suggest structural rearrangements into more elaborate architectures.[235] When compared with those two reagents, which are known to form large reverse micelles, the interfacial activity of a chosen representative of a neutral phase-transfer reagent (TODGA) is less pronounced for much more concentrated ligand solutions.[153] As such, while the previous two ligands have been shown to solubilize large quantities of water and support the formation of water-in-oil microemulsions, the neutral solvent extraction agents typically form only weakly hydrated small micelles when contacted with highly acidic aqueous electrolyte mixtures. Numerous studies have shown that the ratio of water-to-amphiphilic polar group for such aggregated small reverse micelles does not exceed 1.[74,99,147] It is feasible, however, under designed conditions, to host a greater number of water molecules in the micellar core, as demonstrated by Osseo-Asare for the TBP system.[236]

High interfacial activity of the amphiphile may influence the curvature of the liquid–liquid interface so strongly that the dispersion of aqueous microdroplets in the nonaqueous phase occurs, which leads to the formation of stable water-in-oil microemulsions. Such mixtures arrange into large spherical reverse micelles, which solubilize pockets of water, namely, water pools, in the nonaqueous environment. For example, the polar core volume of the inverted HDNNS has been reported to measure roughly 1650 Å^3, which may accommodate ~50 water molecules.[234,237] The transition between the systems that form hydrated reverse micelles described in the previous section to the w/o microemulsion is believed to occur when the water-to-polar core ratio exceeds 3.[63,151] The polar microenvironment inside the reverse micelle is capable of supporting a variety of solutes, creating opportunity for a diverse spectrum of applications, such as micellar catalysis, drug delivery, and nanoparticle synthesis.[238] The interest in w/o microemulsion-mediated metal ion extraction mainly stems from its catalytic influence on the kinetics of the solute transfer across the liquid–liquid boundary.[227] The enhanced interfacial activity of emulsified systems increases the surface area at the liquid–liquid interface, resulting in a faster, more efficient equilibrium.[239]

As mentioned above, Osseo-Asare argued that w/o microemulsions may solubilize solutes of value to metal ion extraction applications.[29] He indicated that a polar aqueous microenvironment could accommodate hydrated metal ions. Since metal ion dehydration is typically the largest thermodynamic barrier to its phase transfer into the nonaqueous solution, the retention of partial hydration would represent a significant energetic incentive to metal extraction into w/o organic phases. In traditional liquid–liquid extraction, waters of hydration challenge the thermodynamic stability of the complex if the coordination ligand does not dehydrate the metal ion completely. Such systems benefit from the addition of the second "dehydrating" reagent—a synergist (as described above)—resulting in a large exothermic push to enhance the lipophilicity of the metal complex.[229]

6.4 BEHAVIOR OF EXTRACTANTS IN SPECIFIC SYSTEMS

6.4.1 NEUTRAL ORGANOPHOSPHOROUS EXTRACTANTS

6.4.1.1 TBP

Organophosphorous compounds—both neutral and acidic ones—of widely varying compositions have been the most versatile and, to date, the most successful reagents for the separation of actinides from used nuclear fuels. They are also at the core of new approaches for chemical partitioning, such as the uranium extraction (UREX) process.[240,241] Of all the organophosphorous extractants, TBP is king. It is arguably the most important extractant molecule in nuclear technology for the recovery of plutonium and uranium in the plutonium and uranium recovery by extraction (PUREX) process, as well as in other hydrometallurgical separations of industrial use. TBP selectively extracts Pu^{4+}, U^{6+}, and Th^{4+} from nitric acid solutions, leaving behind most of the fission products and the transuranic elements.[242] Actinide ion extraction by TBP has been investigated in detail, as has the nature of the structures formed in extractant phases at high solute loadings.

As mentioned above, solvent extraction has received much attention in the development of separation techniques for elements in used nuclear fuel. Several of the extraction reagents used have been shown to, under certain conditions, arrange themselves into aggregates and cause the formation of a separate phase. TBP is an example of this. Because TBP has been so extensively studied and there are numerous publications, including a four-volume set of the science and technology of TBP,[243–246] it is a good system to use as a benchmark, particularly when comparing similar molecules.

It was observed early on[247] that TBP dissolved in kerosene in contact with a concentrated acidic aqueous solution would form third phases, although the exact nature of this phenomenon was not known at that time. TBP has long been known to extract mineral acids, in particular nitric acid, which has been investigated and modeled by several groups reporting various species of $TBP_x[HNO_3]_y$, where the 1:1 $TBP:HNO_3$ species is common for all studies, although higher-order species such as TBP_2HNO_3 and others are included to improve the models.[247–250]

Osseo-Asare presented studies[4,236] on the organic-phase characteristics, including changes in volume, surface tension, conductivity, and viscosity for systems of TBP-diluent-water-acid. In these studies, Osseo-Asare postulated that the hydrated monomer of TBP would dimerize and polymerize, creating elongated chains of TBP monomers that would, under certain conditions, rearrange into spherical aggregates, reverse micelles. These reverse micelles could solubilize larger amounts of water, creating a microemulsion in the organic phase. In the first of these two studies,[236] Osseo-Asare investigated the extraction of HCl by TBP. He found that as HCl is extracted as a function of total HCl concentration, the H_2O uptake and volume increases gradually and goes through three different regimes: (1) First, for low HCl (<1 mmol), the water uptake is flat and volume change is constant; (2) for intermediate HCl (1 < HCl < 10 mmol), there is a linear uptake and change of volume with a certain slope; and, (3) above 10 mmol HCl, the volume change is still linear but with a lower slope, and the water uptake is flat.

In the first region, there is hydration of TBP, and the water concentration in the organic phase corresponds to a 1:1 ratio of H_2O:TBP, attributable to a mono-hydrated TBP molecule. In the second region, there are TBP-H_2O complexes but also hydrated TBP-HCl complexes of the stoichiometry $TBP_y(H_2O)_xHCl$, most likely as an ion pair having water between the P=O and the proton, such as TBP-H_2O...H^+...Cl^-. It was shown that the complex in the second region is of the form $TBP_2(H_2O)_6HCl$. This is formed by the combination of two hydrated TBP mono-mers extracting HCl and four water molecules.

$$2(TBP \cdot H_2O) + HCl + 4H_2O \leftrightarrows (H_2O)_6TBP_2 \, H^+Cl^-$$

Osseo-Asare made the observation that for the second region the HCl does not contribute to the volume change, but the six waters (increase of two per TBP molecule) do. The author further assumes that the TBP-water-H^+ complex has an "open structure" where the Cl^- ion fits without any volume change. Thus, the TBP-water-proton complex is prearranged to accommodate Cl^-. In the third region, there is no more water uptake, but the volume still changes due to the extraction of HCl. A rapid increase in viscosity in this third region, suggesting extensive aggregation in the organic phase, was also reported. It was suggested that the aggregation follows the reaction:

$$2[(H_2O)_6TBP_2 \, H^+Cl^-] + 2H^+ + 2Cl^- \leftrightarrows [(H_2O)_3(TBP)H^+Cl^-]_4$$

The product is a spherical aggregate of 4 TBP molecules surrounding a core of 12 H_2O and 4 HCl molecules.

Osseo-Asare also measured the electrical conductivity of the solutions in the three regions. In the first region, the conductivity was low. Then, in region 2, the conductivity rises because of HCl (H^+ and Cl^- ions) present in the organic phase. In the final region, the conductivity is constant, and finally, it decreases somewhat. This plateau and decrease has been attributed to the formation of reverse micelles with water pools. The interfacial film of the TBP molecules (the nonpolar corona of the reverse micelles) interferes with the ion transport and offsets the increase in conductivity due to the increased HCl extraction. At high concentrations of HCl, the ratio of HCl to TBP is greater than 1:1. This is explained by solubilization of HCl in the droplet of the reverse micelle, but not associated directly with the TBP molecule creating a w/o microemulsion. As such, the acid in the water droplet inside the reverse micelle has the same chemical potential as in the aqueous phase in contact with the microemulsion phase.

Chiarizia and Briand[176] investigated the formation of aggregates of TBP formed as a third phase—a Winsor type III system—in contact with different acidic media.

In these studies, the water and metal uptake in the organic phase was monitored. Water and acid were found to concentrate into the third phase as the organic phase split, whereas the remaining organic-phase was depleted of these molecules. The limiting organic concentration (LOC) of the different mineral acids before third-phase formation was observed to follow the same trend as the amount of water uptake in the organic phase. The interpretation of this phenomenon is that the reverse micelles in the organic phase have strong intermolecular attraction when their cores include high concentrations of water. These van der Waals forces between the reverse micelles make them agglomerate and form a separate phase, resulting in a splitting of the organic phase and the formation of a third phase. This can be seen as a transition from a Winsor type II microemulsion to a Winsor type III. This transition between different types of Winsor microemulsions has been observed for other systems by changing conditions such as temperature or aqueous media.[30,251]

Jiang et al.[25] compared the extraction of inorganic acids by three different neutral phosphorus-containing reagents, a phosphate TBP, a phosphonate DiAMP (di-isoamylmethyl phosphonate), and a phosphine oxide TRPO (C6–C8). In this work, a reference to an earlier study by Foa et al.[252] was made, where an adduct (ion pair) of TBP-HCl is proposed, much like the work by Osseo-Asare.[236] Jiang et al.[25] indicated that the ion pair—the hydronium ion bonding to the phosphoryl oxygen—can be formed in systems of TRPO and DiAMP, and with mineral acids other than HCl. They suggest that when the species of protonated neutral ligand ($R_3P=O-H_2O-H^+$) reach a certain concentration, they may aggregate and solubilize larger amounts of electrolyte, resulting in large changes in the organic-phase volume and large changes in the conductivity. Their NMR data also supported the theory of hydrogen bonding to the P=O group of the TBP, first by water, but when acidity was increased, the H_3O^+ would bind. Their dynamic light-scattering (DLS) data show that when the acid concentration increases, the average radius of aggregation increases. The formation of these aggregates leads to an increase in density, which may lead to phase splitting (third-phase formation). This phenomenon was observed for TBP, TRPO, and DiAMP. The authors do indeed call the third phase a Winsor type III microemulsion, and they claim that the formation of aggregates is related to the formation of molecule-ion species in the organic phase. The tendency to form microemulsions follows the order TRPO > DiAMP > TBP, which follows the order of the electron density of the P=O group.

By far, the most important direct structure technique employed in studies of aggregates in organic solutions of TBP has been SANS, which provides information on the morphology of extractant aggregates at large length scales, typical of aggregation and collective behaviors, including micellization. Through the selective introduction of deuterated organic compounds into the extraction diluent, as well as parallel variations of the aqueous media (H_2O and D_2O), the neutron scattering contrast between ^1H and ^2H can be readily discerned. Detection of the contrast becomes easier as the sizes of the scattering particles increase in the solution, and despite the 10% difference in the strength of the H–O and D–O hydrogen bonds, the morphology of the water-in-oil microemulsion structures is independent of the H_2O–D_2O substitution.[62] In view of the depth and breadth of the chemical separation literature about solute organization and third-phase formation in TBP-based SX, only selected aspects are highlighted here with some specialist applications.

In solutions of TBP in neat n-alkane diluents without contact with aqueous media, TBP molecules self-associate to form dimers.[253] Upon contact with aqueous media containing metal salts, the TBP solvates of the metal salts spontaneously organize in the organic phase in more complex architectures. SANS of TBP solutions loaded with HNO_3 alone or with $HNO_3/UO_2(NO_3)_2$ or $HNO_3/Th(NO_3)_4$ mixtures revealed the presence, both before and after phase splitting, of ellipsoidal aggregates of nanoscopic dimensions.[74,118–120,123,143,145,152,175–177,187,188,193,201] The driving force for the observed aggregation is solvent-mediated interactions between TBP-solvated metal nitrates. It is now generally recognized that TBP in alkane diluents, in contact with aqueous phases containing nitric acid, and uranyl, thorium, zirconium, and plutonium nitrates, forms small reverse micelles containing up to four TBP molecules.[126,145,192,193] Water, nitric acid, and metal nitrates are incorporated into the polar core of the micelles, and these interact through van der Waals forces between their polar cores. The separation of most of the solute particles upon organic-phase collapse in a new third phase takes place when the energy of attraction between the particles in solution becomes about twice the average thermal energy k_BT. Moreover, the splitting of the organic phase does not result in the creation of particles, but rather results from the apparent coalescence of existing particles in the organic phase, superficially resembling a precipitation process. However, observations regarding the thermodynamic reversibility of the phase separation process and the continuous variation of third-phase compositions as a function of solute concentrations imply that, except in one case described below, the process is far more complex than a simple precipitation reaction. The results from a number of investigations have confirmed that when a metal cation is transferred from an aqueous phase into paraffinic diluents with 20% TBP, its effect on promoting third-phase formation, as measured by how rapidly the energy of interaction between reverse-micelle-like particles reaches the $-2k_BT$ critical value, strongly depends on cation charge and size. For example, the extraction of $Zr(NO_3)_4$ from aqueous HNO_3 solutions by 0.73 M TBP in n-octane was investigated under a variety of conditions.[121] The results show that although $Zr(NO_3)_4$ is much less extracted than $Th(NO_3)_4$ and $UO_2(NO_3)_2$, it exhibits a much stronger tendency to promote organic-phase splitting. The Baxter model for hard spheres with surface adhesion interpreted the features of the SANS data in the low scattering range (Q, Å^{-1}) as arising from interactions between small reverse-micelle-like particles containing two or three TBP molecules associated with $Zr(NO_3)_4$, HNO_3, and water. The interaction between the TBP-containing particles was quantified through the potential energy of attraction, $U(r)$, which increased sharply when third-phase formation was approached, reaching $-2.1k_BT$ at the point of phase splitting. In efforts to discern a correlation between the interparticle potential energy of attraction and the nature of the extracted cations, the results for the $Zr(NO_3)_4$ system were compared with those previously obtained for the $Th(NO_3)_4$ and $UO_2(NO_3)_2$ systems. A robust linear relationship was obtained between the derivative of $U(r)$ with respect to the total nitrate concentration in the organic phase and the cation hydration enthalpy or ionization potential.

In the absence of metal ions, the extraction of inorganic mineral acids with different hydration properties by TBP and by its tri-*iso*-butyl and tri-*sec*-butyl isomers was investigated.[176][178] The series of effectiveness with respect to third-phase formation ($HClO_4 > H_2SO_4 > HCl > H_3PO_4 > HNO_3$) was found to correlate with the amount of water present in the organic phase at the critical point of phase splitting. This result reinforces the validity of the reverse-micellar model for third-phase formation for the extraction of inorganic solutes by TBP. According to this model, TBP at high concentrations in *n*-alkanes, in contact with aqueous phases containing inorganic acids and metal salts, forms reverse micelles. The presence of large amounts of water and ionic solutes in the reverse micelles increases the strength of the intermicellar attraction, and thus facilitates third-phase formation. Measurements of third-phase formation for the extraction of $HClO_4$ by TBP dissolved in a series of alkane diluents confirmed that topological parameters, such as the connectivity index, can be used for predicting the critical condition for phase splitting in different diluents. SANS data obtained for the TBP–inorganic acid systems were interpreted using the same model of hard spheres with surface adhesion (Baxter model) that was successfully employed for the interpretation of SANS data on the extraction of metal salts. The results of calculations indicated that the critical value of the energy of attraction between reverse micelles is approximately $-2k_BT$ for all acids, in analogy with the extraction of metal salts. The derivative of the energy of attraction with respect to the organic-phase acid concentration follows the same sequence noted above ($HClO_4 > H_2SO_4 > HCl > H_3PO_4 > HNO_3$), supporting the third-phase formation model based upon reverse-micelle interaction.

Since the main contribution to the thermodynamic drive of micellar organization in nonaqueous environments is the hydration of polar amphiphilic head groups, the weak hydration of TBP micelles explains why they exist only in highly concentrated nonaqueous media.[152,254] Naturally, then, an increased temperature of the system will add to the thermal agitation of micelles, and thereby prevent the formation of the third phase as illustrated by several studies in the past.[152,255] Recent studies on malonamides, diglycolamides, and mixed-ligand systems show similar self-assembly trends in that small reversed micelles form when polar solutes enter the nonaqueous environment.[136,147,149]

6.4.1.2   TRPO (TOPO, etc.)

To further examine the validity of the model described in the previous paragraphs, the extraction of HCl by three other organophosphorus extractants— tri(2-ethylhexyl) phosphate (TEHP), tri-*n*-octyl phosphate (TOP), and tri-*n*-octyl phosphine oxide (TOPO)—in *n*-octane was investigated by liquid–liquid distribution of acid and water as well as SANS.[122] No formation of a third phase was observed with TEHP and TOP, whereas for 0.4 M TOPO, the concentration of HCl at the critical condition was 5.1 M in the equilibrium aqueous phase. For higher aqueous HCl concentrations, the organic phase splits into a light layer and a heavy one that was shown to contain large and strongly interacting reverse micelles. The critical value of $U(r)$ calculated through application of the Baxter sticky sphere model to the SANS data for the TOPO system at the limiting organic condition is *identical* with those, $(-2.0$ to $-2.2)k_BT$, found for third-phase formation in TBP (vide supra) and malonamide systems,[99,117,130,132,136,146,149,150,154,155,157,158,160,161,256–259] pointing to the general validity of the model. In stark contrast, the SX of 12-phosphotungstic acid, $H_3PW_{12}O_{40}$, by 0.73 M TBP in n-octane under conditions exactly comparable to those used for the extraction of conventional inorganic mineral acids reveals an extremely low initial concentration of $H_3PW_{12}O_{40}$ (1.1 mM) at the limiting (critical) organic concentration condition,[260] far lower than the most effective third-phase-forming inorganic acid, namely, $HClO_4$. The results from SANS indicate that the interparticle attraction energy—$U(r)$ calculated through application of the Baxter sticky sphere model to the SANS data at the critical condition—does not approach the $-2k_BT$ value associated with phase splitting in previous studies for other TBP systems. Rather, the collapse of the organic phase upon extraction of $H_3PW_{12}O_{40}$ is assumed to stem from the limited solubility of the heavy and highly polar $H_3PW_{12}O_{40}$–TBP_3 solvate in the alkane diluent, wherein solute effects, namely, interactions between opposite-charged ions, and solvent effects, namely, interactions between ions and neutral solvent molecules, are driven by electrostatic interactions. The correlation—attraction and repulsion—effects are thought to transition from contact ion pairs to the precipitation of a structured fluid more dense than the aqueous phase in ways different from ordinary third-phase formation based on van der Waals forces between polar cores of self-assembled extractant reverse micelles. Whether or not the concepts deduced from the SX of $H_3PW_{12}O_{40}$ by TBP apply to the precipitation of alkylammonium salts of 12-phosphomolybdate, $[PMo_{12}O_{40},]^{3-}$, and third-phase formation in amine extraction systems[261–264] is not known.

6.4.2 AMIDES

6.4.2.1 Monoamides

Efforts to replace TBP with other extractants have focused on the use of N,N-dialkylamides. These systems were first proposed for use in the 1960s, largely through the efforts of Siddall,[265,266] and have since been recently revisited for reprocessing schemes by groups in Italy[267–272] and France.[273–280] Although the mesostructural impact on extraction performance is as yet unknown, monoamides like DOBA (dioctyl butanamide), DO2EBA (dioctyl-2-ethyl-butanamide)— the Japanese extractant of interest for the BAMA (branched alkyl monoamide) process—and D2EHiBA (di(2-ethylhexyl)-i-butanamide)—the French extractant for the GANEX (Group Actinide Extraction) process—are of contemporary interest. Advantages of use include the extractant composition, namely, the four readily incinerable elements C, H, O, and N; the radiolytic and hydrolytic degradation products do not affect performance; the extractants are compatible with existing PUREX plants; and the U–Pu partition does not require redox chemistry.

Although flowsheets were developed and tested, no plant use has been reported. All commercial, production-scale nuclear-fuel reprocessing is still done with TBP. As far as research now stands, the monoamide systems mentioned above are ripe for investigation vis-à-vis structural aspects—from the molecular to the micellar scales—to arrive at an understanding of the basis for extraction selectivity between uranyl(VI) and plutonium(IV).

6.4.2.2 Diamides

The separation of long-lived radionuclides, specifically the extraction of An^{3+} from Ln^{3+} in aqueous nitrate media by SX, is at the core of a significant body of contemporary international research in the field of nuclear hydrometallurgical process technologies.[281–283] For example, one of the processes—dubbed DIAMEX—conceived and developed at the CEA (Commissariat à l'Energie Atomique)[284–288] to deal with this intricate issue employs a malonamide extractant. The one of contemporary interest is DMDOHEMA, and before that, DMDBTDMA (N,N'-dimethyl-N,N'-dibutyltetradecylmalonamide) (see Table 6.1). Malonamides other than DMDOHEMA and DMDBTDMA, including TMMA (N,N,N',N'-tetramethylmalonamide)[289,290] and TEMA (N,N,N',N'-tetraethylmalonamide),[80,291,292] to name but two, have also been studied beforehand[293,294] with either a single Ln or a selection of Ln ions. It has been shown that these reagents in various aliphatic diluents, including n-alkanes and their mixtures, such as hydrogenated tetrapropylene, and under practical conditions of SX, self-assemble into supramolecular entities consisting of extractant molecules and their complexes with trivalent Ln.[99,117,130,150,154,155,160,161,256,258]

The malonamide, DMDBTDMA, was, in a similar fashion to TBP, shown to undergo the formation of aggregates resulting in third-phase formation.[158] It was seen that for extraction of neodymium in a system containing DMDBTDMA in n-dodecane contacted with an aqueous nitric acid phase, a third phase was formed. The uptake of water and neodymium was followed, and it was confirmed that the phase splitting occurred when a large amount of water was extracted. The interfacial tension showed the typical decreasing trends with an increase in diamide concentration, indicating that DMDBTDMA behaves as a surfactant or surface-active molecule. This study included VPO as a method of observing the formation of aggregates in the organic phase before phase splitting. The nature of these aggregates was probed using SANS, which showed them to contain six or seven DMDBTDMA molecules. This is more than what is indicated from slope analysis of conventional solvent extraction studies, and more than seen from crystallographic structures of related diamides, wherein the coordination number and size of a neodymium ion do not allow such large numbers of diamides to bind without steric interferences. These socalled superstochiometric, ligand-to-metal ratios have been observed for other systems.

Except for recent studies with DMDBTDMA[130] and DMDOHEMA,[131] there is no hierarchal disentanglement of the micro-, meso-, and macroscale environments in which the Ln ions are confined. Such metal-to-aggregate metrical knowledge is pivotal to an understanding of the SX method from the Ln–ligand interactions responsible for ion transfer to the morphology of aggregates and their bulk phase behavior. Despite current practice and future prospects of the DIAMEX extraction system, the basis for the effective partitioning of Am^{3+} is not, as yet, fully understood. The solution chemistry becomes even more complicated with the addition of an acidic organophosphorus extractant, such as HDHP (di-n-hexyl phosphoric acid), which is of importance to the separation of An^{3+} from Ln^{3+} in certain adaptations of the GANEX process, to the malonamide-containing diluent.

Structural details from a combination of studies (electrospray ionization–mass spectrometry, VPO, as well as small-angle X-ray and neutron scattering) that fill in gaps in the knowledge about the DIAMEX-SANEX system chemistry responsible for the selectivity of the method and, ultimately, that will drive advances in 4f- and 5f-element separations, show the formation of mixed, reverse-micellar species with a recurrent, aggregate composition of four or five DMDOHEMA and two HDHP molecules before and after extraction of Ln^{3+} and Am^{3+} nitrates from acidic and neutral media.[75,146,149] These mixed, reverse micelles have a diameter in the range of 19–24 Å with a polar core diameter of 10–14 Å. The favorable partitioning of Am^{3+} was attributed to, in part, the supramolecular organization of the Am(III)–extractant complexes in the organic phase. In comparison, the picture that has emerged about the extraction of Ln^{3+} and Am^{3+} with HDHP alone reveals species of approximate composition $M(DHP)_3(HDHP)_3(H_2O)_n$, where the number of water molecules (n) coordinating to the metal is between 0 and 3. This is consistent with the general behavior of dialkyl phosphoric acids[75,146,149,295,296] with trivalent metal ions, M^{3+}. This aggregate species has a spherical morphology with a polar core diameter of ~7 Å and total diameter of 11–15 Å, both of which are dependent upon $[M^{3+}]$. By contrast, the aggregation of DMDOHEMA when used on its own in SX is a progressive phenomenon with respect to its concentration. The average aggregation number of the water- and metal-loaded DMDOHEMA system increases with increasing concentration. An average aggregation number of ~2 is obtained in the 0.2–0.6 M extractant concentration range. Larger aggregates of the reverse-micelle type with a weight-average aggregation number of 9–10 form at higher concentrations.[149,161] The metal-loaded DMDOHEMA aggregates can be considered as interacting spheres with a polar core diameter between ca. 11 and 16 Å, depending on composition, and a total diameter of up to ~25 Å. Through use of systematic variations of extraction conditions and extractant combinations in the diluent n-dodecane, the results from XAS show that the coordination environments of Nd^{3+}, Eu^{3+}, Yb^{3+}, and Am^{3+} in n-dodecane solutions containing 0.3 M HDHP involve $M–O_6$ coordination and distant M–P interactions, whereas the solutions containing 0.7 M DMDOHEMA involve $M–O_8$ coordination, resulting from two bidentate diamide molecules and a combination of monodentate and bidentate coordination of nitrate ions, consistent with SX data.[75] The extended X-ray absorption fine-structure (EXAFS) data for the mixed-HDHP-DMDOHEMA system were shown to be linear combinations of the EXAFS data for the single-extractant complexes with HDHP and DMDOHEMA—a result that does not square with the prediction of the aggregate species unless, by coincidence of averaging, the nearest O environments in terms of M–O distances and O coordination numbers are of unresolvable differences.

Recent research on associative structural phenomena highlights the effects of aqueous acidities on Ln^{3+} extraction.[130–132] The effects of nitric acid on hierarchical structures and extractive properties involving Ce^{3+} and DMDBTDMA, as well as other Ln^{3+} and DMDOHEMA, include significantly improved Ln extraction but phase splitting (third-phase formation) happens more readily than observed in the absence of nitric acid. Using new methods to interpret SAXS data, especially the GIFT method,[129,297–300] and combining this with Ln EXAFS and other spectroscopic techniques, it has proven possible to relate the molecular, supramolecular, and nanoscale organic-phase structures to the macroscopic system behavior. For example, a summary of recent findings is presented in Figure 6.10, where, on the

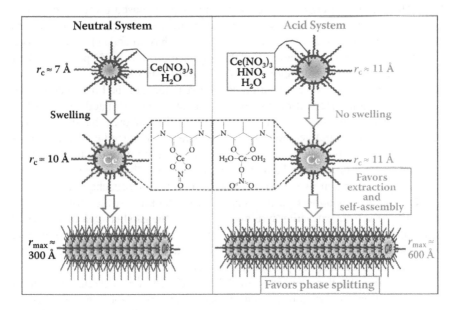

FIGURE 6.10 Graphical illustration of the effect of acid on the extractive properties and micro- and macromolecular organic-phase structures of the Ce^{3+}-DMDBTDMA-n-dodecane system resulting from the extraction of varying Ce^{3+} concentrations from neutral (left panel) and acidic (right panel) aqueous phases of 3 M $LiNO_3$ and 3 M HNO_3, respectively. The micellar core radii, r_c, and the maximum linear extents, r_{max}, of the rodlike aggregates as obtained from the SAXS analyses are shown along with insets depicting the different modes of nitrate coordination to Ce^{3+}, which is essentially bidentate in the neutral organic phases and essentially monodentate in the acidic organic phases, as obtained from Ce L_3-edge EXAFS analyses. The behaviors for the Ln^{3+}-DMDOHEMA-n-dodecane system are exactly comparable.[131] From this type of research approach, it is becoming increasingly evident that interdependent structural hierarchies, from the molecular to the mesoscale, exist in SX organic phases, and that these are intrinsically linked to the macromolecular extractive properties. (From Ellis, R. J., and Antonio, M. R., *Langmuir*, 28(14), 2012, 5987–5998.)[130]

molecular scale, the acidity of the aqueous phase was shown to influence the Ln^{3+} coordination chemistry in the organic phase following extraction from neutral and acidic media, 3 M $LiNO_3$ and 3 M HNO_3, respectively. Specifically, nitrate is essentially monodentate when acid is present and bidentate when it is not. This difference in nitrate coordination was attributed to the presence of water or nitric acid within the reverse-micelle core structures that stabilize the metal complex by either satisfying the inner coordination sphere via direct metal coordination or interacting with the nitrate in the outer coordination sphere, and thus contributes toward improved extraction. In both acidic and neutral organic phases (Figure 6.10), incorporation of Ln^{3+} into the micellar structure and the resulting increased interaction between micellar cores caused the growth of micelle chains, which were identified and monitored using SAXS data interpreted with the GIFT method. In the acidic system, the solution contained sizable chains even without Ce^{3+}. These chains grew as the Ce^{3+} concentration increased toward the critical concentration (concentration at which phase splitting occurs), where interactions between micellar chains were observed. This growth of micelle chains was associated with third-phase formation, and the presence of acid clearly promoted their growth. This multipronged approach of conducting simultaneous investigations into both the coordination and supramolecular structures of the Ln^{3+}-malonamide solvent extraction system afforded new insights into how extracted solutes affect organic-phase assemblies. In particular, the results have shown that the structure of macromolecular architectures formed as a result of micellar coagulation is more ordered than previously thought,[143] and that extraction of solutes stimulates one-dimensional growth of chains that increase in length and begin to interact in a second dimension as the critical concentration is approached. These studies serve as a foundation to a new approach for looking at the structures and properties of solvent extraction systems, and bridge the gap between fundamental soft-matter chemistry and practical separation science, giving new insights into how acid affects coordination chemistry within reverse micelles and the macromolecular architectures that prelude phase transition.

The supramolecular chemistry of DMDOHEMA was correlated with its water extraction properties from neutral media in many of the same manners as done beforehand for the predecessor extractant DMDBTDMA,[99,117,130,132,136,150, 154,155,158,160,256–259] which has been abandoned in practice for its instability to third-phase formation. The alkyl ether substitution ($R = C_2H_4OC_6H_{13}$) for the tetradecyl chain ($R = n\text{-}C_{14}H_{29}$) (see Table 6.1) provides improved extraction and stability so that the organic-phase splitting happens at higher metal concentrations,[258,301] a result that is likely to have its origins in supramolecular chemistry because the polar portions of both DMDOHEMA and DMDBTDMA, especially the carbonyl moieties responsible for bonding with M^{3+} ions, are identical. Investigations of the organization of DMDOHEMA in n-heptane by VPO and by use of combined SANS and SAXS, respectively, revealed two types of aggregates of the reverse-micelle type in the organic phase—one with 4 diamides per aggregate, and an oligomeric structure composed of ca. 10 diamide molecules forming at high extractant concentrations ($\geq$0.8 mol/L).[161] The larger aggregates can extract about five times more water than the DMDOHEMA monomers. As interrogated by SAXS, the evidence for intermediate-range micelle–micelle correlations in n-heptane solutions of varying DMDOHEMA concentrations contacted with a 2.93 M LiNO$_3$ aqueous phase *without metal ions* at 24°C is dramatic. The data of Figure 6.11 show the evolution and structural ripening of the morphology of the DMDOHEMA species as a function of solute concentration. At 0.5 M extractant concentration (black solid curve), the shape and the intensity of the SAXS data are typical of small aggregates of the reverse-micelle type, such as characterized in previous studies with DMDOHEMA in n-dodecane.[149] At concentrations of 0.8 M and higher, the intensities of the SAXS response at low momentum transfer (Q, Å^{-1}) decrease with increasing [DMDOHEMA], and simultaneously, correlation peaks appear at 0.25–0.35 Å^{-1}, whose positions provide an estimate of the average (center of mass to center of mass) intermicellar distance ($2\pi/Q$) of 18–25 Å. The decrease of $I(Q)$ can be interpreted as an increase in the repulsive potential between the reverse micelles. Recently, Bauduin et al.[154] have shown that for the DMDBTDMA and DMDBPMA extractants, a transition occurs from reverse micelles to locally lamellar structures or locally connected cylinders at very high extractant concentrations (>1 M). The appearance of such structures can also explain the decrease in the SAXS signal intensities at high extractant concentrations. Independent research confirms the structure of the reverse micelles formed above the critical micelle concentration and the formation of lamellar-like structures at very high solute concentrations.[130,131] For the present purposes of this review, the qualitative, empirical interpretation of the SAXS data of Figure 6.11

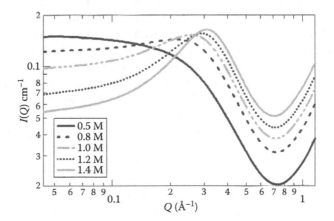

FIGURE 6.11   Log–log plot of the absolute X-ray scattered intensities, $I(Q)$ cm^{-1}, vs. momentum transfer, Q (Å^{-1}), for organic solutions containing various concentrations of DMDOHEMA in n-heptane at 24°C after equilibration with a 2.93 M LiNO$_3$ aqueous phase. The gradual evolution of the increasing correlation peak intensities and position shifting to higher Q values with increasing extractant concentrations indicate a structural ripening and compaction of the mean intermicellar distance. (From Meridiano, Y. et al., *Solvent Extr. Ion Exch.*, 27, 2009, 607–637.)[161]

indicates the presence of two aggregation states: at (and below) a 0.5 M extractant concentration, a spherical reverse-micelle type; and above this, a new aggregation state, the likes of which may be found in the SAXS literature of room temperature ionic liquids.[302,303]

6.4.2.3 Diglycolamides

Similar in molecular structure to the aforementioned diamides, the neutral extractant N,N,N',N'-tetra(n-octyl)-3-oxapentane-1,5-diamide (also known as tetra-n-octyldiglycolamide, or TODGA for short) is also used for the co-extraction of actinides and lanthanides. Studies by Sasaki et al.[255,304] revealed third-phase formation under selected conditions. This finding spawned the investigation by two independent groups at the CEA[153] and Argonne National Laboratory (ANL)[144] of the nature of the aggregates formed by TODGA before the phase splitting occurs. In these investigations, the amount of water and nitric acid extracted was followed by Karl-Fischer and potentiometric titrations, respectively, and it was found that there was a strong correlation between the amount of acid and water solubilized in the organic phase. Measurements of the surface tension also indicate the typical decrease as a function of TODGA concentration, indicating surface-active tendencies; this curve also exhibits the typical break point where the cmc is reached (see Figure 6.7). Metal extraction experiments showed that at around 1 M HNO_3 the nitrate dependency of americium(III) changes from ~3 to ~6, which could be indicative of reverse-micelle formation. SANS and SAXS measurements were carried out on the organic phase, and it was clear that compared to pure solvent, the scattering was intensified, suggesting the presence of aggregates. The study by Nave et al.[153] (directed from CEA) fitted the SANS and SAXS data using Baxter's sticky hard-sphere model and found that this correlated to 4 TODGA molecules in one reverse micelle around each metal. The study by Yaita et al.[144] (directed from ANL) found that modeling the SANS data gave a better fit when cylinders or elongated aggregates were assumed rather than spherical ones. VPO data were correlated with the SANS measurements, indicating that at high nitric acid concentrations the stoichiometric aggregation number reaches very high values, between 6 and 7. A direct correlation was also observed among SANS, VPO, and americium extraction data indicating that the aggregates started growing at the same point that americium extraction unexpectedly increased. The observation was made by Yaita et al.[144] that "despite the propensity of Ln^{3+} and An^{3+} to form complexes with high coordinating numbers (e.g., 8 or 9), the coordination of 4 bi- or tridentate TODGA molecules and 3–6 nitrate molecules by a single trivalent f-element cation is only possible if most of the ligands are monodentate or some of the ligands reside in a second, more distant, coordination sphere." Nave et al.[153] made a similar observation: "When there is a supramolecular organization involving the extractant, a simple molecular speciation cannot explain the experimental parameters such as separation factors or distribution coefficients." These two insightful observations point out one of the problems associated with the formation of reverse-micellar aggregates in the organic phase during solvent extraction (as has been iterated previously in this text). Namely, the possibility to successfully model the response depends on how the system behaves, and it is apparent that these changes will make a model based on classical solvent extraction, with classic thermodynamic equilibrium constants, invalid in the high concentration range where the consequences of aggregation are equal to or outweigh the influences of metal coordination chemistry.

Another illustration of the influence of aggregative behavior on the metal ion extraction using TODGA was reported by Jensen et al.[147,148] The study attributed the enhancement in the extraction of trivalent neodymium by TODGA to a tetramer formation when the concentration of nitric acid reached 0.7 M. According to the study, the small inverted micelles were preorganized to strongly interact with trivalent f-elements. Such evidence of supramolecular influence on the metal extraction is hard to come by in the area of nuclear-fuel processing, where charge-controlled electrostatic bonding dominates in highly acidic aqueous mixtures. The metal ion partitioning is mainly facilitated by neutral solvent extraction reagents in such acidic systems, where the solvation of the inorganic anions in the nonaqueous phase is necessary to maintain charge neutrality for the metal complex.

6.4.3 Acidic Organophosphorous Extractants

6.4.3.1 HDEHP

The importance of extractant aggregation in solvent extraction is perhaps the most apparent for monoacidic organophosphorus extractants, which strongly dimerize in solvents of low polarity ($\varepsilon < 10$).[305] For liquid–liquid distribution systems, where low metal-to-ligand ratios are encountered, the metal ion is often fully dehydrated and coordinated by three phosphoric acid dimers at the aqueous–organic interface, where each dimer dissociates one proton to neutralize the charge of the extracted complex.[306,307] Metal ion complexation by such liquid cation exchangers is typically exothermic, indicating that strong electrostatic interactions between the oxygen-donor ligand and the metal overcome the unfavorable energetic cost of metal dehydration.[146,308,309]

The self-association of monoacidic organophosphorus extractants is possible due to the presence of both the hydrogen bond donating and accepting functionalities. The strength of the dimerization equilibrium is thus dependent on structural features of the extractant, which affect the acidity (proton dissociation) and the basicity (electron density on the phosphoryl group). Kolarik reviewed the physicochemical properties of monoacidic organophosphorus extractants, pointing out that electron-inducing effects by alkyl substituents significantly affect the acid-base properties of such reagents.[305] The disruption of the dimer is necessary to the metal ion extraction. As such, the electron-withdrawing groups weaken the hydrogen bonding during the extractant self-association, and dialkyl phosphoric acids, such as HDEHP, extract metals more strongly than dialkyl phosphinates and dialkyl phosphinic acids. Jensen and Bond illustrated that incorporation of sulfur into the extractant structure of dialkyl phosphinic acids affects the aggregation equilibrium due to the weaker nature of the thiol-supported hydrogen bonding.[310]

In contrast to dimeric phosphinic acid, the thiophosphinic and dithiophosphinic acid monomers dominate the extraction of samarium and europium, yielding the metal complex of lower lipophilicity in the organic phase. Mason et al. studied the depolymerizing effect of an alcohol on the extraction of trivalent lanthanides by HDEHP.[311] When 1-decanol was present in the organic solutions, the extraction of metals decreased four orders of magnitude, demonstrating the importance of dimeric structures of HDEHP.[312] The observed decrease in metal partitioning may again be attributed to the solvation energetics. The metal complex no longer relies on a well-organized structure to shield the polar core from the bulk organic diluent. Although the dielectric constant of decanol presents a nonaqueous environment of more inviting polarity to the metal complex, its thermodynamic stability suffers from the exposed core, substantially decreasing the liquid–liquid distribution of the metal ion.

Owing to the fact that this molecule has one polar end, one hydrophobic end, and structural similarities with AOT, HDEHP has been treated as a surface-active molecule, and several studies exist where this ligand is shown to form reverse micelles in the organic phase.[63,64,251,313] Shioi et al.[251] studied the aggregates formed in the organic phase when the sodium salt of HDEHP was used in alkaline conditions (pH ~ 9). Their SAXS studies indicated that there may be a shift from disc-like to spherical aggregates when the fraction of sodium chloride to sodium-DEHP was varied. The formed microemulsion was compared to a microemulsion formed by using AOT, and it was seen that Na-DEHP gave a stable microemulsion under more aggressive conditions and always resulted in an optically transparent phase compared to using AOT. This could indicate smaller reverse micelles in the case of Na-DEHP than AOT.

Steytler et al.[63,64] examined metal and ammonium salts of HDEHP at alkaline pH and observed aggregates forming in the organic phase (cyclohexane). SANS and viscosity measurements were carried out to gain insight into the shape of the aggregates. For ammonium-DEHP, the aggregates change shape from elongated (rodlike) structures to spherical reverse micelles with the increase of water in the organic phase. For the metal salts of HDEHP, the shape of the aggregates also depended on which metal ion was present. For copper, the addition of water promotes smaller aggregates, while the presence of nickel results in the aggregates going from small spherical to larger rodlike shapes by the addition of water. The aggregates formed by metal salts of manganese, calcium, and cobalt are unaffected by the addition of water, and aluminum-DEHP entities do not display any large aggregate associations.

Neuman et al. investigated HDEHP at elevated pH in a number of different studies.[42,45,52,54,55,313–316] In some of these studies, the group prepared NaDEHP by addition of sodium hydroxide to HDEHP (saponification) and investigated the aggregate characteristics of this compound with and without extractable transition metal ions, e.g., Ni^{2+}, Co^{2+}, and Zn^{2+}. In one of these studies by Neuman et al.,[313] the interfacial characteristics and phase composition of metal extraction systems using HDEHP in n-alkane diluents was investigated. Several analytical tools were used in this study to probe the microemulsion: efficiencies and rates of metal extraction, Karl-Fischer titration for water extraction, VPO measurements for aggregation numbers, and tensiometry for measuring the surface tension between organic and aqueous phases. In addition, several spectrometric techniques, such as Fourier transform infrared (FT-IR), NMR, and fluorescence, were utilized. Finally, light-scattering techniques were used to examine the aggregation formation. All these measurements correlated well, and by increasing the HDEHP concentration, three regions were identified regarding the interactions in the organic phase. One region was called nucleation, which is characterized by a negligible amount of water in the organic phase, and no significant change in the distribution and rate constants. However, a decrease in interfacial tension is noted in this region, suggesting surface-active molecules are present and that some small aggregates may start to form. The next region was called linear growth, where larger aggregates presumably start forming, and the distribution and rate constants increase, as well as the water content. Between regions 1 and 2, there is a break in the decrease in interfacial tension, suggesting that the critical micelle concentration is reached and larger micelles are forming. The third region is called structural reorganization and growth, which is characterized by an abrupt increase in distribution ratios, rate of transfer, water content in the organic phase, and aggregation number. The light-scattering technique used also correlated and showed a small increase in scattering intensity at low HDEHP concentration to show a very large scattering intensity at high HDEHP concentration. These experiments were carried out at a pH between 2.5 and 4, indicating that reverse micelles will form in liquid–liquid extraction experiments using HDEHP at a lower pH than what was seen in previous studies.

Overall, Neuman et al. reported on rodlike micelles formed by linking of metal–DEHP$_2$ complexes, reported for nickel, cobalt, and zinc.[314] These M(DEHP)$_2$ complexes stack linearly with varying aggregation numbers and somewhat varying size, with the Co aggregates being larger than the Ni aggregates. In a later paper, Neuman and Ibrahim[45] show that the water associated with the aggregates of [Ni(DEHP)$_2$(H2O)$_2$]$_x$ was bound on the outside of the aggregates forming what the authors called open-water channels. This observation is in contradiction to other studies of aggregates where the water is often solubilized in droplets of reverse micelles, as described elsewhere in this chapter.

In another paper by the same team,[55] the authors made another discovery defying general considerations for reverse micelles, where they observed water having a negative effect on the micelle size. In this study, the authors were very careful in controlling the water content and to produce optically clean samples for the light-scattering study, which they claimed as one of the keys to their discovery. Their light-scattering data for samples that had been exposed to the atmosphere (wet) show that at 2 mM NaDEHP there is an extensive increase in the scattering as the concentration of NaDEHP increases, indicating the onset of reverse-micelle formation. Below this region, the NaDEHP molecules exist as monomers. Above this concentration, there are large cylindrical aggregates present. The water concentration in these samples was 40 ppm. The molar ratio of water to NaDEHP was 1.2 at 2 mM NaDEHP. They also measured dry samples and found that the dramatic increase in the light scattering started already at 0.13 mM NaDEHP. This means that the cmc value for dry NaDEHP is one order of magnitude lower than the samples that had been exposed to the humid air. This goes against previous theories that it is the hydrated monomers that form aggregates. The dry solutions were exposed to air after the measurement, and it was shown that the aggregates were dissolved or degraded after water was absorbed into the solution. It was also found that the dry micelles were larger than those formed in the wet conditions with a hydrodynamic radius of 50 nm compared to 33 nm. The authors attempted to postulate a possible explanation wherein the sodium ions in the NaDEHP complex must be solubilized in a reverse micelle when it is dry, but if water is present it can help solubilize the sodium ion in the monomeric ligand. Still, the authors make the observation that the true situation is likely more complex. The concept of water serving as a gluing agent is often based on the size observation rather than on the point where the micelles form. In general, if the reverse micelles are already formed, then more water will make them bigger. Thus, the authors suggest that it is better to study the cmc rather than the micelle size.

Ritcey et al. reported on the preferential extraction of Co(II) over Ni(II) by HDEHP when the temperatures were elevated from 25°C to 50°C.[317] Raising the temperature introduces a favorable entropic contribution into the solvent extraction of Co(II) due to its transition from a hydrated octahedral complex with HDEHP to an anhydrous tetrahedral configuration.[318] In highly metal-loaded HDEHP systems, polymerization has been shown to be a gradual phenomenon.[319,320] The progressive changes in the metal complex aggregation were followed using the viscosity measurements. Baes et al. noticed that the increasing organic-phase viscosity of HDEHP in hexane loaded with uranyl ion agreed with the isopiestic equilibration study, indicating the formation of chain polymers when the UO_2^{2+}:HDEHP ratio approached 0.5.[321] Ritcey and Ashbrook determined that simple stoichiometry of the coordination of Co(II) by two monoacidic organophosphorus reagents governs the liquid–liquid distribution of cobalt despite growing polymerization due to ligand loading.[322] The polymer formation was found to be reversible if more ligand or depolymerizing alcohol was added.[319,320]

In the processes investigated for possible implementation in the advanced nuclear-fuel cycle in the United States, one of the final steps is the separation of lanthanides from the remaining actinides in the waste streams. This would be the next step after the co-extraction mentioned above (in, for example, the TRUEX process). One process that was developed in the 1960s at Oak Ridge National Laboratory for this purpose is the TALSPEAK process.[323] This old but still very relevant process chemistry utilizes a combination of several molecules to achieve group separation between actinides and lanthanides. HDEHP is used for its strong extraction properties at low acidity (pH ~ 3) for the trivalent actinides and lanthanides. To hold the actinides back in the aqueous phase, diethyltriaminepentaacetic acid (DTPA) is present in the aqueous phase to selectively form strong complexes with actinides. To buffer the pH, large amounts of lactic acid, HL (1–2 M), are present in the aqueous phase. The combination of these three reagents adds significant complexity to the process and has resulted in contradictory observations and conclusions regarding the presence or absence of ordered structures in the organic phase. Studies of this system have indicated that lactic acid, apart from buffering pH, plays other important roles in this system, increasing the extraction kinetics, phase disengagement rate, and solubility of DTPA.[324] Kolarik and Kuhn[325] investigated the kinetics of the TALSPEAK system and found that the rate of extraction as a function of lactic acid concentration gave unexpected results. Their results were replotted in a recent TALSPEAK review by Nilsson and Nash[326] showing that the rate of transfer as a function of lactic acid concentration can be fitted with a quadratic function, indicating a system that goes from being independent of [HL] to being dependent on it, and finally depending on the square of [HL]. The same review also indicated that the total extraction system with HDEHP, DTPA, and HL cannot be accurately described using known thermodynamic stability constants to calculate extraction trends for a shift in pH. Recent results from Grimes et al.[327] indicate that lactic acid itself partitions into the organic phase in a manner not previously observed, and that the concentration of HDEHP in the organic phase also affects the amount of lactic acid extracted into the organic phase. The observation that water and lactic acid was extracted by HDEHP and studies showing that the kinetics of extraction was improved by the addition of lactic acid may give reason to believe that the extraction system in the TALSPEAK process may be prone to forming reverse micelles. However, no compelling evidence of reverse micelles has been observed; instead, the phenomena may be explained by mixed extracted species of HDEHP:Ln(III):lactate aggregates in a 4:2:2 ratio observed by Marie et al.[328] The improved kinetics may be explained by lactate having an impact on the binding kinetics of DTPA to metal ions observed in the work by Nash et al.[329]

6.4.3.2 Alkylenediphosphonic Acids

Interesting parallels between the aggregative equilibria and metal ion extraction have been noted for alkyl-substituted alkylenediphosphonic acid reagents.[330] Methylenediphosphonic acid (Table 6.1), the first member of series of SX reagents where $-CH_2-$ groups separate two phosphonic acid functionalities, forms very strong complexes with a variety of metal ions.[331] It strongly dimerizes in nonpolar diluents, and similarly to the acidic monofunctional organophosphorus analogs, the hydrogen bonds are broken to accommodate the metal ion during the complexation process.[332] Two dimeric aggregates fully encapsulate divalent and trivalent metals. The strength of the metal complex stems from the thermodynamic stability of the formed six- and eight-member rings when the metal ion is coordinated.[330] The extraction of trivalent americium by 2-ethylhexyl-substituted methylenediphosphonic acid is exothermic, with the positive entropy contribution as the metal dehydration increases the disorder of the system.[333]

The extensive literature on acidic organophosphorous extractants[334] demonstrates that they can organize much like their neutral counterparts in organic solutions. For example, exhaustive prior research on a variety of substituted diphosphonic acid extractants bears this out.[332,335–338] These molecules, such as P,P'-di(2-ethylhexyl) methylene- (H$_2$DEH[MDP]), ethylene- (H$_2$DEH[EDP]), and butylene- (H$_2$DEH[BuDP]) diphosphonic acids, spontaneously organize to form "n-mers" through hydrogen bonding interactions at the hydrophilic ends of the molecules. Detailed investigations of the compositions of organic-phase species formed by SX with these extractant systems (as performed using a variety of techniques, including infrared spectroscopy, VPO, SANS, and distribution methods) reveal that dimers, trimers, and hexamers are common structural motifs. The extraction of Ln^{3+} and An^{3+} as well as alkaline earth and transition metal ions from aqueous nitric acid solutions indicates that the extractant aggregation state strongly affects the metal SX chemistry.[332] The hexameric extractant aggregates are spherical in toluene, adopting a structure comparable to those of reverse micelles, as depicted in Figure 6.1. Under high metal loading conditions, even larger aggregates with rodlike and cylindrical shapes develop; these exhibit different morphological variations with changes of the metal concentrations in the organic phases.[332] Such extended solute organizations are also found for neutral extractants that are rather more complex than TBP and with additional functionality, like CMPO, wherein investigations of structures at various loadings of mineral acids[133] and metals[85,87,88] in the TRUEX process solvent media revealed large cylindrical aggregates.

Impacts of aggregation of this subset of amphiphilic structures on solvent extraction arrive when additional methylene linkers are inserted into the structure. McAlister et al. showed that the increasing separation between the phosphonic acid functionalities results in profound changes in ligand aggregation.[339,340] When two methylene groups are present in the structure of substituted ethylenediphosphonic acid, a predominantly hexameric aggregation dominates. The propylene linker returned the ligand to a dimeric association, whereas the butylene bridge promotes the presence of trimers.[340] Further structural inquiry established that those ligands with an odd number of methylene spacers between the phosphonic acid groups formed dimeric pairs. An even number of –CH$_2$– groups promoted more elaborate aggregate structures. This even-odd effect, as the authors called it, on the aggregative behavior of those alkylenediphosphonate structures left a distinct imprint on the metal ion partitioning behavior. With the exception of methylenediphosphonic acid, where the extraction is dominated by the enhanced stability of the chelate ring formation, the more highly aggregated reagents extracted metal ions such as Am^{3+} more efficiently, compared to those forming dimers.[340] This observation indicated the beneficial impact of the increasing ordering in a nonaqueous environment on the metal ion extraction. Otu et al. reported the thermochemical features of metal extraction (Am^{3+}, UO$_2^{2+}$) by substituted alkylenediphosphonic acids.[333,341] Interestingly, the enthalpic contribution to the extraction by a hexameric ethylenediphosphonic acid was quite small, with the extraction driven by favorable entropy terms. Such a thermochemical signature may be indicative of micellar solubilization of a solute, similar to those found for micellar type phase-transfer reagents such as dinonylnaphthalene sulfonic acid, HDNNS.

Depending on the composition of aqueous and organic phases in a liquid–liquid system, the extra-stoichiometric rearrangements of the metal complexes at high metal loading result in the polymeric or reverse-micellar structures. Jensen et al., for example, reported the presence of dinuclear neodymium complexes with HDEHP in toluene when the metal-to-ligand ratio was 3.[342] Chiarizia and Herlinger studied the aggregation of Ca^{2+}, Al^{3+}, La^{3+}, Nd^{3+}, UO_2^{2+}, Th^{4+}, and Fe^{3+} complexes with dialkyl-substituted alkylenediphosphonic acids in toluene.[330] Large rodlike polymeric formations were observed for iron with methylenediphosphonic acid, and smaller, comprising up to 12 complexing units, for the rest of the metals.[330,332,336] The hexameric morphology of the ethylenediphosphonic acid was found to persist at high metal loading studies, which illustrates the importance of ligand aggregation in the solvation of polar solutes in nonaqueous media.[337] A preformed polar cavity of ethylenediphosphonic acid remained a preferred solvation environment, and structural rearrangements were not necessary. In such weakly hydrated liquid–liquid systems, higher aggregates appear to withstand the solvation demands to a greater degree.

6.4.4 SURFACTANT SYSTEMS AND SYNERGISTIC EXTRACTION SYSTEMS

Studies have been carried out using HDEHP in combination with a sulfonic acid surfactant. Van Dalen and Gerritsma[343] showed a synergistic system where americium and cerium were extracted very effectively by HDEHP from 1 M perchloric acid by the addition of a small amount of HDNNS. Under conventional solvent extraction conditions, HDEHP is used as an acidic chelating agent for the extraction of trivalent metals and would, by itself, display no extraction at 1 M perchloric acid. The authors indicated that the HDNNS formed micelles with an aggregation number of seven HDNNS molecules in the organic phase, although they observed a decrease in water extraction when HDEHP was present, possibly indicating a decrease in micelle formation. Brejza and de Ortiz[19] studied the extraction of aluminum and zinc by HDEHP in kerosene in combination with the surfactant sodium dodecylbenzene sulfonate and n-butanol as a co-surfactant. Their study indicated that zinc extraction was suppressed by the formation of a microemulsion, whereas the extraction of aluminum was promoted. This observation was explained by the amount of hydration for each metal. Zinc is weakly hydrated and is effectively extracted by HDEHP in kerosene, while aluminum is more strongly hydrated and thus facilitated by the formation of a reverse micelle that can solubilize the aluminum ion without first dehydrating it.

Moyer et al.[344] studied the extraction of copper from sulfuric acid by HDDNS in toluene and included a modeling effort to describe extraction behavior by using the SXLSQA software. This model, as well as later adaptations of it, uses Hildebrand solubility parameters to account for nonideality in the organic phase and Pitzer parameters to account for nonideality in the aqueous phase. The general assumption made by the authors when HDDNS (denoted HA below) is used to extract metal ions is that an aggregate of a number (a) of monomers (H_aA_a)

is formed in the organic phase, and this aggregate incorporates copper without changing the aggregate size (aggregate number, a).

$$H_aA_{a(Org)} + Cu^{2+}_{(aq)} \rightleftharpoons CuH_{a-2}A_{a(org)} + 2H^+_{(aq)}$$

Based on a number of assumptions (small Cu loading; constant $[H_2SO_4]$ in the aqueous phase; $[H_2SO_4]$ much larger than $[Cu^{2+}]_{aq}$; large aggregates, where a is a large constant number; and ideal behavior of the phases), slope analysis of the above reaction can be used. Based on these assumptions, the extraction of Cu should not be dependent on the Cu concentration. This means that species containing one Cu ion only (mononuclear species) are considered, and the slope of $\log D$ versus $\log[HDDNS]$ should be 1 (HDDNSs are in aggregates of constant size and monodisperse). It was seen[344] that at high HDDNS concentrations and low loading (low Cu concentration) this was true, but when the concentration of HDDNS decreases, there is a stronger dependency of the Cu extraction on the [HDDNS]. Their study at higher loading of Cu results in an aggregation number, a, equal to 4–7 in the concentration range of 0.01 to 1 M of HDDNS. At lower HDDNS, the extraction mechanism must include the de-aggregation of the larger aggregates to monomers that then complex around the metal ion.

In the same paper, Moyer et al.[344] also studied the uptake of water and found that the aggregates are highly hydrated, as was established for these surfactants earlier.[345] The water concentration was proportional to the HDDNS concentration, which again suggests a constant aggregate size. The number of water, W, per HDDNS was found to be 5.4 ± 0.3.

The de-aggregation equilibrium was necessary to explain the behavior of the extraction curve throughout the entire concentration range of HDDNS, i.e., 0.0001–0.1 M. The model showed that a was equal to 4 or 5 (at lower concentration, $a = 4$ gave a better fit, although $a = 5$ was acceptable). The species most likely are $H_4A_4W_{22}$, $CuH_2A_4W_{22}$, and HAW_2 for the aggregate, metal-containing aggregate, and monomer, respectively. No evidence was found for bimetallic complexes. The authors also found that the aggregation number of HDDNS in toluene is lower than in CCl_4 ($a = 6$), and that the de-aggregation occurs at a lower concentration in CCl_4[346] (aggregates form more easily at lower concentrations). It was suggested that by matching the solubility parameter of the solvent with that of the aggregate, one could estimate the aggregate number. The authors suggested a decrement of 0.2 for the solubility parameter for each increment in the aggregate size. The solubility parameter for monomeric HDDNS was 9.7, and to reach 8.9 for toluene means 9.7 – 0.8 (i.e., 0.2 × 4) and to reach that of CCl_4, 8.55, means 9.7 – 1.2 (i.e., ~0.2 × 6). Although this algorithm seemed to yield reasonable results using a single parameter such as the Hildebrand solubility parameter, which was developed to describe nonpolar solvents, to describe molecules in a system that clearly contains polar (amphiphilic) groups may be questionable. However, the corona of the reverse micelles themselves is nonpolar and might, arguably, be described using this method.

Otu and Westland studied the thermodynamic features of extraction of several divalent (Ca^{2+}, Zn^{2+}) and trivalent (lanthanides, Bi^{3+}, Al^{3+}) metal ions by HDNNS.[347] This liquid cation exchanger has been shown to form reverse-micellar aggregates, typically involving ~10 monomers.[344,345] Such well-defined inverted micelles have been shown to contain water molecules inside their hydrophilic cores, providing a fitting environment to accommodate partially hydrated metal ions, whereas the hydrocarbon chains are extended into the bulk solvent to stabilize the micelle. Such micelles may host a variety of metals without a substantial impact on the enthalpic contribution to the extraction, as the hydration state of the metal cations changes minimally on extraction. The extraction by HDNNS and its sodium salt is governed by the entropic contribution to the Gibbs–Helmholtz relationship, while the ΔH term is fairly constant.[345,347–349] Those features match up with ethylenediphosphonic acid ligand quite well and might suggest that the liquid–liquid distribution mechanism for this hexameric reagent is also micellar. The similarities end there, however, as ethylenediphosphonic acid exhibits a pronounced sensitivity to the charge density of the metal. Chiarizia et al. showed the extraction constants for the partitioning of the lanthanide series using ethylenediphosphonic acid spanning three orders of magnitude.[335] The extraction of lanthanides by HDNNS is much less specific, for example, with a $\Delta(\Delta G)$ of 2 kJ/mol between lanthanum and erbium.[347] Such mass-to-charge recognition suggests that ethylenediphosphonic acid reagent behaves more like monofunctional organophosphorus reagents, such as HDEHP, and less like micellar HDNNS. The lack of molecular recognition based on electrostatics for micellar phase-transfer reagents may be restored through solubilization of metal ion coordination reagents inside the polar core of the micelle. Such molecular templating has been accomplished in highly ordered water-in-oil emulsified systems, and will be discussed later.

While the interest in using w/o microemulsions in metal ion separations also revolves around synergistic effects, the origins of the enhancement are quite different from those typically invoked in traditional SX.[350–353] The addition of a hydrophobic reagent, as Osseo-Asare describes it, offers the secondary electrostatic interaction with the metal ion.[29] The presence of reverse-micellar structures in such systems plays a key catalytic role in this synergistic behavior. A closer examination of several reports of synergistic behavior in w/o microemulsions will demonstrate this premise.

For mixtures of HDNNS and HDEHP, Van Dalen et al., and later, for mixtures of HDNNS and oxime, Keeney and Osseo-Asare, illustrated that the addition of a secondary amphiphile decreases the interfacial activity of HDNNS, indicating that the reagents mutually interact.[234,237] Van Dalen et al. suggested that the core of the sulfonic acid micelle may host up to three HDEHP molecules. Also, their calorimetric investigation for this sequential inclusion process, despite uncertainties resulting from several assumptions, roughly estimated an exothermic outcome.[237] Keeney and Osseo-Asare also estimated that the observed changes in the surface tension result from exothermic formation of a hydrogen bond between sulfonic acid and oxime.[234] Both groups reported the enhancement in the metal ion extraction when HDNNS interacted with a secondary reagent.[237,354]

Van Dalen et al. reported on the synergistic extraction of Am(III) and Ce(III) from 1 M perchloric acid when the molar ratio for HDEHP grew until it reached 6, when micelles were destroyed and metal distribution drastically decreased. Osseo-Asare and Renninger reported that the liquid–liquid distribution of nickel and cobalt by mixtures of HDNNS and LIX-63 goes through a maximum when the organic-phase conditions approach a 1:3:2 M^{2+}:LIX 63:HDNNS ratio.[354] However, due to very low aqueous solubility of oximes, it was postulated that it preferably interacted with the hydrophobic micellar layer.[227] The catalytic function of the HDNNS micelle may be visualized very well at this point. It serves as a platform where the reaction ingredients are preconcentrated (metal ions in the polar core, oxime ligand in the lipophilic layer), allowing for the enhanced outcome. Naturally, the optimal recipe may be diluted, and the effect would diminish. The observed liquid–liquid distribution maxima for Ni(II) and Co(II) has been observed for other metal ions.[355]

The observed synergistic enhancement in metal extraction results from the optimized conditions to support the metal ion complexation reaction at the micellar interface.[227] Accumulation of a hydrophobic agent inside or outside the reverse micelle enhances the kinetics, pushes the equilibrium, and adds to the hydrophobicity of the metal ion complex. This enhanced solubility has been demonstrated by Paatero et al., who found that the presence of Versatic 10 prevents the precipitation of the AOT complexes of copper.[30] Naganawa et al. studied the combination of AOT and CMPO in dodecane.[26] They found that CMPO enhanced the formation of w/o microemulsions in lower-acidity regions, where AOT would preferentially remain in the aqueous phase. This enhanced micellization, illustrating the interaction of both reagents, results in a very large synergistic enhancement of Ln^{3+} extraction.

The thermodynamic study by Arnold and Otu outlined the energy balance for the extraction of the lanthanide family by a mixture of HDNNS and a 2-ethylhexyl phenylphosphonic acid, HEHΦP.[356] The thermochemical features for this mixed-ligand system were compared with those assembled for both ligands individually. As outlined previously, the extraction of trivalent f-elements by a monoacidic dialkyl organophosphorus liquid cation exchanger is enthalpy controlled due to the formation of a stable chelate complex. The entropic contribution is unfavorable due to the ordering of the system. The micellar extraction may be characterized by a fairly constant enthalpic term, whereas the favorable entropy resulting from the partial dehydration of the metal ion drives the extraction. Arnold and Otu demonstrated that the thermochemical features for the extraction of trivalent lanthanides by a ligand mixture (HDNNS + HEHΦP) resembled those observed for the micellar system. The extraction was still controlled by the entropic contribution. The thermodynamic effects of the presence of the liquid cation exchanger, when solubilized by HDNNS micelles, did not match those expected if the two processes were added together. A lower than expected exothermic balance might indicate that the electrostatic interactions between HEHΦP and lanthanides are weaker in microemulsions than without them. A similar study on the mixtures of HDNNS and LIX 63,[227] where Osseo-Asare and Keeney argued for the solubilization of LIX 63 in the hydrophobic layer of the micelle, could be very valuable to delineate this thermochemical discrepancy. Nonetheless, the synergistic effect offered by the presence of w/o microemulsions stems from a micelle acting as a vehicle for the accumulation of the auxiliary hydrophobic agent, which further interacts with the metal ion. As such, the thermochemical features of synergistic contribution must be superimposed on those observed for the micellar mechanism of phase transfer.

6.5 OUTLOOK

An exhaustive amount of physical, chemical, structural, and spectroscopic characterization has provided clues to the fundamental chemistry that lies behind the phenomena of microemulsion formation and performance in chemical separations. The emerging concepts of micellization and microemulsification in separation science and technology complement the conventional metal-centered understanding favored by solvent extraction researchers. Whereas the formation of aggregates has been observed in many different chemical processes used in both industry and analytical lab scales, there is, as yet, little knowledge available about the deliberate and rational design of aggregation for extraction. And so, the control of aggregate nucleation and architecture to achieve improved separations remains a challenge for the present and future, and improving the understanding and ability to tune these phenomena is of great value. To relate solvent organization and solution structures (especially those that exhibit correlation in terms of intermediate, liquid–crystal-like order) with solute partitioning and 4f-/5f-ion (Ln/An) separation factors-will lead to a renaissance in hydrometallurgy. In particular, the myriad processes in the nuclear-fuel cycle are prone to microemulsion phenomena. For nuclear-fuel recycling to achieve international and domestic acceptance, demonstrably viable processes for the recycling and management of the waste byproducts are essential. Toward this end, the design and understanding of solvent organization as well as solute aggregation in process operations is still in its scientific infancy.

6.6 SUMMARY

From the selected systems perspectives brought to light herein, understanding the process of aggregation of solutes in nonaqueous solvents is important in broad areas of chemical separations, technology, and material sciences. Complications, especially in SX, arise when the aggregation leads to the formation of third phases and the precipitation of interfacial cruds. Although exact mechanistic understandings of how hierarchical organic-phase structures produce macroscopic system behaviors is not known, it does not preclude the use of thermodynamically stable, microheterogeneous reverse-micelle and microemulsion systems. Indeed, throughout this review, two pivotal aspects have come to light: (1) the general understanding of aggregation has increased markedly in recent years, and (2) aggregation can be an advantage for extraction strength and selectivity, as well as kinetics. The presumption that particularly effective metal distribution and, possibly, separation factors can arise from, in part, the supramolecular organization of the metal–extractant complexes and their conformational flexibilities in organic phases needs more development. To provide further proof that the organization of amphiphilic extractants in nonpolar diluents mediates the partitioning of solutes in all SX processes of practical importance, research must continue to delineate links among extractants and solutes and solvents at two extremes of length scale in a progressive zoom strategy. In this, at large length scales, typical of aggregation and collective behaviors, including micellization, extractant organization can be probed primarily through use of VPO and scattering techniques, including SAXS and SANS, respectively, as well as by high-energy X-ray scattering (HEXS). At smaller length scales, typical of molecular complexation, the coordination environments of extracted solutes (especially f-element ions) are readily determined through use of chemical, physical, electroanalytical, vibrational, optical, spectrometric, and spectroscopic techniques, including XAS. The combination and correlation of such systematic research results that have begun in this approach, when viewed in the context of the considerable body of international research in the field of separation science, will continue to provide unique entries to information about both short- and long-range interatomic interactions of solvents and solutes with extractants. With such a bottom-up description of the solution chemistry and knowledge of the fundamental interactions that operate in liquid–liquid extraction, researchers will be poised to open new avenues to the understanding and improvement of complex, multifunctional systems for 4f- and 5f-element separation processes by explicit design. We expect that results to be obtained from multipronged, coordinated programs of experimentation and theory in the future will provide new system-wide perspectives on solvent and solute behaviors in SX, and will serve as benchmarking information about the pivotal interfacial chemistry of SX,[22,357–359] reviewed elsewhere in this volume, in terms of kinetics and chemical reactivities, for theorists as they begin to study multifunctional liquid–liquid separation systems. In the end, SX is not all about distribution ratios and separation factors; diffusion phenomena coupled with kinetics at the interface control the mechanisms of SX.

ACKNOWLEDGMENTS

We thank our Argonne colleagues Drs. Ross Ellis and Renato Chiarizia for assistance and insight. The work at Argonne National Laboratory is supported by the U.S. Department of Energy, Office of Science, Office of Basic Energy Sciences, Division of Chemical Sciences, Biosciences and Geosciences, under contract DE-AC02-06CH11357. The work at UC-Irvine is supported by the U.S. Department of Energy through the Nuclear Energy University Program (NEUP) contract 120569. The work at Idaho National Laboratory is supported by the U.S. Department of Energy, Office of Nuclear Energy, under DOE Idaho Operations Office contract DE-AC07-05ID14517.

REFERENCES

1. Ritcey, G. M. 2006. *Solvent extraction: Principles and applications to process metallurgy.* G. M. Ritcey & Associates, Ottawa, Canada.
2. Rydberg, J., Cox, M., Musikas, C., and Choppin, G. R,, eds. 2004. *Solvent extraction principles and practice,* 2nd ed. Marcel Dekker, New York.
3. Tedder, W. D. 1991. Water extraction. In *Science and Technology of Tributyl Phosphate,* Vol. IV, ed. W. W. Schultz, J. D. Narratil, and A. S. Kertes, pp. 35–70. CRC Press, Boca Raton, FL.
4. Osseo-Asare, K. 1991. Aggregation, reversed micelles, and microemulsions in liquid–liquid-extraction—the tri-normal-butyl phosphate-diluent-water-electrolyte system. *Adv. Colloid Interface Sci.,* 37(1–2): 123–173.
5. Israelachvili, J. N., Mitchell, D. J., and Ninham, B. W. 1976. Theory of self-assembly of hydrocarbon amphiphiles into micelles and bilayers. *J. Chem. Soc. Faraday Trans. 2,* 72: 1525–1568.
6. Paatero, E., and Sjöblom, J. 1990. Phase behavior in metal extraction systems. *Hydrometallurgy,* 25(2): 231–256.
7. Winsor, P. A. 1948. Hydrotropy, solubilization, and related emulsification processes. I. *Trans. Faraday Soc.,* 44: 376–382.
8. Winsor, P. A. 1954. *Solvent properties of amphiphilic compounds.* Butterworth, London.
9. Chaiko, D. J. 1992. Partitioning of polymeric plutonium(IV) in Winsor II microemulsion systems. *Sep. Sci. Technol.,* 27(11): 1389–1405.
10. Rees, G. D., and Robinson, B. H. 1993. Microemulsions and organogels: Properties and novel applications. *Adv. Mater.,* 5(9): 608–619.
11. Vijayalakshmi, C. S., Annapragada, A. V., and Gulari, E. 1990. Equilibrium extraction and concentration of multivalent metal ion solutions by using Winsor II microemulsions. *Sep. Sci. Technol.,* 25(6): 711–727.
12. Vijayalakshmi, C. S., and Gulari, E. 1991. An improved model for the extraction of multivalent metals in Winsor II microemusion systems. *Sep. Sci. Technol.,* 26(2): 291–299.
13. Vijayalakshmi, C. S., and Gulari, E. 1992. Extraction of trivalent metals and separation of binary mixtures of metals using Winsor II microemulsion systems. *Sep. Sci. Technol.,* 27(2): 173–198.
14. Watarai, H. 1997. Microemulsions in separation sciences. *J. Chromatogr. A,* 780(1–2): 93–102.
15. Wu, J. G., Zhou, N. F., Shi, N., Zhou, W. J., Gao, H. C., and Xu, G. X. 1997. Extraction and surface chemistry. 2. Microscopic interfacial phenomena in solvent extraction. *Prog. Nat. Sci.,* 7(4): 385–388.
16. Wu, J. G., Zhou, W. J., Zhou, N. F., Gao, H. C., and Xu, G. X. 1997. Extraction and surface chemistry. 1. Microscopic interfacial phenomena in solvent extraction. *Prog. Nat. Sci.,* 7(3): 257–264.
17. Zeng, S., Yang, Y. Z., Zhu, T., Han, J., and Luo, C. H. 2005. Uranium(VI) extraction by Winsor II microemulsion systems using trialkyl phosphine oxide. *J. Radioanal. Nucl. Chem.,* 265(3): 419–421.
18. Friberg, S. E., and Qamheye, K. 1990. When is a microemulsion a microemulsion? In *The structure dynamics and equilibrium properties of colloidal systems,* NATO ASI Series C, Vol. 324, ed. D. M. Bloor and E. Wyn-Jones, pp. 221–231. Kluwer Academic, Dordrecht, The Netherlands.
19. Brejza, E. V., and de Ortiz, E. S. P. 2000. Phenomena affecting the equilibrium of Al(III) and Zn(II) extraction with Winsor II microemulsions. *J. Colloid Interface Sci.,* 227: 244–246.

20. de Castro Dantas, T. N., de Lucena Neto, M. H., Dantas Neto, A. A., Alencar Moura, M. C. P., and Barros Neto, E. L. 2005. New surfactant for gallium and aluminum extraction by microemulsion. *Ind. Eng. Chem. Res.*, 44(17): 6784–6788.

21. de Castro Dantas, T. N., de Lucena Neto, M. H., and Dantas Neto, A. A. 2002. Gallium extraction by microemulsions. *Talanta*, 56: 1089–1097.

22. Diss, R., and Wipff, G. 2005. Lanthanide cation extraction by malonamide ligands: From liquid–liquid interfaces to microemulsions. A molecular dynamics study. *Phys. Chem. Chem. Phys.*, 7(2): 264–272.

23. Fourre, P., Bauer, D., and Lemerle, J. 1983. Microemulsions in the extraction of gallium with 7-(1-ethenyl-3,3,5,5-tetramethylhexyl)-8-quinolinol from aluminate solutions. *Anal. Chem.*, 55(4): 662–667.

24. Hebrant, M., Goetz-Grandmont, G., Brunette, J.-P., and Tondre, C. 2005. Kinetics of complexation of lanthanide ions by 3-methyl-4-acyl-5-pyrazolone derivatives in micellar and microemulsion media. *Colloids Surf. A*, 253: 95–104.

25. Jiang, J. Z., Li, W. H., Gao, H. C., and Wu, J. G. 2003. Extraction of inorganic acids with neutral phosphorus extractants based on a reverse micelle/microemulsion mechanism. *J. Colloid Interf. Sci.*, 268(1): 208–214.

26. Naganawa, H., Suzuki, H., and Tachimori, S. 2000. Cooperative effect of carbamo-ylmethylene phosphine oxide on the extraction of lanthanides(III) to water-in-oil microemulsion from concentrated nitric acid medium. *Phys. Chem. Chem. Phys.*, 2: 3247–3253.

27. Nave, S., Eastoe, J., Heenan, R. K., Steytler, D., and Grillo, I. 2000. What is so special about aerosol-OT? 2. Microemulsion systems. *Langmuir*, 16(23): 8741–8748.

28. Ogino, K., and Abe, M. 1993. Microemulsion formation with some typical surfac-tants. In *Surface and colloid science*, ed. E. Matijevic, pp. 85–123. Plenum Press, New York.

29. Osseo-Asare, K. 1988. Enhanced solvent extraction with water-in-oil microemul-sions. *Sep. Sci. Technol.*, 23(12–13): 1269–1284.

30. Paatero, E., Sjöblom, J., and Datta, S. K. 1990. Microemulsion formation and metal extraction in the system water Aerosol OT extractant isooctane. *J. Colloid Interface Sci.*, 138(2): 388–396.

31. Watarai, H., Ogawa, K., and Suzuki, N. 1993. Formation of fluorescent complexes of europium(III) and samarium(III) with β-diketones and trioctylphosphine oxide in oil-water microemulsions. *Anal. Chim. Acta*, 277: 73–77.

32. Stubenrauch, C., ed. 2009. *Microemulsions. Background, new concepts, applica-tions, perspectives*. Wiley-Blackwell, Chichester, UK.

33. Steed, J., and Atwood, J. 2009. *Supramolecular chemistry*, 2nd ed. Wiley, Chichester, UK.

34. Eicke, H. F. 1980. Aggregation in surfactant solutions—Formation and properties of micelles and microemulsions. *Pure Appl. Chem.*, 52(5): 1349–1357.

35. Eicke, H. F. 1981. Properties of amphiphilic electrolytes in non-polar solvents. *Pure Appl. Chem.*, 53(7): 1417–1424.

36. Testard, F., Zemb, T., Bauduin, P., and Berthon, L. 2010. Third-phase formation in liquid/liquid extraction: A colloidal approach. In *Ion exchange and solvent extraction. A series of advances*, Vol. 19, ed. B. A. Moyer, pp. 381–428. CRC Press: Boca Raton, FL.

37. Kahlweit, M., Strey, R., Haase, D., Kunieda, H., Schmeling, T., Faulhaber, B., Borkovec, M., Eicke, H. F., Busse, G., Eggers, F., Funck, T., Richmann, H., Magid, L., Soderman, O., Stilbs, P., Winkler, J., Dittrich, A., and Jahn, W. 1987. How to study microemulsions. *J. Colloid Interface Sci.*, 118(2): 436–453.

38. Sager, W., and Eicke, H. F. 1991. The experimentalists kit to describe microemul-sions. *Colloids Surf.*, 57(3–4): 343–353.

39. Zalupski, P. R., Chiarizia, R., Jensen, M. P., and Herlinger, A. W. 2006. Metal extraction by sulfur-containing symmetrically-substituted bisphosphonic acids. Part I. P,P'-di(2-ethylhexyl) methylenebisthio-phosphonic acid. *Solvent Extr. Ion Exch.*, 24(3): 331–346.

40. Danesi, P. R., Magini, M., and Scibona, G. 1969. Aggregation of alkylammonium salts by vapor pressure (lowering) and light-scattering studies. In *Solvent extraction research. Proceedings of the Fifth International Conference on Solvent Extraction Chemisty (5th ICSEC)*, Jerusalem, Israel, September 16–18, 1968, ed. A. S. Kertes and Y. Marcus, pp. 185–193. Wiley-Interscience, New York.

41. Eicke, H. F., and Rehak, J. 1976. Formation of water-oil-microemulsions. *Helv. Chim. Acta*, 59(8): 2883–2891.

42. Yu, Z. J., and Neuman, R. D. 1994. Giant rodlike reversed micelles form by sodium bis(2-ethylhexyl) phosphate in *n*-heptane. *Langmuir*, 10(8): 2553–2558.

43. Gaonkar, A. G., Garver, T. M., and Neuman, R. D. 1988. H-1-NMR spectroscopic investigation of reversed micellization in metal organophosphorous surfactant system. *Colloids Surf.*, 30(3–4): 265–273.

44. Ibrahim, T. H., and Neuman, R. D. 2004. Nanostructure of open water-channel reversed micelles. I. H-1 NMR spectroscopy and molecular modeling. *Langmuir*, 20(8): 3114–3122.

45. Neuman, R. D., and Ibrahim, T. H. 1999. Novel structural model of reversed micelles: The open water-channel model. *Langmuir*, 15(1): 10–12.

46. Olsson, U., Shinoda, K., and Lindman, B. 1986. Change of the structure of microemulsions with the hydrophile lipophile balance of nonionic surfactant as revealed by NMR self-diffusion studies. *J. Phys. Chem.*, 90(17): 4083–4088.

47. Zalupski, P. R., Jensen, M. P., Chiarizia, R., Chiarelli, M. P., and Herlinger, A. W. 2006. Acid-base and organic-water distribution equilibria for symmetrically-substituted P,P'-dialkyl alkylenebisphosphonic acids. *Solvent Extr. Ion Exch.*, 24(2): 177–195.

48. Stewart, W. E., and Siddall, T. H., III. 1970. Nuclear magnetic resonance studies of amides. *Chem. Rev.*, 70: 517–551.

49. Barelli, A., and Eicke, H. F. 1986. Electron spin resonance spectroscopic investigation on the exchange kinetics of surfanctants in water AOT isooctane microemulsions. *Langmuir*, 2(6): 780–786.

50. Eicke, H. F., and Christen, H. 1978. Is water critical to formation of micelles in apolar media? *Helv. Chim. Acta*, 61(6): 2258–2263.

51. Podlipskaya, T. Y., Bulavchenko, A. I., and Sheludyakova, L. A. 2011. Properties of water in Triton N-42 reverse micelles during the solubilization of HCl solutions according to FT-IR and photon correlation spectroscopy. *J. Struct. Chem.*, 52(5): 970–979.

52. Neuman, R. D., Jones, M. A., and Zhou, N. F. 1990. Photon-correlation spectroscopy applied to hydrometallurgical solvent-extraction systems. *Colloids Surf.*, 46(1): 45–61.

53. Bu, W., Hou, B. Y., Mihaylov, M., Kuzmenko, I., Lin, B. H., Meron, M., Soderholm, L., Luo, G. M., and Schlossman, M. L. 2011. X-ray fluorescence from a model liquid/liquid solvent extraction system. *J. Appl. Phys.*, 110(10).

54. Yu, Z. J., and Neuman, R. D. 1994. Giant rodlike reversed micelles. *J. Am. Chem. Soc.*, 116(9): 4075–4076.

55. Yu, Z. J., Zhou, N. F., and Neuman, R. D. 1992. The role of water in the formation of reversed micelles—An antimicellization agent. *Langmuir*, 8(8): 1885–1888.

56. Barnes, I. S., Hyde, S. T., Ninham, B. W., Derian, P. J., Drifford, M., and Zemb, T. N. 1988. Small-angle X-ray scattering from ternary microemulsions determines microstructure. *J. Phys. Chem. B*, 92(8): 2286–2293.

57. Ben Azouz, I., Ober, R., Nakache, E., and Williams, C. E. 1992. A small-angle X-ray scattering investigation of the structure of a ternary water-in-oil microemulsion. *Colloids Surf.*, 69(2–3): 87–97.

58. Hilfiker, R., Eicke, H. F., Sager, W., Steeb, C., Hofmeier, U., and Gehrke, R. 1990. Form and structure factors of water AOT oil microemusions from synchrotron SAXS. *Ber. Bunsenges. Phys. Chem.*, 94(6): 677–683.

59. Shioi, A., Harada, M., and Tanabe, M. 1996. X-ray and light scattering from oil-rich microemulsions containing sodium bis(2-ethylhexyl) phosphate. *Langmuir*, 12(13): 3201–3205.

60. Zemb, T. N., Hyde, S. T., Derian, P. J., Barnes, I. S., and Ninham, B. W. 1987. Microstructure from X-ray scattering: The disordered open connected model of microemulsions. *J. Phys. Chem. B*, 91(14): 3814–3820.

61. Szekely, P., Ginsburg, A., Ben-Nun, T., and Raviv, U. 2010. Solution X-ray scattering form factors of supramolecular self-assembled structures. *Langmuir*, 26(16): 13110–13129.

62. Huang, J. S., Sung, J., and Wu, X. L. 1989. The effect of H_2O and D_2O on a water-in-oil microemulsion. *J. Colloid Interface Sci.*, 132(1): 34–42.

63. Steytler, D. C., Jenta, T. R., Robinson, B. H., Eastoe, J., and Heenan, R. K. 1996. Structure of reversed micelles formed by metal salts of bis(ethylhexyl) phosphoric acid. *Langmuir*, 12(6): 1483–1489.

64. Steytler, D. C., Sargeant, D. L., Welsh, G. E., Robinson, B. H., and Heenan, R. K. 1996. Ammonium bis(ethylhexyl) phosphate: A new surfactant for microemulsions. *Langmuir*, 12(22): 5312–5318.

65. Pilsl, H., Hoffmann, H., Hofmann, S., Kalus, J., Kencono, A. W., Lindner, P., and Ulbricht, W. 1993. Shape investigation of mixed micelles by small-angle neutron scattering. *J. Phys. Chem.*, 97(11): 2745–2754.

66. Chen, S. H. 1986. Small angle neutron scattering studies of the structure and interaction in micellar and microemulsion systems. *Annu. Rev. Phys. Chem.*, 37: 351–399.

67. Eicke, H. F., Geiger, S., Sauer, F. A., and Thomas, H. 1986. Dielectric study of fractal clusters formed by aqueous nanodroplets in apolar media. *Ber. Bunsenges. Phys. Chem.*, 90(10): 872–876.

68. Eicke, H. F., Hopmann, R. F. W., and Christen, H. 1975. Kinetics of conformational change during micelle formation in apolar media. *Ber. Bunsenges. Phys. Chem.*, 79(8): 667–673.

69. Eicke, H. F., and Shepherd, J. C. 1974. Dielectric properties of apolar micelle solutions containing solubilized water. *Helv. Chim. Acta*, 57(7): 1951–1963.

70. Eicke, H. F., Hilfiker, R., and Thomas, H. 1985. Probing order phenomena in macrofluids by pulsed electro-optical Kerr effect measurements. *Chem. Phys. Lett.*, 120(3): 272–275.

71. Eicke, H. F., and Markovic, Z. 1981. Temperature-dependent coalescence in water-oil microemulsions and phase transitions to lyotropic mesophases. *J. Colloid Interface Sci.*, 79(1): 151–158.

72. Eicke, H.F. 1980. Surfactants in nonpolar solvents: Aggregation and micellization. In *Topics in current chemistry: Micelles*, Vol. 87, ed. F. L. Boschke, pp. 85–145. Springer-Verlag, Berlin.

73. Burattini, E., Dangelo, P., Giglio, E., and Pavel, N. V. 1991. EXAFS study of probe molecules in micellar solutions. *J. Phys. Chem.*, 95(20): 7880–7886.

74. Chiarizia, R., Jensen, M. P., Borkowski, M., Ferraro, J. R., Thiyagarajan, P., and Littrell, K. C. 2003. Third phase formation revisited: The U(VI), HNO_3–TBP, *n*-dodecane system. *Solvent Extr. Ion Exch.*, 21(1): 1–27.

75. Gannaz, B., Antonio, M. R., Chiarizia, R., Hill, C., and Cote, G. 2006. Structural study of trivalent lanthanide and actinide complexes formed upon solvent extraction. *Dalton Trans.*, 38: 4553–4562.

76. Longo, A., Portale, G., Bras, W., Giannici, F., Ruggirello, A. M., and Liveril, V. T. 2007. Structural characterization of frozen n-heptane solutions of metal-containing reverse micelles. *Langmuir*, 23(23): 11482–11487.

77. Ma, G., Yan, W. F., Hu, T. D., Chen, J., Yan, C. H., Gao, H. C., Wu, J. G., and Xu, G. X. 1999. FTIR and EXAFS investigations of microstructures of gold solvent extraction: Hydrogen bonding between modifier and Au(CN)$_2^-$. *Phys. Chem. Chem. Phys.*, 1(22): 5215–5221.

78. Sun, Y., Yang, Z. L., Zhang, L., Zhou, N. F., Weng, S. F., and Wu, J. G. 2003. The interaction of Co^{2+} ions and sodium deoxycholate micelles. *J. Mol. Struct.*, 655(2): 321–330.

79. Yaita, T., Narita, H., Suzuki, S., Tachimori, S., Motohashi, H., and Shiwaku, H. 1999. Structural study of lanthanides(III) in aqueous nitrate and chloride solutions by EXAFS. *J. Radioanal. Nucl. Chem.*, 239(2): 371–375.

80. Den Auwer, C., Charbonnel, M. C., Drew, M. G. B., Grigoriev, M., Hudson, M. J., Iveson, P. B., Madic, C., Nierlich, M., Presson, M. T., Revel, R., Russell, M. L., and Thuery, P. 2000. Crystallographic, X-ray absorption, and IR studies of solid- and solution-state structures of tris(nitrato) N,N,N',N'-tetraethylmalonamide complexes of lanthanides. Comparison with the americium complex. *Inorg. Chem.*, 39(7): 1487–1495.

81. Den Auwer, C., Charbonnel, M. C., Presson, M. T., Madic, C., and Guillaumont, R. 1998. XAS study of actinide solvent extraction compounds. II. UO$_2$(NO$_3$)$_2$L$_2$ (with L = tri-isobutylphosphate, tri-n-butylphosphate, trimethylphosphate and triphenylphosphate). *Polyhedron*, 17(25–26): 4507–4517.

82. Antonio, M. R., and Soderholm, L. 2006. X-ray absorption spectroscopy of the actinides. In *The chemistry of the actinide and transactinide elements*, Vol. 5, 3rd ed., ed. L. R. Moiss, N. M. Edelstein, and J. Fuger, pp. 3086–3198. Springer, Dordrecht, The Netherlands.

83. Miyake, C., Hirose, M., Yoneda, Y., and Sano, M. 1990. The third phase of extraction process in fuel reprocessing. (II) EXAFS study of zirconium monobutylphosphate and zirconium dibutylphosphate. *J. Nucl. Sci. Technol.*, 27(3): 256–261.

84. Burns, J. H. 1983. X-ray diffraction studies of the structures of organic-phase solvent-extraction complexes. In *International Solvent Extraction Conference, AICHE ISEC '83*, Denver, CO, August 26–September 2, 1983, pp. 363–364. American Institute of Chemical Engineering.

85. Diamond, H., Thiyagarajan, P., and Horwitz, E. P. 1990. Small-angle neutron scattering studies of praseodymium-CMPO polymerization. *Solvent Extr. Ion Exch.*, 8(3): 503–513.

86. Thiyagarajan, P., Diamond, H., Danesi, P. R., and Horwitz, E. P. 1987. Small-angle neutron scattering studies of cobalt(II) organo-phosphorus polymers in deuteriobenzene. *Inorg. Chem.*, 26(25): 4209–4212.

87. Thiyagarajan, P., Diamond, H., and Horwitz, E. P. 1988. Small-angle neutron scattering studies of the aggregation of Pr(NO$_3$)$_3$–CMPO and PrCl$_3$–CMPO complexes in organic solvents. *J. Appl. Crystallogr.*, 21: 848–852.

88. Thiyagarajan, P., Diamond, H., Soderholm, L., Horwitz, E. P., Toth, L. M., and Felker, L. K. 1990. Plutonium(IV) polymers in aqueous and organic media. *Inorg. Chem.*, 29(10): 1902–1907.

89. Lohithakshan, K. V., Aswal, V. K., and Aggarwal, S. K. 2008. Small angle neutron scattering study of U(VI) third phase formation in HNO$_3$/DHDECMP-n-dodecane system. *Pramana J. Phys.*, 71(5): 985–989.

90. Molina, P. G., Silber, J. J., Correa, N. M., and Sereno, L. 2007. Electrochemistry in AOT reverse micelles. A powerful technique to characterize organized media. *J. Phys. Chem. C*, 111(11): 4269–4276.

91. Charlton, I. D., and Doherty, A. P. 2000. Simultaneous observation of attractive interaction, depletion forces, and "sticky" encounters between AOT reverse micelles in isooctane using microelectrode voltammetry. *J. Phys. Chem. B*, 104(33): 8061–8067.

92. Charlton, I. D., and Doherty, A. P. 2000. Voltammetry as a tool for monitoring micellar structural evolution? *Anal. Chem.*, 72(4): 687–695.

93. Charlton, I. D., and Doherty, A. P. 1999. Electrochemistry in true reverse micelles. *Electrochem. Commun.*, 1(5): 176–179.

94. Kazarinov, V. F., ed. 1987. *The interface structure and electrochemical processes at the boundary between two immiscible liquids*. Springer-Verlag, Berlin.

95. Kitatsuji, Y., Yoshida, Z., Kudo, H., and Kihara, S. 2002. Transfer of actinide ion at the interface between aqueous and nitrobenzene solutions studied by controlled-potential electrolysis at the interface. *J. Electroanal. Chem.*, 520(1–2): 133–144.

96. Rusling, J. F. 1997. Molecular aspects of electron transfer at electrodes in micellar solutions. *Colloid Surf. A Physicochem. Eng. Asp.*, 123: 81–88.

97. Osakai, T., and Shinohara, A. 2008. Electrochemical aspects of the reverse micelle extraction of proteins. *Anal. Sci.*, 24(7): 901–906.

98. Ellis, R. J., and Antonio, M. R. 2012. Redox chemistry of third phases formed in the cerium/nitric acid/malonamide-*n*-dodecane solvent extraction system. *ChemPlusChem*, 77(1): 41–47.

99. Erlinger, C., Belloni, L., Zemb, T., and Madic, C. 1999. Attractive interactions between reverse aggregates and phase separation in concentrated malonamide extractant solutions. *Langmuir*, 15(7): 2290–2300.

100. Subbuthai, S., Ananthanarayanan, R., Sahoo, P., Rao, A. N., and Rao, R. V. S. 2012. Feasibility studies for the detection of third phase during reprocessing of fast reactor fuel. *J. Radioanal. Nucl. Chem.*, 292(2): 879–883.

101. Scholz, F., Schroder, U., and Gulaboski, R. 2005. *Electrochemistry of immobilized particles and droplets*. Springer, Berlin.

102. Beeby, A., Clarkson, I. M., Eastoe, J., Faulkner, S., and Warne, B. 1997. Lanthanide-containing reversed micelles: A structural and luminescence study. *Langmuir*, 13(22): 5816–5819.

103. Burrows, H. D., and Tapia, M. J. 2002. Lanthanide ion binding in AOT/water/isooctane microemulsions. *Langmuir*, 18(17): 6706–6708.

104. Mwalupindi, A. G., Blyshak, L. A., Ndou, T. T., and Warner, I. M. 1991. Sensitized room-temperature luminescence in reverse micelles using lanthanide counterions as acceptors. *Anal. Chem.*, 63(13): 1328–1332.

105. Mwalupindi, A. G., Ndou, T. T., and Warner, I. M. 1992. Characterization of select organic analytes in reverse micelles using lanthanide counterions as acceptors. *Anal. Chem.*, 64(17): 1840–1844.

106. Eicke, H. F., Shepherd, J. C. W., and Steinemann, A. 1976. Exchange of solubilized water and aqueous-electrolyte solutions between micelles in apolar media. *J. Colloid Interface Sci.*, 56(1): 168–176.

107. Guo, F. Q., Li, H. F., Zhang, Z. F., Meng, S. L., and Li, D. Q. 2009. Synthesis of REF_3 (RE = Nd, Tb) nanoparticles via a solvent extraction route. *Mater. Res. Bull.*, 44(7): 1565–1568.

108. Guo, F. Q., Zhang, Z. F., Li, H. F., Meng, S. L., and Li, D. Q. 2010. A solvent extraction route for CaF_2 hollow spheres. *Chem. Commun.*, 46(43): 8237–8239.

109. Sato, H., and Komasawa, I. 2000. Preparation of concentrated ultrafine particles in reverse micellar systems using extractant-metal ion complex as a metal ion source. *J. Chem. Eng. Jpn.*, 33(2): 262–266.

110. Fu, X., Yu, L. L., Lin, Y. S., Zhu, H. T., Wang, H. T., and Zhou, X. D. 2004. Structure adjustment of mesoporous ZrO_2 prepared with the middle phase formed in extraction systems. *Solvent Extr. Ion Exch.*, 22(5): 885–895.

111. Hu, Z. S., Hu, X. P., Cui, W., Wang, D. B., and Fu, X. 1999. Three phase extraction study II—TBP-kerosene/H_2SO_4-TiOSO$_4$ system and the preparation of ultrafine powder of TiO_2. *Colloid Surf. A Physicochem. Eng. Asp.*, 155(2–3): 383–393.

112. Yang, C. F., Hong, B., and Chen, J. Y. 1996. Production of ultrafine ZrO_2 and Y-doped ZrO_2 powders by solvent extraction from solutions of perchloric and nitric acid with tri-n-butyl phosphate in kerosene. *Powder Technol.*, 89(2): 149–155.

113. Masson, S., Holliman, P., Kalaji, M., and Kluson, P. 2009. The production of nanoparticulate ceria using reverse micelle sol gel techniques. *J. Mater. Chem.*, 19(21): 3517–3522.

114. Sager, W. F. C. 1998. Controlled formation of nanoparticles from microemulsions. *Curr. Opin. Colloid Interface Sci.*, 3(3): 276–283.

115. Baba, Y., Kubota, F., Kamiya, N., and Goto, M. 2011. Recent advances in extraction and separation of rare-earth metals using ionic liquids. *J. Chem. Eng. Jpn.*, 44(10): 679–685.

116. Glatter, O. 1979. The interpretation of real-space information from small-angle scattering experiments. *J. Appl. Crystallogr.*, 12: 166–175.

117. Erlinger, C., Gazeau, D., Zemb, T., Madic, C., Lefrancois, L., Hebrant, M., and Tondre, C. 1998. Effect of nitric acid extraction on phase behavior, microstructure and interactions between primary aggregates in the system dimethyldibutyltetradecylmalonamide (DMDBTDMA) n-dodecane water: A phase analysis and small angle X-ray scattering (SAXS) characterisation study. *Solvent Extr. Ion Exch.*, 16(3): 707–738.

118. Chiarizia, R., Briand, A., Jensen, M. P., and Thiyagarajan, P. 2008. SANS study of reverse micelles formed upon the extraction of inorganic acids by TBP in n-octane. *Solvent Extr. Ion Exch.*, 26(4): 333–359.

119. Chiarizia, R., Jensen, M. P., Borkowski, M., Ferraro, J. R., Thiyagarajan, P., and Littrell, K. C. 2003. SANS study of third phase formation in the U(VI)–HNO$_3$/TBP–n-dodecane system. *Sep. Sci. Technol.*, 38(12–13): 3313–3331.

120. Chiarizia, R., Jensen, M. P., Borkowski, M., Thiyagarajan, P., and Littrell, K. C. 2004. Interpretation of third phase formation in the Th(IV)–HNO$_3$, TBP–n-octane system with Baxter's "sticky spheres" model. *Solvent Extr. Ion Exch.*, 22(3): 325–351.

121. Chiarizia, R., Jensen, M. P., Rickert, P. G., Kolarik, Z., Borkowski, M., and Thiyagarajan, P. 2004. Extraction of zirconium nitrate by TBP in n-octane: Influence of cation type on third phase formation according to the "sticky spheres" model. *Langmuir*, 20(25): 10798–10808.

122. Chiarizia, R., Stepinski, D., and Antonio, M. R. 2010. SANS study of HCl extraction by selected neutral organophosphorus compounds in n-octane. *Sep. Sci. Technol.*, 45: 1668–1678.

123. Chiarizia, R., Nash, K. L., Jensen, M. P., Thiyagarajan, P., and Littrell, K. C. 2003. Application of the Baxter model for hard spheres with surface adhesion to SANS data for the U(VI)–HNO$_3$, TBP–n-dodecane system. *Langmuir*, 19(23): 9592–9599.

124. Baxter, R. J. 1968. Percus-Yevick equation of hard spheres with surface adhesion. *J. Chem. Phys.*, 49: 2770–2774.

125. Menon, S. V. G., Kelkar, V. K., and Manohar, C. 1991. Application of Baxter's model to the theory of cloud points of nonionic surfactant solutions. *Phys. Rev. A*, 43: 1130–1133.

126. Plaue, J., Gelis, A., and Czerwinski, K. 2006. Actinide third phase formation in 1.1 M TBP/nitric acid/alkane diluent systems. *Sep. Sci. Technol.*, 41: 2065–2074.
127. Glatter, O. 1980. Evaluation of small-angle scattering data from lamellar and cylindrical particles by the indirect transformation method. *J. Appl. Crystallogr.*, 13: 577–584.
128. Fritz, G., and Bergmann, A. 2004. Interpretation of small-angle scattering data of inhomogeneous ellipsoids. *J. Appl. Crystallogr.*, 37: 815–822.
129. Glatter, O., Fritz, G., Lindner, H., Brunner-Popela, J., Mittelbach, R., Strey, R., and Egelhaaf, S. U. 2000. Nonionic micelles near the critical point: Micellar growth and attractive interaction. *Langmuir*, 16: 8692–8701.
130. Ellis, R. J., and Antonio, M. R. 2012. Coordination structures and supramolecular architectures in a cerium(III)–malonamide solvent extraction system. *Langmuir*, 28(14): 5987–5998.
131. Ellis, R. J., Meridiano, Y., Chiarizia, R., Berthon, L., Muller, J., Couston, L., and Antonio, M. R. 2013. Periodic behavior of lanthanide coordination within reverse micelles. *Chemistry—A European Journal*, DOI: 10.1002/chem.201202880.
132. Ellis, R. J., D'Amico, L., Chiarizia, R., and Antonio, M. R. 2012. Solvent extraction of cerium(III) using an aliphatic malonamide: The role of acid in organic phase behaviors. *Sep. Sci. Technol.*, 47(14–15): 2007–2014.
133. Ellis, R. J., Audras, M., and Antonio, M. R. 2012. Mesoscopic aspects of phase transitions in a solvent extraction system. *Langmuir*, 28(44): 15498–15504.
134. Bellini, T., Mantegazza, F., Piazza, R., and Degiorgio, V. 1989. Stretched-exponential relaxation of electric birefringence in a polydisperse colloidal solution. *Europhys. Lett.*, 10(5): 499–503.
135. Chevalier, Y., and Zemb, T. 1990. The structure of micelles and microemulsions. *Rep. Prog. Phys.*, 53(3): 279–371.
136. Testard, F., Bauduin, P., Martinet, L., Abecassis, B., Berthon, L., Madic, C., and Zemb, T. 2008. Self-assembling properties of malonamide extractants used in separation processes. *Radiochim. Acta*, 96(4–5): 265–272.
137. Sharma, S. C., Shrestha, R. G., Shrestha, L. K., and Aramaki, K. 2009. Viscoelastic wormlike micelles in mixed nonionic fluorocarbon surfactants and structural transition induced by oils. *J. Phys. Chem. B*, 113: 1615–1622.
138. Shrestha, L. K., Sato, T., Dulle, M., Glatter, O., and Aramaki, K. 2010. Effect of lipophilic tail architecture and solvent engineering on the structure of trehalose-based nonionic surfactant reverse micelles. *J. Phys. Chem. B*, 114: 12008–12017.
139. Shrestha, L. K., Sato, T.-A., Shrestha, R. G., Hill, J., Ariga, K., and Aramaki, K. 2011. Structure and rheology of reverse micelles in dipentaerythrityl tri-(12-hydroxystearate)/oil systems. *Phys. Chem. Chem. Phys.*, 13: 4911–4918.
140. Shrestha, L. K., Shrestha, R. G., and Aramaki, K. 2011. Intrinsic parameters for the structure control of nonionic reverse micelles in styrene: SAXS and rheometry studies. *Langmuir*, 27: 5862–5873.
141. Shrestha, L. K., Shrestha, R. G., Oyama, K., Matsuzawa, M., and Aramaki, K. 2010. Structure of diglycerol polyisostearate nonionic surfactant micelles in nonpolar oil hexadecane: A SAXS study. *J. Oleo Sci.*, 59: 339–350.
142. Shrestha, L. K., Yamamoto, M., Arima, S., and Aramaki, K. 2011. Charge-free reverse wormlike micelles in nonaqueous media. *Langmuir*, 27: 2340–2348.
143. Chiarizia, R., Thiyagarajan, P., Jensen, M. P., Borkowski, M., and Littrell, K. C. 2003. Third phase formation in TBP solvent extraction systems as a result of interaction between reverse micelles. In *Learning and Solution Purification*, Vol. 1 of the Proceedings of Hydrometallurgy 2003, ed. C. Young, A. Alfantazi, C. Anderson, A. James, D. Dreisinger, and B. Harris, pp. 917–928. Minerals, Metals and Materials Society, Warrendale, PA.

144. Yaita, T., Herlinger, A. W., Thiyagarajan, P., and Jensen, M. P. 2004. Influence of extractant aggregation on the extraction of trivalent f-element cations by a tetraalkyl-diglycolamide. *Solvent Extr. Ion Exch.*, 22(4): 553–571.
145. Chiarizia, R., Jensen, M. P., Borkowski, M., and Nash, K. L. 2006. A new interpretation of third-phase formation in the solvent extraction of actinides by TBP. In *Separations for the nuclear fuel cycle in the 21st century*, Vol. 933, ed. G. J. Lumetta, K. L. Nash, S. B. Clark, and J. I. Friese, pp. 135–150. ACS Symposium Series. Washington, DC.
146. Gannaz, B., Chiarizia, R., Antonio, M. R., Hill, C., and Cote, G. 2007. Extraction of lanthanides(III) and Am(III) by mixtures of malonamide and dialkylphosphoric acid. *Solvent Extr. Ion Exch.*, 25(3): 313–337.
147. Jensen, M. P., Yaita, T., and Chiarizia, R. 2007. Reverse-micelle formation in the partitioning of trivalent f-element cations by biphasic systems containing a tetraal-kyldiglycolamide. *Langmuir*, 23(9): 4765–4774.
148. Jensen, M. P., Yaita, T., and Chiarizia, R. 2008. Extractant aggregation as a mechanism of metal ion selectivity. In *Solvent extraction: Fundamentals to industrial applications. Proceedings of the 18th International Solvent Extraction Conference, ISEC'08*, Vol. 2, ed. B. A. Moyer, pp. 1029–1034. Canadian Institute of Mining, Metallurgy and Petroleum, Montreal.
149. Antonio, M. R., Chiarizia, R., Gannaz, B., Berthon, L., Zorz, N., Hill, C., and Cote, G. 2008. Aggregation in solvent extraction systems containing a malonamide, a dial-kylphosphoric acid and their mixtures. *Sep. Sci. Technol.*, 43(9–10): 2572–2605.
150. Lefrancois, L., Delpuech, J. J., Hebrant, M., Chrisment, J., and Tondre, C. 2001. Aggregation and protonation phenomena in third phase formation: An NMR study of the quaternary malonamide/dodecane/nitric acid/water system. *J. Phys. Chem. B*, 105(13): 2551–2564.
151. Zhou, N. F., and Wu, J. G. 2003. Review on aggregation of acid extractants in solvent extraction of metal ions: Remark on the general model. *Prog. Nat. Sci.*, 13(1): 1–12.
152. Nave, S., Mandin, C., Martinet, L., Berthon, L., Testard, F., Madic, C., and Zemb, T. 2004. Supramolecular organisation of tri-*n*-butyl phosphate in organic diluent on approaching third phase transition. *Phys. Chem. Chem. Phys.*, 6(4): 799–808.
153. Nave, S., Modolo, G., Madic, C., and Testard, F. 2004. Aggregation properties of N,N,N',N'-tetraoctyl-3-oxapentanediamide (TODGA) in *n*-dodecane. *Solvent Extr. Ion Exch.*, 22(4): 527–551.
154. Bauduin, P., Testard, F., Berthon, L., and Zemb, T. 2007. Relation between the hydrophile/hydrophobe ratio of malonamide extractants and the stability of the organic phase: Investigation at high extractant concentrations. *Phys. Chem. Chem. Phys.*, 9(28): 3776–3785.
155. Berthon, L., Martinet, L., Testard, F., Madic, C., and Zemb, T. 2007. Solvent penetration and sterical stabilization of reverse aggregates based on the DIAMEX process extracting molecules: Consequences for the third phase formation. *Solvent Extr. Ion Exch.*, 25(5): 545–576.
156. Guo, F. Q., Li, H. F., Zhang, Z. F., Meng, S. L., and Li, D. Q. 2008. Reversed micelle formation in a model liquid–liquid extraction system. *J. Colloid Interface Sci.*, 322(2): 605–610.
157. Abecassis, B., Testard, F., Zemb, T., Berthon, L., and Madic, C. 2003. Effect of *n*-octanol on the structure at the supramolecular scale of concentrated dimethyldioc-tylhexylethoxymalonamide extractant solutions. *Langmuir*, 19(17): 6638–6644.
158. Martinet, L., Berthon, L., Peineau, N., Madic, C., and Zemb, T. 2002. Aggregation of the solvates formed in liquid–liquid extraction of metallic cations by malonamides: Study of the Nd(NO$_3$)$_3$/H$_2$O/LiNO$_3$/DMDBTDMA/dodecane system. In *Proceedings*

of the International Solvent Extraction Conference, ISEC 2002, Vol. 2, ed. K. C. Sole, P. M. Cole, J. S. Preston, and D. J. Robinson, pp. 1161–1167. South African Institute of Mining and Metallurgy, Johannesburg, South Africa.

159. Dozol, H., and Berthon, C. 2006. Applications of experiments DOSY to systems incorporated in organic phase. *C. R. Chim.*, 9(3–4): 556–563.

160. Dozol, H., and Berthon, C. 2007. Characterisation of the supramolecular structure of malonamides by application of pulsed field gradients in NMR spectroscopy. *Phys. Chem. Chem. Phys.*, 9(37): 5162–5170.

161. Meridiano, Y., Berthon, L., Crozes, X., Sorel, C., Dannus, P., Antonio, M. R., Chiarizia, R., and Zemb, T. 2009. Aggregation in organic solutions of malonamides: Consequences for water extraction. *Solvent Extr. Ion Exch.*, 27: 607–637.

162. Kertes, A. S., and Gutmann, H. 1976. Surfactants in organic solvents: The physical chemistry of aggregation and micellization. In *Surface and colloid science*, Vol. 8, ed. E. Matijevic, pp. 193–295. John Wiley, New York.

163. Kon-no, K. 1993. Properties and applications of reversed micelles. In *Surface and colloid science*, Vol. 15, ed. E. Matijevic, pp. 125–151. Plenum, New York.

164. Ward, A. J. I., and du Reau, C. 1993. Surfactant association in nonaqueous media. In *Surface and colloid science*, Vol. 15, ed. E. Matijevic, pp. 153–196. Plenum, New York.

165. Marchelli, R., Dossena, A., Redenti, E., Corradini, R., Casnati, G., Pellizzetti, E., and Gasparini, G. M. 1991. Highly selective and micellar extraction of uranyl and alkaline-earth cations. In *New separation chemistry techniques for radioactive waste and other specific applications*, ed. L. Cecille, M. Casarci, and L. Pietrelli, pp. 33–3. Elsevier, London.

166. Danesi, P. R. 1992. Solvent extraction kinetics. In *Principles and practices of solvent extraction*, ed. J. Rydberg, C. Musikas, and G. R. Choppin, pp. 157–207. Marcel Dekker, New York.

167. Ballesteros-Gomez, A., and Rubio, S. 2012. Environment-responsive alkanol-based supramolecular solvents: Characterization and potential as restricted access property and mixed-mode extractants. *Anal. Chem.*, 84(1): 342–349.

168. Ballesteros-Gomez, A., Sicilia, M. D., and Rubio, S. 2010. Supramolecular solvents in the extraction of organic compounds. A review. *Anal. Chim. Acta*, 677(2): 108–130.

169. Rassat, S. D., Sukamto, J. H., Orth, R. J., Lilga, M. A., and Hallen, R. T. 1999. Development of an electrically switched ion exchange process for selective ion separations. *Sep. Purif. Technol.*, 15(3): 207–222.

170. Lilga, M. A., Orth, R. J., Sukamto, J. P. H., Rassat, S. D., Genders, J. D., and Gopal, R. 2001. Cesium separation using electrically switched ion exchange. *Sep. Purif. Technol.*, 24(3): 451–466.

171. Chen, W., and Xia, X. H. 2007. Highly stable nickel hexacyanoferrate nanotubes for electrically switched ion exchange. *Adv. Funct. Mater.*, 17(15): 2943–2948.

172. Yamaguchi, M., Kojima, D., Goto, M., Nakashio, F., and Nagao, S. 2000. Effect of applied voltage on the extractive separation of rare earth metals by D2EHPA in the presence of an oil-soluble complexing agent. In *Solvent extraction for the 21st century. Proceedings of the International Solvent Extraction Conference, ISEC'99*, Barcelona, Spain, Vol. 2, ed. M. Cox, M. Hidalgo, and M. Valiente, pp. 865–870. Society of Chemical Industry (SCI), London.

173. Okugaki, T., Kitatsuji, Y., Kasuno, M., Yoshizumi, A., Kubota, H., Shibafuji, Y., Maeda, K., Yoshida, Z., and Kihara, S. 2009. Development of high performance electrochemical solvent extraction method. *J. Electroanal. Chem.*, 629(1–2): 50–56.

174. Blain, J. F., Kikindai, T., and Gourisse, D. 1969. A study of third phase formation in trilaurylamine solutions. In *Solvent extraction research. Proceedings of the Fifth International Conference on Solvent Extraction Chemisty (5th ICSEC)*, Jerusalem, September 16–18, 1968, ed. A. S. Kertes and Y. Marcus, pp. 201–209. Wiley-Interscience, New York.

175. Borkowski, M., Chiarizia, R., Jensen, M. P., Ferraro, J. R., Thiyagarajan, P., and Littrell, K. C. 2003. SANS study of third phase formation in the Th(IV)–HNO₃/TBP-n-octane system. *Sep. Sci. Technol.*, 38(12–13): 3333–3351.

176. Chiarizia, R., and Briand, A. 2007. Third phase formation in the extraction of inorganic acids by TBP in n-octane. *Solvent Extr. Ion Exch.*, 25(3): 351–371.

177. Chiarizia, R., Rickert, P. G., Stepinski, D., Thiyagarajan, P., and Littrell, K. C. 2006. SANS study of third phase formation in the HCl–TBP-n-octane system. *Solvent Extr. Ion Exch.*, 24(2): 125–148.

178. Chiarizia, R., Stepinski, D. C., and Thiyagarajan, P. 2006. SANS study of third phase formation in the extraction of HCl by TBP isomers in n-octane. *Sep. Sci. Technol.*, 41(10): 2075–2095.

179. Gourisse, D. 1966. Méchanisme d'extraction de l'acide nitrique et de l'aeu parles solutions organiques d'alcoylamines tertiaires. Report CEA-R-3005.

180. Kedari, C. S., Coll, T., Fortuny, A., and Sastre, A. 2005. Third phase formation in the solvent extraction system Ir(IV)—Cyanex 923. *Solvent Extr. Ion Exch.*, 23(4): 545–559.

181. Kertes, A. S., and Habousha, Y. E. 1963. The third phase formation in extraction systems involving amine hydrochlorides. *J. Inorg. Nucl. Chem.*, 25: 1531–1533.

182. Kolarik, Z. 1979. The formation of a third phase in the extraction of plutonium(IV), uranium(IV) and thorium(IV) nitrates with tributyl phosphate in alkane diluents. In *Proceedings of the International Solvent Extraction Conference, ISEC'77*, Toronto, September 9–16, 1977, Vol. 1, ed. B. H. Lucas, G. M. Ritooy, and Herman W. Smith, pp. 178–187. Canadian Institute of Mining and Metallurgy, Montreal.

183. Kumar, S., and Koganti, S. B. 2003. Speciation studies in third phase formation: U(IV), Pu(IV), and Th(IV) third phases in TBP systems. *Solvent Extr. Ion Exch.*, 21(4): 547–557.

184. Kumar, S., and Koganti, S. B. 2003. An extended Setschenow model for Pu(IV) third phase formation in 20% Tri-n-butyl phosphate based nuclear solvent extraction system. *Solvent Extr. Ion Exch.*, 21(3): 423–433.

185. Kumar, S., Mudali, U. K., and Natarajan, R. 2011. Speciation studies of nitric acid and U(VI) third phases in n,n-dialkyl amides/n-dodecane systems. *J. Radioanal. Nucl. Chem.*, 289(3): 717–720.

186. Lefrancois, L., F, B., Noel, D., and C, T. 2001. Third phase formation: A new predictive approach. In *Solvent extraction for the 21st century. Proceedings of the International Solvent Extraction Conference, ISEC'99*, Barcelona, Spain, Vol. 1, ed. M. Cox, M. Hidalgo, and M. Valentine, pp. 637–641. Society of Chemical Industry (SCI), London.

187. Mallick, S., Suresh, A., Srinivasan, T. G., and Rao, P. R. V. 2010. Comparative studies on third-phase formation in the extraction of thorium nitrate by tri-n-butyl phosphate and tri-n-amyl phosphate in straight chain alkane diluents. *Solvent Extr. Ion Exch.*, 28(4): 459–481.

188. Mallick, S., Suresh, A., Srinivasan, T. G., and Rao, P. R. V. 2011. Third phase formation in the extraction of nitric acid and metal ions by octyl(phenyl)-N,N-diisobutyl carbamoyl methyl phosphine oxide (OΦCMPO) based solvents. *Desalin. Water Treat.*, 25(1–3): 216–225.

189. Nakashima, T., and Kolarik, Z. 1983. The formation of a third phase in the simultaneous extraction of actinide(IV) and uranyl nitrates by tributyl phosphate in dodecane. *Solvent Extr. Ion Exch.*, 1: 497–513.

190. Osseo-Asare, K. 1999. Third phase formation in solvent extraction: A microemulsion model. In *Metal separation technologies beyond 2000: Integrating novel chemistry with processing*, ed. K. C. Liddell and D. J. Chaiko, pp. 339–346. Minerals, Metals and Materials Society, Warrendale, PA.
191. Osseo-Asare, K. 2002. Microemulsions and third phase formation. In *Proceedings of the International Solvent Extraction Conference, ISEC 2002*, ed. K. C. Sole, P. M. Cole, J. S. Preston, and D. J. Robinson, pp. 118–124. South African Institute of Mining and Metallurgy, Johannesburg, South Africa.
192. Plaue, J., Gelis, A., and Czerwinski, K. 2006. Plutonium third phase formation in the 30% TBP/nitric acid/hydrogenated polypropylene tetramer system. *Solvent Extr. Ion Exch.*, 24(3): 271–282.
193. Plaue, J., Gelis, A., Czerwinski, K., Thiyagarajan, P., and Chiarizia, R. 2006. Small–angle neutron scattering study of plutonium third phase formation in 30% TBP/ HNO₃/alkane diluent systems. *Solvent Extr. Ion Exch.*, 24(3): 283–298.
194. Rao, P. R. V., and Kolarik, Z. 1996. A review of third phase formation in extraction of actinides by neutral organophosphorus extractants. *Solvent Extr. Ion Exch.*, 14(6): 955–993.
195. Rao, P. R. V., Srinivasan, T. G., and Suresh, A. 2010. Third phase formation in the extraction of thorium nitrate by trialkyl phosphates. *IOP Conf. Ser. Mater. Sci. Eng.*, 9(1).
196. Ravi, J., Prathibha, T., Venkatesan, K. A., Antony, M. P., Srinivasan, T. G., and Rao, P. R. V. 2012. Third phase formation of neodymium (III) and nitric acid in unsymmetrical N,N-di-2-ethylhexyl-N',N'-dioctyldiglycolamide. *Sep. Purif. Technol.*, 85: 96–100.
197. Srinivasan, T. G., Ahmed, M. K., Shakila, A. M., Dhamodaran, R., Vasudeva, R. P. R., and Mathews, C. K. 1986. Third phase formation in the extraction of plutonium by tri-*n*-butyl phosphate. *Radiochim. Acta*, 40: 151–154.
198. Srinivasan, T. G., Vijayasaradhi, S., Dhamodaran, R., Suresh, A., and Rao, P. R. V. 1998. Third phase formation in extraction of thorium nitrate by mixtures of trialkyl phosphates. *Solvent Extr. Ion Exch.*, 16(4): 1001–1011.
199. Suresh, A., Deivanayaki, R., Srinivasan, T. G., and Rao, P. R. V. 2006. Third phase formation in the extraction of Th(NO₃)₄ by tri-2-methyl butyl phosphate from nitric acid media. *Radiochim. Acta*, 94(6–7): 319–324.
200. Suresh, A., Rao, C., Sabharwal, K. N., Srinivasan, T. G., and Rao, P. R. V. 1999. Third phase formation in the extraction of Nd(III) by octyl(phenyl)-N,N-diisobutyl carbamoyl methyl phosphine oxide (O Φ CMPO). *Solvent Extr. Ion Exch.*, 17(1): 73–86.
201. Suresh, A., Srinivasan, T. G., and Rao, P. R. V. 2009. Parameters influencing third-phase formation in the extraction of Th(NO₃)₄ by some trialkyl phosphates. *Solvent Extr. Ion Exch.*, 27(2): 132–158.
202. Vidyalakshmi, V., Subramanian, M. S., Srinivasan, T. G., and Rao, P. R. V. 2001. Effect of extractant structure on third phase formation in the extraction of uranium and nitric acid by N,N-dialkyl amides. *Solvent Extr. Ion Exch.*, 19(1): 37–49.
203. Enderby, J.E., and Neilson, G. W. 1981. The structure of electrolyte solutions. *Rep. Prog. Phys.*, 44(6): 593–653.
204. Marcus, Y., and Hefter, G. 2006. Ion pairing. *Chem. Rev.*, 106: 4585–4621.
205. Merkling, P. J., Munoz-Paez, A., and Marcos, E. S. 2002. Exploring the capabilities of X-ray absorption spectroscopy for determining the structure of electrolyte solutions: Computed spectra for Cr³⁺ or Rh³⁺ in water based on molecular dynamics. *J. Am. Chem. Soc.*, 124(36): 10911–10920.

206. Yamaguchi, T. 1990. Diffraction and X-ray absorption studies of electrolyte-solutions. *Pure Appl. Chem.*, 62(12): 2251–2258.
207. Skanthakumar, S., and Soderholm, L. 2006. Studying actinide correlations in solution using high energy X-ray scattering. In *MRS Symposium Proceedings*, Vol. 893 (Actinides 2005—Basic Science, Applications and Technology), pp. 411–416.
208. Soderholm, L., Skanthakumar, S., and Wilson, R. E. 2009. Structures and energetics of erbium chloride complexes in aqueous solution. *J. Phys. Chem. A*, 113: 6391–6397.
209. Wilson, R. E., Skanthakumar, S., and Soderholm, L. 2007. The structures of polynuclear Th(IV) hydrolysis products. In *MRS Symposium Proceedings*, Vol. 986 (Actinides 2006—Basic Science, Applications and Technology), pp. 183–188.
210. Marcus, Y., and Kertes, A. S. 1969. *Ion exchange and solvent extraction of metal complexes.* Wiley-Interscience, London.
211. Miles, J. H. 1990. Separation of plutonium and uranium. In *Science and technology of tributyl phosphate*, Vol. III, ed. W. W. Schulz, L. L. Burger, and J. D. Navratil, pp. 11–54. CRC Press, Boca Raton, FL.
212. Dam, H. H., Reinhoudt, D. N., and Verboom, W. 2007. Multicoordinate ligands for actinide/lanthanide separations. *Chem. Soc. Rev.*, 36(2): 367–377.
213. Dam, H. H., Beijleveld, H., Reinhoudt, D. N., and Verboom, W. 2008. In the pursuit for better actinide ligands: An efficient strategy for their discovery. *J. Am. Chem. Soc.*, 130(16): 5542–5551.
214. Bulavchenko, A. I., Podlipskaya, T. Y., Batishcheva, E. K., and Torgov, V. G. 2000. The study of the $PtCl_6^2$ concentration with reversed micelles as a function of the acid-salt content of the feed. *J. Phys. Chem. B*, 104(20): 4821–4826.
215. Soderholm, L., Almond, P. M., Skanthakumar, S., Wilson, R. E., and Burns, P. C. 2008. The structure of the plutonium oxide nanocluster $[Pu_{38}O_{56}Cl_{54}(H_2O)_8]^{14-}$. *Angew. Chem. Int. Ed.*, 47(2): 298–302.
216. Wilson, R. E., Skanthakumar, S., and Soderholm, L. 2011. Separation of plutonium oxide nanoparticles and colloids. *Angew. Chem. Int. Ed.*, 50(47): 11234–11237.
217. du Preez, J. G. H., Sumter, N. M., Viviers, C., Rohwer, H. E., and van Brecht, B. J. A. M. 1996. Coordination chemistry and the development of metal ion specific separating agents. In *Emerging separation technologies for metals II*, ed. R. G. Bautista, pp. 265–278. Minerals, Metals and Materials Society, Warrendale, PA.
218. Laing, M. 1994. Solvent extraction of metals is coordination chemistry. In *Coordination chemistry. A century of progress*, Vol. 565, ed. G. B. Kauffman, pp. 382–394. ACS Symposium Series 565. Washington, DC.
219. Zolotov, Y. A. 1994. Coordination chemistry in the solovent extraction of metals. Developments from Russian laboratories. In *Coordination chemistry. A century of progress*, Vol. 565, ed. G. B. Kauffman, pp. 395–403. ACS Symposium Series 565. Washington, DC.
220. Lumetta, G. J., Rapko, B. M., Garza, P. A., Hay, B. P., Gilbertson, R. D., Weakley, T. J. R., and Hutchison, J. E. 2002. Deliberate design of ligand architecture yields dramatic enhancement of metal ion affinity. *J. Am. Chem. Soc.*, 124(20): 5644–5645.
221. Koma, Y., Watanabe, M., Nemoto, S., and Tanaka, Y. 1998. A counter current experiment for the separation of trivalent actinides and lanthanides by the SETFICS process. *Solvent Extr. Ion Exch.*, 16(6): 1357–1367.
222. Weaver, B., Kappelmann, F. A., and Topp, A. C. 1953. Quantity separation of rare earths by liquid–liquid extraction. I. The first kilogram of gadolinium oxide. *J. Am. Chem. Soc.*, 75: 3943–3945.
223. Singleterry, C. R. 1955. Micelle formation and solubilization in nonaqueous solvents. *J. Am. Oil Chem. Soc.*, 32: 446–452.

224. Ruckenstein, E., and Nagarajan, R. 1980. Aggregation of amphiphiles in nonaqueous media. *J. Phys. Chem.*, 84: 1349–1358.
225. Cote, G., and Szymanowski, J. 1992. Processing of interfacial tension data in solvent extraction studies. Interfacial properties of various acidic organophosphorus extractants. *J. Chem. Tech. Biotechnol.*, 54: 319–329.
226. Marcus, Y. 1969. The law of mass-action versus non-ideal behaviour in distribution equilibria. *Pure Appl. Chem.*, 20(1): 85–92.
227. Osseo-Asare, K., and Keeney, M. E. 1980. Sulfonic acids: Catalysts for the liquid–liquid extraction of metals. *Sep. Sci. Technol.*, 15(4): 999–1011.
228. Osseo-Asare, K. 1996. Microemulsions: Compartmentalized fluids for metal separations and materials synthesis. In *Proceedings of a Symposium on Emerging Separation Technologies for Metals II*, Kona, HI, June 16–21, 1996, ed. R. G. Bautista, pp. 159–169. Minerals, Metals and Materials Society, Warrendale, PA.
229. Marcus, Y., and Kertes, A. S. 1969. Synergistic extraction. In *Ion exchange and solvent extraction of metal complexes*, pp. 815–858. Wiley-Interscience, London.
230. Cote, G. 2003. The supramolecular speciation: A key for improved understanding and modelling of chemical reactivity in complex systems. *Radiochim. Acta*, 91(11): 639–643.
231. Leveque, A., and Helgorsky, J. 1979. The recovery of gallium from Bayer process aluminate solutions by liquid–liquid extraction. In *Proceedings of the International Solvent Extraction Conference, ISEC'77*, Toronto, September 9–16, 1977, Vol. 1, ed. B. H. Lucas, G. M. Ritcey, and H. W. Smith, pp. 439–442. Canadian Institute of Mining and Metallurgy, Montreal.
232. Helgorsky, J., and Leveque, A. 1979. Liquid/liquid extraction of gallium contained in basic aqueous solutions. Patent BR7808230A. Rhone-Poulenc Industries, Paris.
233. De, T. K., and Maitra, A. 1995. Solution behavior of Aerosol OT in non-polar solvents. *Adv. Colloid Interface Sci.*, 59: 95–193.
234. Keeney, M. E., and Osseo-Asare, K. 1982. Molecular interaction in a mixed α-hydroxyoxime-sulfonic acid solvent extraction system. *Polyhedron*, 1(5): 453–455.
235. Gaonkar, A. G., and Neuman, R. D. 1987. Interfacial activity, extractant selectivity, and reversed micellization in hydrometallurgical liquid/liquid extraction systems. *J. Colloid Interf. Sci.*, 119(1): 251–261.
236. Osseo-Asare, K. 1990. Volume changes and distribution of hydrochloric acid and water in the tri-n-butyl phosphate–H₂O–HCl liquid–liquid system: A reversed micellar phenomenological model. *Colloids Surf.*, 50: 373–392.
237. Van Dalen, A., Gerritsma, K. W., and Wijkstra, J. 1974. Inclusion of some organic acids in micelles of dinonyl naphthalene sulfonic acid. *J. Colloid Interf. Sci.*, 48(1): 127–133.
238. Osseo-Asare, K. 1999. Microemulsion-mediated synthesis of nanosize oxide materials. In *Handbook of microemulsion science and technology*, ed. P. Kumar and K. L. Mittal, pp. 549–603. Marcel Dekker, New York.
239. Nyman, B. G., Hummelstedt, L. 1974. Use of liquid cation exchange for separation of nickel(II) and cobalt(II) with simultaneous concentration of nickel sulphate. In *Fundamental and applied aspects of solvent extraction in industry. Proceedings of the International Solvent Extraction Conference (ISEC'74)*, Vol. 1, ed. G. V. Jeffreys, pp. 669–684. Society of Chemical Industry (SCI), Lyon, France.
240. Laidler, J. J., Burris, L., Collins, E. D., Duguid, J., Henry, R. N., Hill, J., Karell, E. J., McDeavitt, S. M., Thompson, M., Williamson, M. A., and Willit, J. L. 2001. Chemical partitioning technologies for an ATW system. *Prog. Nucl. Energy*, 38(1–2): 65–79.

241. Regalbuto, M. C. 2011. Alternative separation and extraction: UREX+ processes for actinide and targeted fission product recovery. In *Advanced separation techniques for nuclear fuel reprocessing and radioactive waste treatment*, ed. K. L. Nash and G. J. Lumetta, pp. 176–200. Woodhead Publishing, Cambridge.

242. Naylor, A., and Wilson, P. D. 1983. Recovery of uranium and plutonium from irradiated nuclear fuel. In *Handbook of solvent extraction*, ed. T. C. Lo and M. H. I. Baird, pp. 783–798. Wiley, New York.

243. Schulz, W. W., and Navratil, J. D. 1984. *Science and Technology of Tributyl Phosphate, Vol. I, Synthesis, Properties, Reactions, and Analysis*. CRC Press, Boca Raton, FL.

244. Schulz, W. W., and Navratil, J. D. 1987. *Science and Technology of Tributyl Phosphate, Vol. II, Selected Technical and Industrial Uses*. CRC Press, Boca Raton, FL.

245. Schulz, W. W., Burger, L. L., and Navratil, J. D. 1990. *Science and Technology of Tributyl Phosphate, Vol. III, Applications of Tributyl Phosphate in Nuclear Fuel Processing*. CRC Press, Boca Raton, FL.

246. Schulz, W. W., Navratil, J. D., and Kertes, A. S. 1991. *Science and Technology of Tributyl Phosphate, Vol. IV, Extraction of Water and Acids*. CRC Press, Boca Raton, FL.

247. Alcock, K., Grimley, S. S., Healy, T. V., Kennedy, J., and McKay, H. A. C. 1956. Extraction of nitrates by tributyl phosphate (TBP). I. System TBP + diluent + H_2O + HNO_3. *Trans. Faraday Soc.*, 52: 39–47.

248. Blaylock, C. R., and Tedder, D. W. 1989. Competitive equilibria in the system: Water, nitric acid, tri-n-butyl phosphate, and Amsco 125–82. *Solvent Extr. Ion Exch.*, 7(2): 249–271.

249. Chaiko, D. J., Fredrickson, D. R., Reichley-Yinger, L., and Vandegrift, G. F. 1988. Thermodynamic modeling of chemical equilibria in metal extraction. *Sep. Sci. Technol.*, 23(12–13): 1435–1451.

250. Naganawa, H., and Tachimori, S. 1997. Complex formation between tributyl phosphate and nitric acid and the hydration of the complexes in dodecane. *Bull. Chem. Soc. Jpn.*, 70(4): 809–819.

251. Shioi, A., Harada, M., and Matsumoto, K. 1991. Phase equilibrium of sodium bis(2-ethylhexyl)phosphate/water/n-heptane/sodium chloride microemulsion. *J. Phys. Chem.*, 95: 7495–7502.

252. Foa, E., Rosintal, N., and Marcus, Y. 1961. Three phase formation in the system hydrochloric acid–water–tributyl phosphate–diluent. *J. Inorg. Nucl. Chem.*, 23: 109–114.

253. Motokawa, R., Suzuki, S., Ogawa, H., Antonio, M. R., and Yaita, T. 2012. Microscopic structures of tri-n-butyl phosphate/n-octane mixtures by X-ray and neutron scattering in a wide q range. *J. Phys. Chem. B*, 116(4): 1319–1327.

254. Shen, X., Gao, H., and Wang, X. 1999. What makes the solubilization of water in reversed micelles exothermic or endothermic? A titration calorimetry investigation. *Phys. Chem. Chem. Phys.*, 1: 463–469.

255. Tachimori, S., Sasaki, Y., and Suzuki, S. 2002. Modification of TODGA–n-dodecane solvent with a monoamide for high loading of lanthanides(III) and actinides(III). *Solvent Extr. Ion Exch.*, 20(6): 687–699.

256. Berthon, L., Testard, F., Martinet, L., Zemb, T., and Madic, C. 2010. Influence of the extracted solute on the aggregation of malonamide extractant in organic phases: Consequences for phase stability. *C. R. Chim.*, 13(10): 1326–1334.

257. Charbonnel, M. C., Daldon, M., Berthon, C., Presson, M. T., Madic, C., and Moulin, C. 2001. Extraction of lanthanides(III) and actinides(III) by N,N'-substituted

malonamides: Thermodynamic and kinetic data. In *Solvent extraction for the 21st century. Proceedings of the International Solvent Extraction Conference, ISEC'99*, Barcelona, Spain, Vol. 2, ed. M. Cox, M. Hidalgo, and M. Valentine, pp. 1333–1338. Society of Chemical Industry (SCI), London.

258. Charbonnel, M. C., Flandin, J. L., Giroux, S., Presson, M. T., Madic, C., and Morel, J. P. 2002. Extraction of Ln(III) and Am(III) from nitrate media by malonamides and polydentate N-bearing ligands. In *Proceedings of the International Solvent Extraction Conference, ISEC 2002*, Vol. 2, ed. K. C. Sole, P. M. Cole, J. S. Preston, and D. J. Robinson, pp. 1154–1160. South African Institute of Mining and Metallurgy, Johannesburg, South Africa.

259. Spjuth, L., Liljenzin, J. O., Skalberg, M., Hudson, M. J., Chan, G. Y. S., Drew, M. G. B., Feaviour, M., Iveson, P. B., and Madic, C. 1997. Extraction of actinides and lanthanides from nitric acid solution by malonamides. *Radiochim. Acta*, 78: 39–46.

260. Antonio, M. R., Chiarizia, R., and Jaffrennou, F. 2010. Third-phase formation in the extraction of phosphotungstic acid by TBP in *n*-octane. *Sep. Sci. Technol.*, 45(12–13): 1689–1698.

261. Moyer, B. A. 1987. Alkane-insoluble trialkylammonium double salts involving the dodecamolybdophosphate anion. 3. Nature of a liquid third phase. *Solvent Extr. Ion Exch.*, 5(1): 195–203.

262. Moyer, B. A. 1988. Trialkylammonium mixed salts in amine extraction systems. Infrared study of the salts $(R_3NH)_3PMo_{12}O_{40}$ and $(R_3NH)Cl$ and mixed salt $(R_3NH)_3PMo_{12}O_{40}$–$3(R_3NH)Cl$. *Solvent Extr. Ion Exch.*, 6(1): 1–37.

263. Moyer, B. A., Baltich, F. J., and Andrews, K. C. 1991. Interfacial precipitates containing dodecamolybdophosphate and dodecamolybdoarsenate anions in tertiary amine solvent extraction. *Hydrometallurgy*, 27(1): 113–122.

264. Moyer, B. A., and McSowell, W. J. 1987. Alkane-insoluble trialkylammonium double salts involving the dodecamolybdophosphate anion. 2. Effect of amine structure on third-phase formation. *Sep. Sci. Technol.*, 22(2–3): 417–445.

265. Siddall, T. H., III. 1960. Effects of structure of N,N-disubstituted amides on their extraction of actinide and zirconium nitrates and of nitric acid. *J. Phys. Chem.*, 64: 1863–1866.

266. Siddall, T. H., III. 1961. *Application of amides as extractants*. U.S. AEC Report DP 541. E. I. du Pont de Nemours & Co., Aiken, SC.

267. Cantale, C., Casarci, M., De, S. A., Gasparini, G. M., Nardi, L., and Salluzzo, A. 1991. N,N-Dialkylaliphatic amides as extractant of platinum group metals. In *New separation chemistry techniques for radioactive waste and other specific applications*, ed. L. Cecille, M. Casarci, and L. Pietrelli, pp. 57–63. Elsevier, London.

268. Carlini, D., Casarci, M., Gasparini, G. M., Grossi, G., and Moccia, A. 1984. Bench scale demonstration of a coprocessing operation with an amidic solvent. In *Fuel reprocessing and waste management, Proceedings of American Nuclear Society International Topical Meeting*, Jackson, WY, August 26–29, 1984, pp. 485–492. American Nuclear Society, LaGrange, IL.

269. Dembinski, W., and Gasparini, G. M. 1980. *The extraction of uranium(VI) from nitric acid solutions with N,N-di-n-butyl octanamide*. Technical Report CNEN-RT/CHI-(80)14. Comitato Nazionale per l'Energia Nucleare.

270. Gasparini, G. M., and Grossi, G. 1980. Application of N,N-dialkyl aliphatic amides in the separation of some actinides. *Sep. Sci. Technol.*, 15: 825–844.

271. Gasparini, G. M., and Grossi, G. 1986. Long chain disubstituted aliphatic amides as extracting agents in industrial applications of solvent extraction. *Solvent Extr. Ion Exch.*, 4(6): 1233–1271.

272. Gasparini, G. M., and Torcini, S. 1980. *N,N-dialkyl substituted aliphatic amides as extractants for actinides and fission products: Preparation and extractive properties of N,N-di-n-butyl-2-ethylhexanamide and N,N-di-n-hexyloctanamide*. Technical Report CNEN-RT/CHI(80)6. Comitato Nazionale per l'Energia Nucleare.

273. Condamines, N., and Musikas, C. 1988. The extraction by *N,N-dialkylamides*. I. Nitric and other inorganic acids. *Solvent Extr. Ion Exch.*, 6: 1007–1034.

274. Condamines, N., and Musikas, C. 1992. The extraction by *N,N*-dialkylamides. II. Extraction of actinide cations. *Solvent Extr. Ion Exch.*, 10: 69–100.

275. Descouls, N., and Musikas, C. 1986. Mechanism of extraction of uranium(IV) ions by *N,N*-dialkylamides. *J. Less Common Met.*, 122: 265–274.

276. Musikas, C. 1987. Solvent extraction for the chemical separations of the 5f elements. *Inorg. Chim. Acta*, 140: 197–206.

277. Musikas, C. 1988. Potentiality of nonorganophosphorus extractants in chemical separations of actinides. *Sep. Sci. Technol.*, 23: 1211–1226.

278. Musikas, C. 1997. Completely incinerable extractants for the nuclear industry. A review. *Miner. Process. Extr. Metall. Rev.*, 17: 109–142.

279. Musikas, C., Condamines, N., and Cuillerdier, C. 1992. Advance in actinides separations by solvent extraction. Research and applications. *Process Metall.*, 7A: 417–422.

280. Musikas, C., Morisseau, J. C., Hoel, P., and Guillaume, B. 1987. Actinide extractants for the nuclear industry of the future. *Inst. Chem. Eng. Symp. Ser.*, 103: 51–65.

281. Cuillerdier, C., Musikas, C., Hoel, P., Nigond, L., and Vitart, X. 1991. Malonamides as new extractants for nuclear waste solutions. *Sep. Sci. Technol.*, 26(9): 1229–1244.

282. Baron, P., Berthon, L., Charbonnel, M. C., and Nicol, C. 1997. State of progress of Diamex process. In *Challenge towards second nuclear era with advanced fuel cycles. Proceedings of the International Conference on Future Nuclear Systems, GLOBAL'97*, Yokohama, Japan, October 5–10, 1997, Vol. 1, pp. 366–370.

283. Charbonnel, M. C., and Berthon, L. 1998. *Optimisation de la molécule extractante pour le procédé DIAMEX*. Report CEA-R-5801.

284. Serrano-Purroy, D., Baron, P., Christiansen, B., Glatz, J. P., Madic, C., Malmbeck, R., and Modolo, G. 2005. First demonstration of a centrifugal solvent extraction process for minor actinides from a concentrated spent fuel solution. *Sep. Purif. Technol.*, 45(2): 157–162.

285. Serrano-Purroy, D., Baron, P., Christiansen, B., Malmbeck, R., Sorel, C., and Glatz, J. P. 2005. Recovery of minor actinides from HLLW using the DIAMEX process. *Radiochim. Acta*, 93(6): 351–355.

286. Serrano-Purroy, D., Christiansen, B., Glatz, J. P., Malmbeck, R., and Modolo, G. 2005. Towards a DIAMEX process using high active concentrate. Production of genuine solutions. *Radiochim. Acta*, 93(6): 357–361.

287. Modolo, G., Vijgen, H., Serrano-Purroy, D., Christiansen, B., Malmbeck, R., Sorel, C., and Baron, P. 2007. DIAMEX counter-current extraction process for recovery of trivalent actinides from simulated high active concentrate. *Sep. Sci. Technol.*, 42(3): 439–452.

288. Sorel, C., Montuir, M., Espinoux, D., Lorrain, B., and Baron, P. 2008. Technical feasibility of the DIAMEX process. In *Solvent extraction: Fundamentals to industrial applications. Proceedings of the International Solvent Extraction Conference ISEC 2008*, Vol. 1, ed. B. A. Moyer, pp. 715–719. Canadian Institute of Mining, Metallurgy and Petroleum: Montreal, Canada.

289. Castellano, E. E., and Becker, R. W. 1981. Structural studies of addition compounds of lanthanides with organic ligands. 2. Lanthanum hexafluorophosphate and *N,N,N',N'*-tetramethylmalonamide. *Acta Crystallogr. B Struct. Commun.*, 37(Nov.): 1998–2001.

290. Castellano, E. E., and Becker, R. W. 1981. Structural studies of addition compounds of lanthanides with organic ligands. 1. Structures of samarium and erbium hexafluorophosphates with N,N,N',N'-tetramethylmalonamide. *Acta Crystallogr. Sect. B Struct. Commun.*, 37(Jan.): 61–67.

291. Fedosseev, A. M., Grigoriev, M. S., Charushnikova, I. A., Budantseva, N. A., Starikova, Z. A., and Moisy, P. 2008. Synthesis, crystal structure and some properties of new perrhenate and pertechnetate complexes of Nd^{3+} and Am^{3+} with 2,6-bis(tetramethylfurano)-1,2,4-triazin-3-yl)-pyridine, tris(2-pyridylmethyl)amine and N,N'-tetraethylmalonamide. *Polyhedron*, 27(8): 2007–2014.

292. Thuéry, P., Nierlich, M., Charbonnel, M.-C., Den Auwer, C., and Dognon, J.-P. 1999. Complexation of lanthanum(III) nitrate by N,N',N,N'-tetraethylmalonamide: Crystal structure of three polymorphic forms. *Polyhedron*, 18(27): 3599–3603.

293. Good, M. L., and Siddall, T. H., III. 1968. Metal chelates of some simple diamides. *J. Inorg. Nucl. Chem.*, 30: 2679–2687.

294. Siddall, T. H., III, and Good, M. L. 1967. Proton magnetic resonance studies and extraction properties of some simple diamides. *J. Inorg. Nucl. Chem.*, 29: 149–158.

295. Hannel, T. S., Otu, E. O., and Jensen, M. P. 2007. Thermochemistry of the extraction of bismuth(III) with bis(2-ethylhexyl) phosphoric and 2-ethyhexylphenylphosphonic acids. *Solvent Extr. Ion Exch.*, 25(2): 241–256.

296. Horwitz, E. P., McAlister, D. R., and Dietz, M. L. 2006. Extraction chromatography versus solvent extraction: How similar are they? *Sep. Sci. Technol.*, 41(10): 2163–2182.

297. Brunner-Popela, J., and Glatter, O. 1997. Small-angle scattering of interacting particles. 1. Basic principles of a global evaluation technique. *J. Appl. Crystallogr.*, 30: 431–442.

298. Brunner-Popela, J., Mittelbach, R., Strey, R., Schubert, K. V., Kaler, E. W., and Glatter, O. 1999. Small-angle scattering of interacting particles. III. D_2O-$C_{12}E_5$ mixtures and microemulsions with *n*-octane. *J. Chem. Phys.*, 110(21): 10623–10632.

299. Glatter, O. 2002. The inverse scattering problem in small-angle scattering. In *Neutrons, X-rays and light*, ed. P. Lindner and T. Zemb, pp. 73–102. Elsevier Science, Amsterdam.

300. Weyerich, B., Brunner-Popela, J., and Glatter, O. 1999. Small-angle scattering of interacting particles. II. Generalized indirect Fourier transformation under consideration of the effective structure factor for polydisperse systems. *J. Appl. Crystallogr.*, 32: 197–209.

301. Bisel, I., Nicol, C., Charbonnel, M. C., Blanc, P., Baron, P., and Belnet, F. 1999. Inactive DIAMEX test with the optimized extraction agent DMDOHEMA. In *Proceedings of the Fifth International Information Exchange Meeting on Actinide and Fission Product Partitioning and Transmutation*, SKC-CEN, Mol, Belgium, November 25–27, 1998, Session II, Paper 5.

302. Firestone, M. A., Dzielawa, J. A., Zapol, P., Curtiss, L. A., Seifert, S., and Dietz, M. L. 2002. Lyotropic liquid-crystalline gel formation in a room-temperature ionic liquid. *Langmuir*, 18: 7258–7260.

303. Triolo, A., Russina, O., Bleif, H. J., and Di Cola, E. 2007. Nanoscale segregation in room temperature ionic liquids. *J. Phys. Chem. B*, 111(18): 4641–4644.

304. Sasaki, Y., Sugo, Y., Suzuki, S., and Tachimori, S. 2001. The novel extractants, diglycolamides, for the extraction of lanthanides and actinides in HNO_3–*n*-dodecane system. *Solvent Extr. Ion Exch.*, 19: 91–103.

305. Kolarik, Z. 2010. Review: Dissociation, self-association, and partition of monoacidic organophosphorus extractants. *Solvent Extr. Ion Exch.*, 28: 707–763.

306. Beitz, J. V., and Sullivan, J. C. 1989. Laser-induced fluorescence studies of europium-extractant complexes in organic phases. *J. Less Common Metals*, 148: 159–166.
307. Kolarik, Z., and Pankova, H. 1966. Acidic organophosphorus extractants. I. Extraction of lanthanides by means of dialkyl phosphoric acids—Effect of structure and size of alkyl group. *J. Inorg. Nucl. Chem.*, 28: 2325–2333.
308. Fidelis, I. 1971. Influence of temperature on the extraction of lanthanides in the HDEHP–HNO$_3$ system. In *Proceedings of the International Solvent Extraction Conference*, Vol. 2, ed. J. G. Gregory, B. Evans, and P. C. Weston, pp. 1004–1007. Society of Chemical Industry (SCI), The Hague, Netherlands.
309. Zalupski, P. R., and Nash, K. L. 2008. Two-phase calorimetry. I. Studies on the thermodynamics of lanthanide extraction by bis(2-ethylhexyl) phosphoric acid. *Solvent Extr. Ion Exch.*, 26: 514–533.
310. Jensen, M. P., and Bond, A. H. 2002. Influence of aggregation on the extraction of trivalent lanthanide and actinide cations by purified Cyanex 272, Cyanex 301, and Cyanex 302. *Radiochim. Acta*, 90: 205–209.
311. Mason, G. W., Lewey, S., and Peppard, D. F. 1964. Extraction of metallic cations by mono-acidic orthophosphate esters in a monomerizing diluent. *J. Inorg. Nucl. Chem.*, 26: 2271–2284.
312. Danesi, P. R., and Vandegrift, G. F. 1981. Kinetics and mechanism of the interfacial mass transfer of Eu^{3+} and Am^{3+} in the system bis(2-ethylhexyl) phosphate-n-dodecane-NaCl-HCl-water. *J. Phys. Chem.*, 85: 3646–3651.
313. Neuman, R. D., Zhou, N. F., Wu, J. G., Jones, M. A., Gaonkar, A. G., Park, S. J., and Agrawal, M. L. 1990. General-model for aggregation of metal-extractant complexes in acidic organophosphorus solvent-extraction systems. *Sep. Sci. Technol.*, 25(13–15): 1655–1674.
314. Neuman, R. D., Yu, Z.-J., and Ibrahim, T. 1996. Role of reversed micelles of acidic organophosphorus extractants in cobalt/nickel separation—A position paper. In *Proceedings of International Solvent Extraction Conference, ISEC'96*, Vol. 1, ed. D. C. Shallcross, R. Paimin, and L. M. Prvcic, pp. 135–140. University of Melbourne, Melbourne, Australia.
315. Neuman, R. D., and Park, S. J. 1992. Characterization of association microstructures in hydrometallurgical nickel extraction by di(2-ethylhexyl)phosphoric acid. *J. Colloid Interface Sci.*, 152(1): 41–53.
316. Yu, Z. J., and Neuman, R. D. 1995. Reversed micellar solution-to-bicontinuous microemulsion transition in sodium bis(2-ethylhexyl) phosphate n-heptane water-system. *Langmuir*, 11(4): 1081–1086.
317. Ritcey, G. M., Ashbrook, A. W., and Lucas, B. H. 1975. Development of a solvent extraction process for the separation of cobalt from nickel. *CIM Bull.*, 68(753): 111–123.
318. Barnes, J. E., Setchfield, J. H., and Williams, G. O. R. 1976. Solvent extraction with di(2-ethylhexyl)phosphoric acid; a correlation between selectivity and the structure of the complex. *J. Inorg. Nucl. Chem.*, 38: 1065–1067.
319. Grimm, R., and Kolarik, Z. 1974. Acidic organophosphorus extractants. XIX. Extraction of Cu(II), Co(II), Ni(II), Zn(II) and Cd(II) by di(2-ethylhexyl) phosphoric acid. *J. Inorg. Nucl. Chem.*, 36: 189–192.
320. Preston, J. S. 1982. Solvent extraction of cobalt and nickel by organophosphorus acids. I. Comparison of phosphoric, phosphonic and phosphinic acid systems. *Hydrometallurgy*, 9: 115–133.
321. Baes, C. F., Jr., Zingaro, R. A., and Coleman, C. F. 1958. The extraction of uranium(VI) from acid perchlorate solutions by di-(2-ethylhexyl)-phosphoric acid in n-hexane. *J. Phys. Chem.*, 62(2): 129–136.

322. Ritcey, G. M., and Ashbrook, A. W. 1984. *Solvent extraction principles and applications to process metallurgy.* Elsevier, Amsterdam.

323. Weaver, B., and Kappelman, F. A. 1968. Preferential extraction of lanthanides over trivalent actinides by monoacidic organophosphates from carboxylic acids and from mixtures of carboxylic and aminopolyacetic acids. *J. Inorg. Nucl. Chem.*, 30(1): 263–272.

324. Persson, G., Wingefors, S., Liljenzin, J. O., and Svantesson, I. 1984. The Cth-Process for HLLW treatment. 2. Hot test. *Radiochim. Acta*, 35(3): 163–172.

325. Kolarik, Z., and Kuhn, W. 1974. Acidic organophosphorus extractants. XXI. Kinetics and equilibriums of extraction of europium(III) by di(2-ethylhexyl) phosphoric acid from complexing media. In *Fundamental and applied aspects of solvent extraction in industry. Proceedings of the International Solvent Extraction Conference (ISEC'74)*, Vol. 3, ed. G. V. Jeffreys, pp. 2593–2606. Society of Chemical Industry (SCI), Lyon, France.

326. Nilsson, M., and Nash, K. L. 2007. Review article: A review of the development and operational characteristics of the TALSPEAK process. *Solvent Extr. Ion Exch.*, 25(6): 665–701.

327. Grimes, T. S., Nilsson, M. A., and Nash, K. L. 2010. Lactic acid partitioning in TALSPEAK extraction systems. *Sep. Sci. Technol.*, 45(12–13): 1725–1732.

328. Marie, C., Hiscox, B., and Nash, K. L. 2012. Characterization of HDEHP-lanthanide complexes formed in a non-polar organic phase using P-31 NMR and ESI-MS. *Dalton Trans.*, 41(3): 1054–1064.

329. Nash, K. L., Brigham, D., Shehee, T. C., and Martin, A. 2012. The kinetics of lanthanide complexation by EDTA and DTPA in lactate media. *Dalton Trans.*, 41(48): 14547–14556.

330. Chiarizia, R., and Herlinger, A. W. 2004. Symmetrical *P,P'*-disubstituted esters of alkylenediphosphonic acids as reagents for metal solvent extraction. In *Ion exchange and solvent extraction*, Vol. 17, ed. B. A. Moyer, pp. 85–164. Marcel Dekker, New York.

331. Nash, K. L., Horwitz, E. P. 1990. Stability constants for europium(III) complexes with substituted methane diphosphonic acids in acid solutions. *Inorg. Chim. Acta*, 169: 245–252.

332. Chiarizia, R., Barrans, R. E., Ferraro, J. R., Herlinger, A. W., and McAlister, D. R. 2001. Aggregation of dialkyl–substituted diphosphonic acids and its effect on metal ion extraction. *Sep. Sci. Technol.*, 36(5–6): 687–708.

333. Otu, E. O., and Chiarizia, R. 2001. Temperature effects in the extraction of metal ions by dialkyl-substituted diphosphonic acids. I. The Am(III) case. *Solvent Extr. Ion Exch.*, 19(5): 885–904.

334. Kolarik, Z. 1982. Critical evaluation of some equilibrium constants involving acidic organophosphours extractants. *Pure Appl. Chem.*, 54(12): 2593–2674.

335. Chiarizia, R., McAlister, D. R., and Herlinger, A. W. 2005. Trivalent actinide and lanthanide separations by dialkyl-substituted diphosphonic acids. *Sep. Sci. Technol.*, 40(1–3): 69–90.

336. Chiarizia, R., Urban, V., Thiyagarajan, P., and Herlinger, A. W. 1998. Aggregation of *P,P'*-di(2-ethylhexyl) methanediphosphonic acid and its Fe(III) complexes. *Solvent Extr. Ion Exch.*, 16(5): 1257–1278.

337. Chiarizia, R., Urban, V., Thiyagarajan, P., and Herlinger, A. W. 1999. SANS study of aggregation of the complexes formed by selected metal cations and *P,P'*-di(2-ethylhexyl) ethane- and butane-diphosphonic acids. *Solvent Extr. Ion Exch.*, 17(5): 1171–1194.

338. Chiarizia, R., Urban, V., Thiyagarajan, P., and Herlinger, A. W. 1999. Aggregation of complexes formed in the extraction of selected metal cations by *P,P'*-di(2-ethylhexyl) methanediphosphonic acid. *Solvent Extr. Ion Exch.*, 17(1): 113–132.

339. Chiarizia, R., McAlister, D. R., Herlinger, A. W., and Bond, A. H. 2001. Solvent extraction by dialkyl-substituted diphosphonic acids in a depolymerizing diluent. I. Alkaline earth cations. *Solvent Extr. Ion Exch.*, 19(2): 261–276.

340. McAlister, D. R., Dietz, M. L., Chiarizia, R., Zalupski, P. R., and Herlinger, A. W. 2002. Metal extraction by silyl-substituted diphosphonic acids. II. Effect of alkylene bridge length on aggregation and metal ion extraction behavior. *Sep. Sci. Technol.*, 37(10): 2289–2315.

341. Otu, E. O., Chiarizia, R. 2001. Thermodynamics of the extraction of metal ions by dialkyl-substituted diphosphonic acids. II. The U(VI) and Sr(II) case. *Solvent Extr. Ion Exch.*, 19(6): 1017–1036.

342. Jensen, M. P., Chiarizia, R., and Urban, V. 2001. Investigation of the aggregation of the neodymium complexes of dialkylphosphoric, -oxothiophosphinic, and -dithiophosphinic acids in toluene. *Solvent Extr. Ion Exch.*, 19(5): 865–884.

343. Van Dalen, A., and Gerritsma, K. W. 1971. Solvent extraction and inclusion compound formation with dinonylnaphthalenesulfonic acid. In *Science and technology of solvent extraction as applied to all branches of industry. Solvent extraction: Proceedings of the International Solvent Extraction Conference ISEC'71*, Vol. 2, ed. J. G. Gregory, pp. 1096–1100. Society of Chemical Industry, London.

344. Moyer, B. A., Baes, C. F., Case, G. N., Lumetta, G. J., and Wilson, N. M. 1993. Equilibrium-analysis of aggregation behavior in the solvent-extraction of Cu(II) from sulfuric-acid by didodecylnaphthalene sulfonic-acid. *Sep. Sci. Technol.*, 28(1–3): 81–113.

345. Markovits, G. Y., and Choppin, G. R. 1973. Solvent extraction with sulfonic acids. In *Ion exchange and solvent extraction*, Vol. 3, ed. J. A. Marinsky and Y. Marcus, pp. 51–81. Marcel Dekker, New York.

346. Lumetta, G. J., Moyer, B. A., Johnson, P. A., and Wilson, N. M. 1991. Extraction of zinc ion by didodecylnaphthalenesulfonic acid (HDDNS) in carbon tetrachloride: The role of aggregation. *Solvent Extr. Ion Exch.*, 9(1): 155–176.

347. Otu, E. O., and Westland, A. D. 1990. The thermodynamics of extraction of some lanthanide and other ions by dinonylnaphthalenesulfonic acid. *Solvent Extr. Ion Exch.*, 8(6): 827–842.

348. Raieh, M. A., and Aly, H. F. 1980. Thermodynamics of dinonylnaphthalene sulfonic acid (HD). Extraction of trivalent americium, curium and californium. *J. Radioanal. Chem.*, 57(1): 17–21.

349. Danesi, P. R., Chiarizia, R., Raieh, M. A., Scibona, G. 1975. Enthalpy and entropy variations in the liquid cation exchange of some lanthanide ions by dinonylnaphthalenesulfonic acid (HD) and bis(2-ethylhexyl)phosphoric acid (HDEHP). *J. Inorg. Nucl. Chem.*, 37: 1489–1493.

350. Caceci, M., Choppin, G. R., and Liu, Q. 1985. Calorimetric studies of the synergistic effect. The reaction of UO_2^{2+}, Th^{4+} and Nd^{3+} TTA complexes with TBP and TOPO in benzene. *Solvent Extr. Ion Exch.*, 3(5): 605–621.

351. Kertes, A. S., and Kassierer, E. F. 1972. Thermochemistry of lanthanide complexes in the thenoyltrifluoroacetone-bipyridyl systems. *Inorg. Chem.*, 11(9): 2108–2111.

352. Modolo, G., and Nabet, S. 2005. Thermodynamic study on the synergistic mixture of bis(chlorophenyl)dithiophosphinic acid and tris(2-ethylhexyl)phosphate for separation of actinides(III) from lanthanides(III). *Solvent Extr. Ion Exch.*, 23: 359–373.

353. Pai, S. A., Mathur, J. N., Khopkar, P. K., and Subramanian, M. S. 1977. Thermodynamics of synergistic estraction of europium(III) with thenoyltrifluoroacetone and tributyl phosphate in various diluents. *J. Inorg. Nucl. Chem.*, 39(7): 1209–1211.

354. Osseo-Asare, K., and Renninger, D. R. 1984. Synergic extraction of nickel and cobalt by LIX63-dinonylnaphthalene sulfonic acid mixtures. *Hydrometallurgy*, 13: 45–62.
355. Dias Lay, M. d. L., Perez de Ortiz, E. S., Gruentges, K. 1996. Aluminium and iron extraction by DNNSA and DNNSA/DEHPA reverse micelles. In *Proceedings of International Solvent Extraction Conference, ISEC'96*, Vol. 1, ed. D. Shallcross, R. Paimin, and L. M. Prvcic, pp. 409–414. University of Melbourne, Melbourne, Australia.
356. Arnold, J. M., and Otu, E. O. 2004. Effect of temperature on the synergistic extraction of some lanthanides by 2-ethylhexyl phenylphosphonic acid and micelles of dinonyl naphthalene sulfonic acid. *Solvent Extr. Ion Exch.*, 22(3): 415–428.
357. Baaden, M., Berny, F., Madic, C., Schurhammer, R., and Wipff, G. 2003. Theoretical studies on lanthanide cation extraction by picolinamides: Ligand-cation interactions and interfacial behavior. *Solvent Extr. Ion Exch.*, 21(2): 199–220.
358. Baaden, M., Burgard, M., and Wipff, G. 2001. TBP at the water-oil interface: The effect of TBP concentration and water acidity investigated by molecular dynamics simulations. *J. Phys. Chem. B*, 105(45): 11131–11141.
359. Szymanowski, J. 2000. Kinetics and interfacial phenomena. *Solvent Extr. Ion Exch.*, 18(4): 729–751.

Index

Note: Page numbers ending in "e" refer to equations. Page numbers ending in "f" refer to figures. Page numbers ending in "t" refer to tables.

Milton Keynes UK
Ingram Content Group UK Ltd.
UKHW020023071024
449327UK00032B/2903